D0758803

Quality of Life Impairment in Schizophrenia, Mood and Anxiety Disorders

Quality of Life Impairment in Schizophrenia, Mood and Anxiety Disorders

New Perspectives on Research and Treatment

Edited by

Michael S. Ritsner
Associate Professor, Department of Psychiatry, Faculty of Medicine,
Technion - Israel Institute of Technology, Haifa
Director, Acute Department, Sha'ar Menashe
Mental Health Center, Israel
Associate Editor, Quality of Life Research,
Amsterdam, The Netherlands

and

A. George Awad
Professor Emeritus, Department of Psychiatry and the Institute of Medical Science University of Toronto Chief of Psychiatry, Humber River Regional Hospital, Toronto, Ontario President, The International Society of CNS Clinical Trials and Methodology, Canada

A C.I.P. Catalogue record for this book is available from the Library of Congress.

ISBN 978-1-4020-5777-9 (HB)
ISBN 978-1-4020-5779-3 (e-book)

Published by Springer,
P.O. Box 17, 3300 AA Dordrecht, The Netherlands.
www.springer.com

Printed on acid-free paper

CONTENTS

CONTRIBUTORS

Roee Admon, Functional Brain Imaging Unit, Wohl Institute for Advanced Imaging TASMC, Medical Science, Faculty of Medicine, Tel Aviv University, Israel; roeeadmon@yahoo.com

A. George Awad, MB, Bch, PhD, FRCPC, Professor Emeritus, Department of Psychiatry and the Institute of Medical Science University of Toronto, Chief of Psychiatry, Humber River Regional Hospital, Canada; gawad@hrrh.on.ca

Maria-Teresa Bascarán, MD, Department of Psychiatry, University of Oviedo, Spain; cuca.b@apdo.es

Andreas Bechdolf, MD, Consultant Psychiatrist, Cologne Early Recognition and Intervention Centre for mental crisis – FETZ, Department of Psychiatry and Psychotherapy, University of Cologne, Kerpenerstr. 6250937 Köln, Germany; andreas.bechdolf@uk-koeln.de

Marcelo T. Berlim, MD, MSc, Depressive Disorders Program and McGill Group for Suicide Studies, Douglas Hospital Research Centre, McGill University, Montréal, Quebec, Canada; marcelo.berlim@douglas.mcgill.ca

Julio Bobes, MD, PhD, Department of Psychiatry, University of Oviedo, Spain; bobes@uniovi.es

Maria-Teresa Bobes-Bascarán, PhD, Department of Psychiatry, University of Oviedo, Spain; bobes@ctv.es

Manuel Bousoño, MD, PhD, Department of Psychiatry, University of Oviedo, Spain; bousono@uniovi.es

Monika Bullinger, MD, Professor, University Medical Centre of Hamburg Eppendorf Institute and Policlinic for Medical Psychology, Hamburg, Germany; bullinge@uke.uni-hamburg.de

Mojca Z. Dernovsek, MD, PhD, Assistant Professor, University Psychiatric Hospital, Institute of Public Health of Republic of Slovenia, Ljubljana, Slovenia; Mojca-zvezdana.dernovsek@guest.arnes.si

Mona Eklund, Professor, Department of Health Sciences, Division of Occupational Therapy, Lund University, Sweden; mona.eklund@med.lu.se

Sherrill Evans PhD, Senior Lecturer in Social Work and Social Care; Centre for Carework Research, Department of Applied Social Sciences, University of Wales, Swansea; S.Evans@swansea.ac.uk

Marcelo P.A. Fleck, MD, PhD, Professor, Department of Psychiatry and Forensic Medicine; Head, Mood Disorders Program Hospital de Clínicas de Porto Alegre, Brazil

M.-P. García-Portilla, MD, PhD, Associate Professor, Department of Psychiatry, University of Oviedo, Spain; albert@uniovi.es

Anatoly Gibel, MD, Senior Psychiatrists, Acute Department, Sha'ar Menashe Mental Health Center, Israel; gibel@shaar-menashe.org.il

Graeme Hawthorne, A/Professor, Department of Psychiatry, The University of Melbourne, Australia; graemeeh@unimelb.edu.au

Talma Hendler MD, PhD, Senior lecturer, Psychology Department and Faculty of Medicine, Tel Aviv University, Director, Functional Brain Imaging Unit, Wohl Institute for Advanced Imaging, Tel Aviv Sourasky Medical Center, Israel; talma@tasmc.health.gov.il

Joachim Klosterkötter, MD, Professor of Psychiatry, Chair, Director, Department of Psychiatry, and Psychotherapy, University of Cologne, Kerpenerstr. 6250937 Köln, Germany; joachim.klosterkoetter@uk-koeln.de

Margaret A. Koury, MS, Research Project Assistant, Cedars-Sinai Health System, Department of Psychiatry, Los Angeles, USA; margaret.koury@cshs.org

Raymond W. Lam, Professor and Head, Division of Clinical Neuroscience, Department of Psychiatry, University of British Columbia, Vancouver, Canada

Shira Louria, MSW, Research assistant at the Roudebush VA Medical Center and a doctoral candidate at the University of Indianapolis, School of Psychological Science, USA

Paul H. Lysaker, Ph.D., Staff Psychologist at the Roudebush VA Medical Center and Assistant Professor of Clinical Psychology in the Department of Psychiatry at the Indiana University School of Medicine, Department of Psychiatry, Indianapolis, USA; plysaker@iupui.edu

Marianna Mazza, MD, PhD, Institute of Psychiatry, Catholic University of Sacred Heart, Rome, Italy; mariannamazza@hotmail.com or marianna.mazza@rm.unicatt.it

Salvatore Mazza, MD, Professor and Head, Neurophysiopathology Unit, Department of Neurosciences, Catholic University of Sacred Heart, Rome, Italy

W. Vaughn McCall, MD, MPH, Professor and Chair, Wake Forest University School of Medicine, Medical Center Blvd., USA

Erin E. Michalak, PhD, Assistant Professor, MSFHR Scholar & CIHR New Investigator, Mood Disorders Centre, Department of Psychiatry, University of British Columbia, Vancouver, Canada; emichala@interchange.ubc.ca

Greg Murray, PhD, Senior Lecturer and Clinical Psychologist, Faculty of Life and Social Sciences, Swinburne University of Technology, Hawthorn, Australia; GWMurray@groupwise.swin.edu.au

Dieter Naber, MD, Professor, University Medical Centre of Hamburg Eppendorf, Department of Psychiatry, Centre of Psychosocial Medicine, Hamburg, Germany

David Papo, Functional Brain Imaging Unit, Wohl Institute for Advanced Imaging TASMC, Israel; david.papo@chups.jussieu.fr

Valentina Prevolnik-Rupel, BA, MSc, Ministry of Health of Republic of Slovenia, Ljubljana, Slovenia

Ralf Pukrop, PhD, ScMD, Associate Professor, Section of Experimental and Clinical Psychology, Department of Psychiatry and Psychotherapy, University of Cologne, Kerpenerstr. 6250937 Köln Germany; ralf.pukrop@uk-koeln.de

Mark Hyman Rapaport, MD, Chairman, Department of Psychiatry, The Polier Endowed Chair in Schizophrenia, and Related Disorders, Cedars-Sinai Medical Center, Department of Psychiatry, Los Angeles, USA; mark.rapaport@cshs.org

Michael S. Ritsner, MD, PhD, Associate Professor, Department of Psychiatry, The Rappaport Faculty of Medicine, Technion - Israel Institute of Technology; and Director, Acute Department, Sha'ar Menashe Mental Health Center, Israel; ritsner@shaar-menashe.org.il

Peter B. Rosenquist, MD, Associate Professor, Department of Psychiatry and Behavioral Medicine, Wake Forest University School of Medicine, Medical Center Blvd., USA; rosenqui@wfubmc.edu

Pilar-Alejandra Sáiz, MD, PhD, Department of Psychiatry, University of Oviedo, Spain; frank@uniovi.es

Silke Schmidt, PhD, University Medical Centre of Hamburg Eppendorf Institute and Policlinic for Medical Psychology, Hamburg, Germany

Rok Tavcar, MD, PhD, Chair of Psychiatry, University Psychiatric Hospital, University of Ljubljana, School of Medicine, Ljubljana, Slovenia

Verena Veith, Clinical Psychologist, Department of Psychiatry and Psychotherapy, University of Cologne, Kerpenerstr. 6250937 Köln, Germany; verena.veithf@gmx.de

Lakshmi N.P. Voruganti, Associate Professor, MD, MS, PhD, McMaster University, Hamilton, ON, Canada; vorugl@mcmaster.ca

Allan H. Young, Professor and LEEF Chair in Depression Research, Associate Director of the Institute of Mental Health, Department of Psychiatry, University of British Columbia, Vancouver, Canada; allan.young@ubc.ca

FOREWORD

Over the past few decades health-related quality of life (HRQL) has emerged as the new image of medicine viewed from a psychosocial perspective. The concept of Quality of Life has attracted a good deal of interest, not only from a clinical perspective but also from psychosocial, health economics as well as cultural aspects. More recently, the neurobiological brain substrates that modulate many aspects of subjective experiences, which is relevant to quality of life such as affect, mood, cognition, pleasure, reward responses as well as feeling of wellbeing and satisfaction has been explored and elucidated.

Such increased interest in HRQL is highlighted by the large number of recent publications. Over the past 10 years at least 350 papers were published describing aspects of HRQL in the psychiatric and mental field. Among them 78% dealt with HRQL in schizophrenia and schizoaffective disorders, 21% with major depression, 14% with anxiety disorders and 4% with bipolar disorder. It is gratifying to witness the enhanced interest and popularity in HRQL research and publications, yet the field continues to suffer from conceptual inconsistencies as well as a good deal of methodological limitations. It is worrisome that we still do not have a clear understanding of the concept itself as applied to mental health and illness as well as the complex array of its determinants. Similarly there is a lag in the application of quality of life data in improving clinical practice.

For some time the field has been mostly concerned about measurement of quality of life with less interest devoted in developing testable conceptual models for different disorders, at different stages of illness and for specific populations. Fortunately this state of affairs has been changing over the past decade with the introduction of a number of theoretical and conceptual models, refining methodology and looking beyond the assessment of quality of life. Though quality of life by itself constitutes an important outcome in the management of psychiatric disorders, one can construe quality of life also as a mediator of other important outcomes.

Historically and traditionally quality of life has always been approached as mostly a psychosocial construct, yet the recent advances in neuroscience has added important knowledge about the genesis of such constructs such as pleasure, satisfaction as well as the broad aspects of affective states, insight and cognitive abilities; all are relevant to the person's ability to correctly interpret their feelings and perceptions of their quality of life. In this context, this book contributed to by some of the most active researchers and clinicians in the field; provides new perspectives not only clarifying some of the ongoing controversies but also proposing new models and different perspectives. The book is organized in three major sections: Key Methodological Issues, Quality of Life Impairment in Severe Mental Disorders and Treatment and Rehabilitation Issues. Though the selection of diverse topics has been mostly guided by the current state of the art and the issues that require further discussion, one can not assume that these are the only important issues at this time. Similarly as in publications contributed to by diverse authors from diverse orientations and academic backgrounds it is inevitable to avoid differences of approaches and opinions as well as some overlap. Indeed we consider such diversity as a measure of strength for this book. We also believe that this book probably is the first of its kind to go beyond the psychosocial aspects of quality of life and delve into the neurobiological basis for emotions, feelings and satisfaction.

We sincerely hope that this book will further knowledge in the complex field of quality of life as well as to be of interest to a broad spectrum of readers including clinicians, researchers and policy makers in the fields of mental health and rehabilitation. We certainly believe that enhancement of quality of life of our patients is in the best interest of patients, their families, clinicians and the society as a whole.

Michael S Ritsner
A. George Awad
Editors, 2006

PART I

KEY METHODOLOGICAL ISSUES

CHAPTER 1

THE DISTRESS/PROTECTION VULNERABILITY MODEL OF QUALITY OF LIFE IMPAIRMENT SYNDROME

Current evidence and future research directions

MICHAEL S. RITSNER

Department of Psychiatry, The Rappaport Faculty of Medicine, Technion - Israel Institute of Technology and Acute Department, Sha'ar Menashe Mental Health Center, Israel

Abstract: In this chapter the author describes the conceptual basis of the health related quality of life (HRQL) impairment syndrome in severe mental disorders (SMD) such as schizophrenia, schizoaffective, mood and anxiety disorders. He presents the evidence for its validity, and identifies some likely directions for future research and development. Based on the author's and his team research contributions and complementary theoretical considerations, the author explores four issues in this chapter: the quality of life concept, interpreting HRQL findings, conceptualizing HRQL impairment in the framework of the Distress/Protection Vulnerability Model (DPV), and implications for future research. Evidence for the concept's validity is assessed, followed by a discussion of the possible evolvement of the concept, to encompass biologic domains. Finally, there is a review of the research implications of the HRQL impairment concept and the DPV model followed by a discussion of some major areas of unresolved questions for future research

Keywords: Health related quality of life, Schizophrenia, Schizoaffective, Mood disorders, Distress/Protection Vulnerability model

DEFINING QUALITY OF LIFE

There are numerous definitions for "Quality of life" (QOL) however, there is no satisfactory definition as yet using both subjective and objective aspects of quality. A number of existing definitions lack clarity and create confusion depending on whether the approach used is subjective or objective. Broadly speaking, the subjective approach centers on issues such as life satisfaction, satisfaction with defined needs, happiness, self-realization and growth. On the other hand the objective approach focus on living conditions, economic and social indicators. Calman[1] suggests that QOL measures the gap, at a particular point in time between the hopes and expectations of the individual and the individual's present experiences. According to Maslow's theory in order to reach self-actualization it is

M.S. Ritsner and A.G. Awad (eds.), Quality of Life Impairment in Schizophrenia, Mood and Anxiety Disorders, 3–19.

necessary to go through a hierarchical needs process[2]. His theory uses the concept of human needs as a basis for development of happiness and true being. Besides such well-known needs as food, sex, and social relations, Maslow talks about a more abstract need to realize oneself. He defines five needs, organized hierarchically: physiological, safety, belonging, love, self-esteem and the need for self-actualization[3].

HEALTH RELATED QUALITY OF LIFE

Since many valued aspects of life, like income, freedom and quality of the environment, etc are not usually considered "health related", the term health-related quality of life (HRQL) came to refer to the physical, psychological, and social domains of health. Those domains are seen as distinct areas that are influenced by the person's experiences, beliefs, expectations, and perceptions[4]. One common element across the various definitions of HRQL is the individual's subjective sense of well-being. It is often postulated that HRQL and subjective well-being are determined by various dimensions of physical, social, and role functioning. In other words, HRQL includes dimensions of physical and social functioning: mental health and general health perceptions including such important concepts as energy, fatigue, pain, and cognitive functioning[5].

HRQL is multidimensional in the sense that the subjects may simultaneously evaluate several dimensions to arrive at an overall judgment. Two persons with the same mental health status may have different HRQL levels since elements such as differences in personality and illness related factors influence a person's perception of health and satisfaction with life. Perceptions of HRQL are based on a cognitive process, which involves identifying the relevant domains comprising QOL, determining which domains are self relevant to one and integrating separate domain assessments into an overall quality of life assessment[6]. Each domain of health has many components that need to be measured. Because of this multidimensionality, there is an almost infinite number of states of health, all with differing qualities.

HRQL is a heterogeneous concept, as reflected in the different perceptions of this construct by psychiatrists and their patients. Such differences are obviously dependant on whether observer rated or self-reporting instruments are used. HRQL differs somewhat from subjective well-being, in that the latter concerns itself primarily with affective states, both positive and negative. A HRQL assessment is much broader and although affect-laden, represents the subjective evaluation of oneself in the context one's social and physical world[7]. Self-reported and observer-rated HRQL data provide distinct types of information, and appear to have different indicators for HRQL[8,9].

The discrepancy between subjective and objective measures of HRQL signifies a genuine difference rather than an anomaly related to the patient's psychiatric condition[10]. Indeed, Kaiser et al.[11] found that psychopathology was the only robust predictor of subjective HRQL but overall the impact seemed to be moderate and did not affect the patient's subjective HRQL ratings. Correlations between subjective and objective measures of HRQL among severely mentally ill patients ranged from very low to insignificant[12,13]. There are several limitations in the interpretation of self reported measures of HRQL, namely, self-report bias, the lack of universally accepted measures, the lack of reliability and validity data for many of the scales, and difficulty in generalizing findings from the various instruments. Observer-rated instruments include mostly negative and deficit symptom items. There is a general consensus regarding the importance of using both self-reported and observer-rated measures of HRQL.

MODEL FOR CLINICAL TRIALS

Despite the increasing importance of quality of life in the mental health field, the theoretical conceptualization of the construct remains poorly developed. The rationale for a HRQL assessment in psychiatric research should be outlined in an analytic model that tests the relationship between predictors and response variables.

In 1997, Awad and associates[14] reported on the development of a *conceptual integrative model* of HRQL in medicated persons with schizophrenia. According to that model, HRQL is the outcome of interaction between the three major determinants (symptoms, side effects, and psychosocial performance) and with several modulators such as personality characteristics, premorbid adjustment, values and attitudes toward health and illness, resources and their availability. Testing validity of this model indicated that the severity of symptoms was the main predictor of HRQL, explaining 32% of its variance, while neuroleptic side effects explained 17%. The contribution of psychosocial indicators and modulators, however, was not significant. Since this model continues to be the only model that is specific to the effects of medication and, as such, is the most applicable to clinical trials with medications, broadening the model to make it applicable to other social or vocational interventions seemed to be warranted.

MEDIATIONAL MODEL

In 1998, Zissi, Barry, and Cochrane[15] proposed a model, which links subjective HRQL with self-related constructs. They tested this model on a group of 54 long-term psychiatric hostel residents in Greece using a modified version of Lehman's Quality of Life Interview. The results indicated that perceived improvements in lifestyle, greater autonomy and positive self-concept were significantly and directly associated with better quality of life. In contrast, no direct relationship was found between objective indicators and subjective quality of life. The authors concluded that the extended mediational model of HRQL for individuals with long-term mental health problems appears to have important implications for the planning and delivery of mental health programmes. However, this model needs further development, testing and validation.

DISTRESS/PROTECTION MODEL

In 2000, Ritsner and associates[16] proposed a *Distress/Protection Model* of HRQL in severe mental disorders. The model, initially based on findings from a sample of 210 inpatients consecutively admitted to closed, open and rehabilitation wards, postulated that subjective HRQL is an outcome of the interaction of an array of distress factors, on the one hand, and protection factors, on the other. It suggests that satisfaction with HRQL decreases when distress factors outweigh protection factors, and vice versa. The data included measures of satisfaction with general and domain-specific HRQL such as physical health, subjective feelings, leisure activities, social relationships, general activities, medication, as well as severity of psychopathology, adverse events, psychological distress, expressed emotions, personality traits, self-constructs, coping styles, and perceived social support. The model includes a greater number of quality of life domains and would, therefore, allow for easy integration of HRQL items into one of the distress or protection factor categories and could provide practitioners with an easily recognized outline for use in their clinical work.

In order to validate the Distress/Protection model, two kinds of multivariate analyses were conducted using cross-sectional and longitudinal data. First, an exploratory factor analysis confirmed the main postulate of the model regarding distress and protection factors underlying the general HRQL ($n = 339$ patients; Table 1). The first and second factors (named 'distress') with a harmful effect on the general HRQL included severity of emotional distress, somatization, illness, symptoms and side effects, insight and emotionally oriented coping. The third factor (named 'protection') was constructed using task and avoidance coping styles, self-efficacy, self-esteem, and social support. HRQL scores associated significantly with first and third factors. Two distress and protection factors accounted for 42.4%, 26% and 27.9% of the total variance of the 29 variables, respectively. The factor structure did not change when we used data obtained from a follow up sample ($n = 199$).

Next, the hierarchical multiple regression analysis was applied in order to predict general Q-LES-Q_{index} scores for schizophrenia patients (Table 2). In the first step, when distress and demographic variables were simultaneously entered into the regression analysis, the *distress factors' model* accounted for 40% of the total variance in Q-LES-Q_{index} scores. This model included two distress variables, three PANSS factors and age accounted for 16.4%, 10.1%, and 2%, respectively. The *protective factors' model* consisting of 5 predictors (self-esteem, self-efficacy, two coping styles and social support) explained 46% of the total variance. Finally, the *combined distress/protection model* that fit the data best ($R^2 = 0.58$), revealed that the protective pattern (self-esteem, self-efficacy, avoidance coping styles and social support) accounted for 18.2% of the variance, while distress patterns (emotional distress, somatization, negative and depressive symptoms) explained only 11.6%. When patients were re-examined about 16 months later, similar analysis revealed that the *combined distress/protection model* that fits the data best ($R^2 = 0.72$), include two protection factors (self-efficacy, and social support) and two distress factors (emotional distress and depressive symptoms) factors.

Table 1. Factor loadings after Varimax rotation of variable values among 339 patients with severe mental disorders[1]

Variables		Factor 1 (eigenvalue=5.43) *'distress'*	Factor 2 (eigenvalue=3.32) *'symptoms'*	Factor 3 (eigenvalue=3.57) *'protective'*
General quality of life(Q-LES-Q_{index})[2]		**0.4695**	−0.0453	**0.6102**
Illness severity (CGI)[3]		−0.1389	0.6777	−0.0759
General functioning (GAF)[4]		0.0115	**−0.4232**	0.0792
Symptoms:	Negative symptoms	−0.1106	**0.6339**	−0.2542
(PANSS)[5]	Positive symptoms	−0.1044	**0.7225**	0.0384
	Activation symptoms	−0.1674	**0.7497**	−0.1244
	Dysphoric mood	**−0.5793**	0.1760	−0.0726
	Autistic preoccupations	−0.1397	0.7352	−0.1508
Insight:	Observer-rated[6]	−0.3455	**−0.4445**	−0.0428
	Self-report[7]	**−0.4266**	−0.2409	−0.2237
Side effects:	Number of Adverse Symptoms	**−0.5886**	0.1613	0.0533
(DSAS)[8]	Mental Distress Index	**−0.5945**	0.1326	0.0457
	Somatic Distress Index	**−0.5525**	0.0013	0.0297
Emotional distress:	Obssesiveness	**−0.6272**	0.0126	−0.3312
(TBDI)[9]	Hostility	**−0.4526**	0.1127	−0.0657
	Sensitivity	**−0.6554**	0.0112	−0.2570
	Depression	**−0.7068**	−0.0822	−0.3454
	Anxiety	**−0.6938**	0.09601	−0.2459
	Paranoid ideation	**−0.5173**	0.1203	−0.2150
Somatization (BSI)[10]		**−0.6225**	0.0855	−0.1112
Self-esteem (RSES)[11]		0.3212	0.0769	**0.4585**
Self-efficacy (GSES)[12]		0.2507	0.0600	**0.6268**
Coping styles:	Task oriented coping	−0.0108	0.1222	**0.6893**
(CISS)[13]	Emotion oriented coping	**−0.5873**	0.0088	0.1181
	Avoidance coping	−0.1014	0.0250	**0.6758**
Social support:	Family support	−0.0209	−0.1757	**0.5228**
(MSPSS)[14]	Friend support	0.1333	−0.0619	**0.4971**
	Other significant support	−0.0262	−0.1575	**0.5497**
Factors' contribution (%)		**42.4%**	**26.0%**	**27.9%**

[1]Variables with an absolute loading greater than the amount set in the minimum loading option (≥ 0.40) were selected; expressed emotion, age, education, age of onset and illness duration were removed to avoid augmenting scores.
[2]Q-LES-Q- the Quality of Life Enjoyment and Life Satisfaction Questionnaire; [3]CGI- the Clinical Global Impression scale; [4]GAF - the Global Assessment of Functioning Scale; [5]PANSS - the Positive and Negative Syndromes Scale; [6]ITAQ – the Insight and Treatment Attitudes Questionnaire; [7]IS – the Insight Self-report Scale; [8]DSAS - the Distress Scale for Adverse Symptoms; [9]TBDI - the Talbieh Brief Distress Inventory; [10]BSI - the Brief Symptom Inventory; [11]RSES - the Rosenberg Self-Esteem scale; [12]GSES - the General Self-Efficacy Scale; [13]CISS - the Coping Inventory for Stressful Situations; [14]MSPSS - the Multidimensional Scale of Perceived Social Support (for references see Appendix, Chapter 10).

Table 2. Summary of hierarchical multiple regression analysis for predicting general quality of life ($Q-LES-Q_{index}$) in schizophrenia

Model	Initial sample of inpatients (n=237)				Follow up sample (n=148)			
	β	P	Partial R^2 (%)	Model's properties	β	P	Partial R^2 (%)	Model's properties
Distress factors' model								
Emotional distress	−0.41	0.001	14.2	$R^2=0.40$	−0.68	0.001	50.3	$R^2=0.55$
Somatization	−0.15	0.025	2.2	adj. $R^2=0.38$	–	–	–	adj. $R^2=0.54$
Negative factor	−0.29	0.001	5.0	F=25.5	–	–	–	F=87.1
Dysphoric mood	−0.15	0.009	2.9	df=6,236	–	–	–	df=3,147
Autistic preoccupations	0.20	0.023	2.2	$p<0.001$	–	–	–	$P<0.001$
Number of adverse symptoms	–	–	–		−0.13	0.018	4.0	
Expressed emotion	–	–	–		−0.14	0.013	4.4	
Age at examination	−0.11	0.032	2.0		–	–	–	
Protective factors' model								
Self-efficacy	0.21	0.001	4.4	$R^2=0.46$	0.30	0.001	12.8	$R^2=0.65$
Self-esteem	0.28	0.001	7.8	$P<0.0001$	0.45	0.001	25.1	adj. $R^2=0.64$
Emotion oriented coping	−0.16	0.005	3.3	adj. $R^2=0.45$	–	–	–	F=87.6
Avoidance coping	0.18	0.006	3.2	df=5,236	–	–	–	df=3,147
Social support	0.21	0.001	5.9	F=39.9	0.21	0.001	8.9	$P<0.001$
Distress/protection model								
Emotional distress	−0.21	0.002	4.9		−0.51	0.001	32.8	$R^2=0.72$
Somatization	−0.14	0.011	2.5	$R^2=0.58$	–	–	–	adj. $R^2=0.71$
Negative symptoms	−0.08	0.023	1.2	adj. $R^2=0.56$	–	–	–	F=92.7
Depressive symptoms	−0.13	0.005	3.0	F=38.4,	−0.13	0.008	4.8	df=4,147
Self-efficacy	0.21	0.001	5.3	df=8,236	0.23	0.001	10.2	$p<001$
Self-esteem	0.15	0.012	2.6	$p<0.001$	–	–	–	
Avoidance coping	0.19	0.002	3.6		–	–	–	
Social support	0.20	0.009	6.7		0.1	0.001	9.1	

Partial R^2 reflects the percentage of variation in the dependent variable ($Q-LES-Q_{index}$) explained by each independent variable adjusted to the effects of the rest independent variables. The higher R^2 value, the greater contribution of independent variable to the model

Since 2000, the Distress/Protection model has been extensively used to compare HRQL impairment among patients with severe mental disorders[16–18], to examine the role of side effects[19], to test mediating effects of coping styles[20], to search longitudinal predictors of general and domain-specific quality of life[21–23], to explore association of HRQL impairment with suicide behavior[24], temperament factors[25,26], and sleep quality[27], and to examine the impact of antipsychotic agents[28]. In addition, findings of other research groups also highlighted the importance of addressing psychosocial issues and their interrelationships in the structures of HRQL that have supported this model[29–34].

MAIN FEATURES OF HRQL IMPAIRMENT

A number of empirical findings of HRQL impairments are reviewed[18,35–39], with the evidence leads to the suggestion that a current profile of HRQL disturbances in severe mental disorders cannot be covered by the "outcome" hypothesis that views poor HRQL level in patients with schizophrenia, schizoaffective, mood and anxiety disorders as a result of the progressive course of the illness.

- Subjects who met criteria for vulnerability to schizophrenia ("schizotaxia") received significantly lower total ratings on the Quality of Life Scale[40] including the interpersonal relations subscale. They also showed significantly higher psychological distress scores (SCL-90), and demonstrated particular elevations on the obsessive-compulsive, anxiety, and hostility subscales (but not other subscales, such as depression, paranoia, and psychoticism)[41,42].
- Despite the absence of psychotic symptoms, individuals with prodromal symptoms (ultra-high-risk) for schizophrenia experience significant HRQL impairments in a manner parallel to that observed in patients with established psychotic illness[43,44].
- There is an association between poor premorbid adjustment and poor HRQL levels in schizophrenia[45–47].
- Poor HRQL is associated with long duration of untreated first-episode schizophrenia patients[45,46,48].
- There are significant differences in HRQL levels between patients with severe mental disorders[18].
    1. The most notable deficit in social relationships is observed in schizophrenia, while in subjective feelings it appears more frequently in mood-disordered patients.
    2. Predictors of changes in HRQL impairment over time, among schizophrenia patients, are distinct from those associated with schizoaffective disorders.
    3. Severe impairment in general HRQL ($Q\text{-}LES\text{-}Q_{index}$) was observed among 33% of patients with schizoaffective disorders, 49% with schizophrenia, and 53% with mood disorders (46% of patients with severe mental disorders).
    4. Rapaport and associates[49] reported that the proportion of patients with severely reduced HRQL varied according to diagnoses: 63% of the study recruits with major depressive disorder, 59% with PTSD and 85% of those with chronic

or double depression had severe impairment, compared with 20-26% of those with panic disorder, obsessive-compulsive disorder and social phobia.

- Taken together, HRQL findings underscore the relatively stable character of HRQL disturbances with mild fluctuations in the general and domain-specific quality of life scores throughout the course of schizophrenia.
  1. Subjective HRQL appears to be lower in first-admission schizophrenia patients than in long-term chronic patients[50].
  2. Although most researchers report some positive changes in HRQL domains after one-year follow-up[51,52], others did not find significant change in the subjective HRQL of schizophrenia patients followed up after 7-10 year periods[53,54].
  3. Our findings suggest that most of the patients with severe mental disorders (63%) remained dissatisfied over a 16 month follow-up period, while 10% were satisfied with general HRQL. Improvement in HRQL was reported by 16% of the patients, whereas worsening in HRQL was reported by 11% of the patients[18].
- HRQL of patients with severe mental disorders was associated with stress related factors such as personality traits, self-esteem, and self-efficacy, emotion-oriented coping, and psychological distress[16,33],[55–57] rather than with illness related symptoms, and side effects[16,21]. Illness-specific symptoms explained a relatively small proportion of the variance in HRQL scores among people with depression or anxiety disorders[49].
- Temperament traits, which are not necessarily part of the deterioration process of the illness, are significantly associated with the HRQL in schizophrenia[26].

CONCEPTUALIZATION OF HRQL IMPAIRMENT

HRQL has not been fully explained by the "outcome" hypothesis. However, this approach is still prominent in most HRQL psychiatric publications. These reports can be divided into two categories: correlation studies and intervention studies.

Correlation studies search for an association between various measures of psychopathology and HRQL domains. The vast majority of those studies view HRQL as a strongly illness related measure and therefore focus mainly on clinical correlates. A major obstacle to research in this field is the multiplicity of illness related factors that conceivably can influence the quality of life of individuals with severe mental disorders.

Intervention studies address the question of how well HRQL measures function as a valid and sensitive outcome indicator of efficacy of treatment interventions (antipsychotic agents, rehabilitation programmes) and of mental health services. Findings from cross-sectional, comparative, repeated-measures and randomised empirical studies have shown that any changes or lack of improvement in patients' HRQL are related to various treatment regimens and treatment programs, an issue that deserves serious consideration in medical practice. One should keep in mind that, in these studies, the confounding effect of numerous stress related factors that influence HRQL structure has not been recognized.

A scientifically valid hypothesis should be able to explain the main observations and be, at least, consistent with the rest of the observations. The major problems of the "outcome" hypothesis of HRQL impairment are the lack of a strong association with most illness related variables and a lack of direct evidence of progressive alteration of HRQL over time in patients with severe mental disorders. Furthermore, HRQL impairment was observed among persons with vulnerability to schizophrenia (schizotaxia) and before manifestation of first episode of illness. They are influenced by poor premorbid adjustment and by many stress related factors. Therefore, recent HRQL findings have emphasized the concept of "HRQL impairment" to replace the "outcome" hypothesis.

According to the HRQL impairment concept, the quality of life deficit syndrome refers to the vulnerability to illness, and, consequently, should be viewed as a definitive expression or a particular syndrome of severe mental disorders, like psychopathology or cognitive impairment. The stress-vulnerability model postulates the vulnerability to illness as stable, enduring, and largely attributable to genetic and environmental factors[58–60]. Greater vulnerability is associated with higher risk for developing schizophrenia, but the actual expression of this predisposition depends on a host of personal and environmental factors, some of which are noxious, while others are protective. It is the interaction of vulnerability, stressors and protective factors that influences both the onset and the course of the disorder.

DISTRESS/PROTECTION VULNERABILITY MODEL

According to "HRQL impairment concept", the Distress/Protection Vulnerability Model (DPV model) [previously formulated as the Distress/Protection Model], suggested that:

1. HRQL impairment is a particular syndrome observed in the most psychiatric and somatic disorders. [The term *syndrome* derives from the Greek and means literally "run together," as the features do. In medicine, the term syndrome is the association of several recognizable features, sings, or characteristics that often occur together. It is most often used when the reason that the features occur together (the pathophysiology of the syndrome) has not yet been discovered. In recent decades the term has been used outside of medicine to refer to a combination of phenomena seen in association (from Wikipedia: http://en.wikipedia.org/wiki/syndrome)].
2. This syndrome is an outcome of the interaction of an array of distressing factors, on the one hand, and putative stress process protective factors, on the other hand. HRQL impairment increases if distressing factors overweigh protective factors, and vice versa.
3. There are primary and secondary factors. Primary or vulnerability related factors are those usually considered inborn or personal characteristics, while secondary factors are related to the illness and the environment[22]. Such primary factors

as harm avoidance, high levels of neuroticism, poor coping skills, elevated emotional distress, emotion-oriented coping, and weak self-constructs[61–63] might lower the vulnerability threshold, and, consequently, result in severe HRQL impairment. Secondary factors influence HRQL impairment via primary factors.
4. HRQL impairment syndrome is characterized on the basis of underlying neurobiology and that may lead to improved understanding of severe mental disorders and to making more effective treatment decisions.

Factors influencing HRQL impairment syndrome in schizophrenia according to the DPV model are summarized in Figure 1 (see also Chapter 10, Table 8).

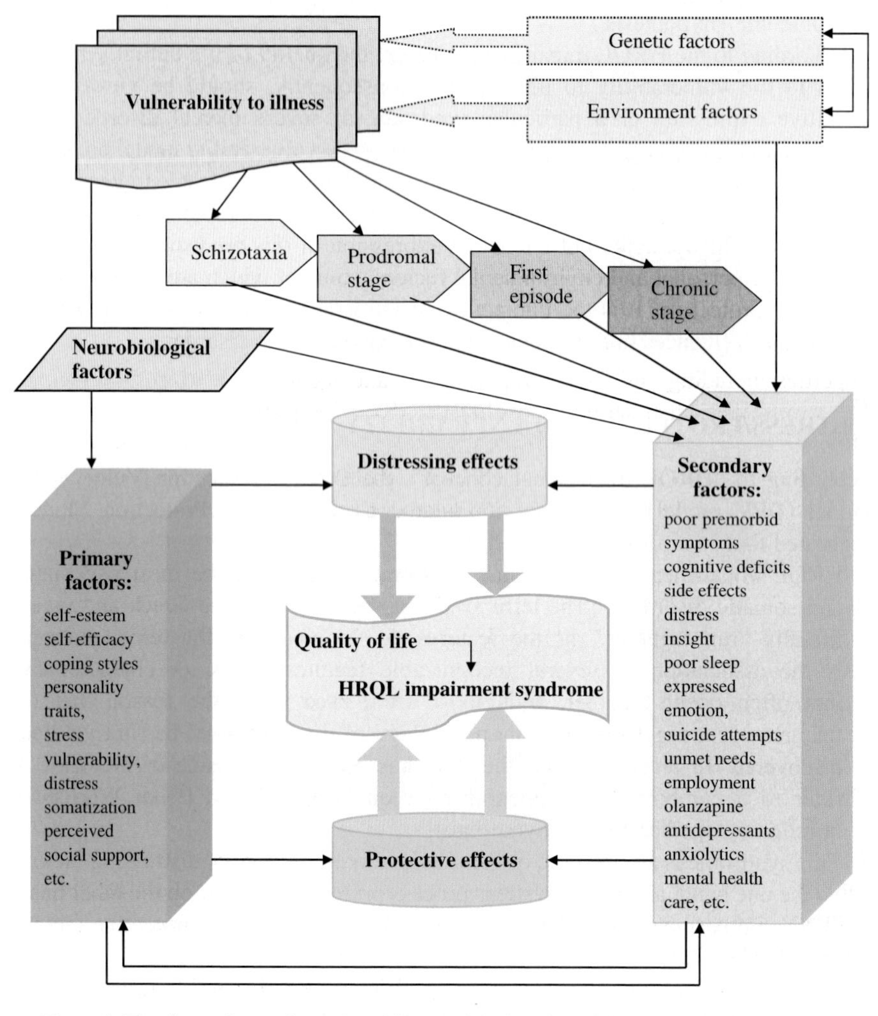

Figure 1. The distress/protection vulnerability model of quality of life impairment syndrome in schizophrenia

FUTURE NEUROBIOLOGICAL RESEARCH DIRECTIONS

Researchers have to pay particular attention to biological correlates and/or brain imaging correlates of HRQL impairment syndrome in severe mental disorders. What neurobiological factors influence HRQL impairment? Is HRQL impairment a neurobiological syndrome? There are no previous HRQL studies to answer this question. Following are a number of issues that require further studies.

Broad-based HRQL studies including stress related biological indicators with and without antipsychotic treatment are warranted in the future. Significantly, advances have been made in understanding of the role of hyperactivity of hypothalamic–pituitary–adrenal (HPA) axis[64], impaired emotional reactivity to daily life stresses[65], and altered hormonal stress-response[66] resulting in florid emotional distress and depressive symptoms. More research is needed to explore the role of emotional distress and depressive symptoms in development of HRQL impairment in patients with SMD.

Advances have been made in understanding the neuroactive role of neurosteroids such as dehydroepiandrosterone (DHEA) and its sulfate conjugate (DHEAS) [together abbreviated DHEA(S)] that mediate several neurotransmitter systems coupled to ion channels, such as gamma-aminobutyric acid ($GABA_A$), N-methyl-D-aspartate (NMDA) and sigma receptors. They modulate neuronal excitability and plasticity, in addition to their neuroprotective properties[67–70]. Since DHEA has memory-enhancing effects[71], it has been hypothesized that DHEA has neuroprotective effects on cognition[72]. These neurosteroids exhibit a variety of properties including anti-stress properties[73,74] and reduction of neurodegeneration [67,75]. It was found that serum cortisol/DHEA(S) ratios were significantly higher in schizophrenia patients than in healthy control subjects[76], and that these elevated ratios may serve as markers of biological mechanisms that are involved in responsivity of schizophrenia patients to antipsychotic treatment[77]. Although there is evidence that the level of neurosteroids is correlated with feelings of "well-being" and enjoyment of "leisure" activities[78], our short-term DHEA supplementation study does not support DHEA's value as an effective adjunct in the treatment of symptoms, side effects and quality of life impairment in schizophrenia, while suggesting that DHEA improves sustained attention, visual, and motor skills[79]. A long-term, large-scale study with a broader dose range is warranted to further investigate DHEA's role in the management of schizophrenia.

Vasopressin increases anxiety, stress and aggressiveness[80], and appears to have a strongly negative effect on HRQL and likely increase HRQL impairment. Oxytocin on the other hand, may have a protective effect on HRQL, since it has been found to comprise a prominent factor in maternal care, social attachment and affiliation in mammals[81]. Further research is necessary to clarify the relationship between these hormones and HRQL impairment.

Researchers should consider neuroimaging findings like perception of emotions and affective style that may influence of HRQL levels. Recent animal data and functional neuroimaging studies in human indicate that emotion perception may be dependent upon the functioning of two neural systems: a ventral system, including the amygdala, insula, ventral striatum, and ventral regions of the anterior cingulate gyrus and prefrontal cortex[82]. Wang and associates[83] reported neuroimaging data showing that stress inducing negative emotions and vigilance implicating the ventral right prefrontal cortex play a key role in the central stress response. Furthermore, patients with schizophrenia exhibit impaired neural responses to emotionally salient stimuli in the right ventral striatum, supporting a role for this structure in the pathophysiology of the illness[84]. Yet, a resilient affective style and well being are associated with high levels of left prefrontal activation, effective modulation of activation in the amygdala and fast recovery in response to negative and stressful events[85].

Several controlled brain imaging studies have shown hippocampal abnormalities in psychiatric disorders, including posttraumatic stress disorder (PTSD), major depressive disorder (MDD), and borderline personality disorder (BPD)[86]. Current literature suggests that individuals with PTSD, MDD, and BPD may suffer hippocampal atrophy as a result of stressors associated with these disorders. Herman et al.[87] summarize current understanding of the relationship between limbic structures and control of ACTH and glucocorticoid release, focusing on the hippocampus, medial prefrontal cortex and amygdala. In general, the hippocampus and anterior cingulate/prelimbic cortex inhibit stress-induced HPA activation. Well-designed studies will be needed to explore the relationship between HRQL and hippocampal atrophy in mental disorders.

Research on the neural systems underlying emotion, over the past two decades, has implicated the amygdala in conditions such as anxiety, autism, stress, phobias, fear and other emotional states[88]. Abnormalities in emotion processing and in structure of the amygdala have consistently been documented in schizophrenia[89–91]. van Rijn S et al.[92] propose that amygdala abnormalities are an endophenotype in schizophrenia and may account for subtle emotional processing deficits that have been described in these high-risk groups. Low basal levels of amygdala activation, in conjunction with situationally appropriate responding, effective top-down regulation and rapid recovery, characterize a pattern that is consistent with high levels of well-being[85]. This suggestion should be tested in future HRQL studies.

Given the well-documented relationship between temperament (i.e. core personality traits) and HRQL, research investigating the neurobiological substrates that underlie temperament is potentially key to our understanding of the biological basis of HRQL impairment in mental disorders. Recently, Whittle et al.[93] have reported that specific areas of the prefrontal cortex and limbic structures (including the amygdala, hippocampus and nucleus accumbens) are key regions associated with three fundamental dimensions of temperament. Introversion/extraversion and neuroticism describe individual personality differences in emotional responding across a range of situations and may contribute to a predisposition for psychiatric disorders. Recent neuroimaging research has begun to provide evidence that neuroticism and introversion/extraversion have specific functional and structural neural correlates. Previous studies in healthy adults have reported an association between neuroticism, introversion/extraversion, and the activity of the prefrontal cortex and amygdala. Studies of individuals with psychopathological states have also indicated that anatomic variations in these brain areas may relate to extraversion and neuroticism. Wright et al.[94] observed that the thickness of specific prefrontal cortex regions correlates with measures of extraversion and neuroticism in healthy subjects ($n = 28$). It would be interesting to include self-reported as well as observer-rated HRQL measures into the design of such studies.

Thus we may conclude that HRQL impairment syndrome in severe mental disorders is likely related to neurobiological alterations that could include changes in the signaling, information encoding, plasticity, and neurochemical properties of neurons or glia. Integration of the quality of life and neurobiological investigations may provide new vistas on the HRQL impairment syndrome in mental disorders.

CONCLUSIONS

The HRQL impairment syndrome occurs before the first psychotic episode and persists throughout the course of the illness. It involves every aspect of quality of life and has an important impact on long-term social and occupational outcomes. Improvement of the HRQL impairment by antipsychotic treatment can be due indirectly to the improvement of therapeutic profiles of the newer antipsychotic agents (eg, higher efficacy on mood and negative symptoms, fewer side effects, less anticholinergic effects) or directly to effects on cerebral functioning (eg, by restoring dopamine prefrontal activity). However, further research is needed regarding the therapeutic effects of the newer antipsychotic drugs on the HRQL impairment. Although newer medications may improve the HRQL impairment, they do not normalize the HRQL impairment in severe mental disorders. In addition, various nonpharmacologic, psychological interventions have been used in the rehabilitation of patients with the HRQL impairment syndrome.

In summary, this chapter discusses the concept of mental health-related quality of life impairment and proposes a new model: the Distress/Protection Vulnerability Model. The definition of HRQL is explored and several studies based on the HRQL "outcome" hypothesis and related models are reviewed. New findings are presented, which can help delineate the HRQL impairment and the Distress/Protection Vulnerability Model, while emphasizing the factors influencing HRQL in severe mental disorders such as schizophrenia, schizoaffective and mood disorders. Finally, there is a brief presentation of selected highlights to illustrate various future research directions.

ACKNOWLEDGEMENTS

I wish to express gratitude to Anatoly Gibel, MD, Yael Ratner, MD, and Stella Lulinski for the dedicated help, as well as, to Galina Ritsner, MD for the considerable support.

REFERENCES

1. Calman KC. Quality of life in cancer patients–an hypothesis. J Med Ethics. 1984; 10:124–127.
2. Maslow, AH. (1962) *Toward a Psychology of Being*. Van Nostrand, New York.
3. Wolf JRLM. Client needs and quality of life. Psychiatric Rehabilitation Journal, 1997; 20:16–24.
4. Testa MA, Simonson DC. Assessment of quality-of-life outcomes. N Engl J Med, 1996; 334:835–840.
5. Wilson IB, Cleary PD. Linking clinical variables with health-related quality of life. A conceptual model of patient outcomes. JAMA, 1995; 273: 59–65.
6. Smith KW, Avis NE, Assmann SF. Distinguishing between quality of life and health status in quality of life research: A meta-analysis. Quality of Life Research, 1999; 8: 447–459.
7. Orley J, Saxena S, Herrman H. Quality of life and mental illness. Reflections from the perspective of the WHOQOL. Br J Psychiatry, 1998; 172:291–293.
8. Fitzgerald PB, Williams CL, Corteling N, et al. Subject and observer-rated quality of life in schizophrenia. Acta Psychiatrica Scandinavica, 2001; 103:387–392.
9. Ruggeri M, Bisoffi G, Fontecedro L, Warner R. Subjective and objective dimensions of quality of life in psychiatric patients: a factor analytical approach: The South Verona Outcome Project 4. British Journal of Psychiatry, 2001; 178: 268–275.
10. Khatri N, Romney DM, Pelletier G Validity of self-reports about quality of life among patients with schizophrenia. Psychiatric Services, 2001; 52: 534–535.
11. Kaiser W, Priebe S, Barr W, Hoffmann K, Isermann M, Roder-Wanner UU, Huxley P. Profiles of subjective quality of life in schizophrenic in- and out-patient samples. Psychiatry Research, 1997; 66: 153–166.
12. Trauer T, Duckmanton RA, Chiu E. A study of the quality of life of the severely mentally ill. International Journal of Socical Psychiatry, 1998; 44:79–91.
13. Lobana A, Mattoo SK, Basu D, Gupta N. Quality of life in schizophrenia in India: comparison of three approaches. Acta Psychiatrica Scandinavica, 2001; 104: 51–55.
14. Awad AG, Voruganti LNP, Heselgrave RJ. A Conceptual Model of Quality of Life in Schizophrenia: Description and Preliminary Clinical Validation. Quality of Life Res 1997; 6: 21–26.
15. Zissi A, Barry MM, Cochrane R. A mediational model of quality of life for individuals with severe mental health problems. Psychological Medicine, 1998; 28: 1221–1230.
16. Ritsner M, Modai I, Endicott J, et al. Differences in quality of life domains and psychopathologic and psychosocial factors in psychiatric patients. J Clin Psychiatry 2000, 61: 880–889.
17. Ritsner M, Kurs R, Gibel A, Hirschmann S, Shinkarenko E, Ratner Y. Predictors of quality of life in major psychoses: a naturalistic follow-up study. Journal of Clinical Psychiatry, 2003; 64: 308–315.
18. Ritsner, M, Kurs, R. Quality-of-Life Impairment in Severe Mental Illness: Focus on Schizoaffective Disorders. In: Schizoaffective Disorder: New Research. Ed. William H. Murray, NOVA Publishers, NY, 2006; pp. 69–107.
19. Ritsner M, Ponizovsky A, Endicott J, et al. The impact of side-effects of antipsychotic agents on life satisfaction of schizophrenia patients: a naturalistic study. Eur Neuropsychopharmacol 2002; 12: 31–38.
20. Ritsner M, Ben-Avi I, Ponizovsky A, Timinsky I, Bistrov E, Modai I. Quality of life and coping with schizophrenia symptoms. Qual Life Res., 2003; 12: 1–9.
21. Ritsner M. Predicting changes in domain-specific quality of life of schizophrenia patients. J Nerv Ment Dis., 2003; 191: 287–294.
22. Ritsner M, Gibel A, Ratner Y. Determinants of Changes in Perceived Quality of Life in the Course of Schizophrenia. Qual Life Res. 2006; 15:515–526.
23. Ritsner M, Modai I, Kurs R, et al. Subjective Quality of Life Measurements in Severe Mental Health Patients: Measuring Quality of Life of Psychiatric Patients: Comparison Two Questionnaires. Quality Life Research, 2002; 11: 553–561.

24. Ponizovsky AM, Grinshpoon A, Levav I, Ritsner MS. Life satisfaction and suicidal attempts among persons with schizophrenia. Comprehensive Psychiatry, 2003; 44: 442–447.
25. Ritsner, M, Farkas, H, Gibel, A. Satisfaction with quality of life varies with temperament types of patients with schizophrenia. Journal of Nervous and Mental Disease, 2003; 191: 668–674.
26. Kurs R, Farkas H, Ritsner M. Quality of life and temperament factors in schizophrenia: comparative study of patients, their siblings and controls. Quality of Life Research, 2005;14: 433–440.
27. Ritsner M, Kurs R, Ponizovsky A, Hadjez J. Perceived quality of life in schizophrenia: relationships to sleep quality. Quality of Life Research, 2004;13: 783–791.
28. Ritsner, M, Perelroyzen, G, Kurs, R, Ratner, Y, Jabarin, M, Gibel, A. Quality of life outcomes in schizophrenia patients treated with atypical and typical antipsychotic agents: A naturalistic comparative study. International Clinical Psychopharmacology, 2004; 24: 582–591.
29. Diener E, Diener M. Cross-cultural correlates of life satisfaction and self-esteem. J Pers Soc Psychol., 1995; 68:653–663.
30. Barry MM. Well-being and life satisfaction as components of quality of life in mental disorders. In. eds. H Katsching, H Freeman, and N Sartorius, Quality of Life in Mental Disorders. Chichester: John Wiley & Sons, 1997.
31. Hansson L, Middelboe T, Merinder L, Bjarnason O, Bengtsson-Tops A, NilssonL, Sandlund M, Sourander A, Sorgaard KW, Vinding H.Predictors of subjective quality of life in schizophrenic patients living in the community. A Nordic multicentre study. : International Journal of Social Psychiatry, 1999; 45: 247–258.
32. Diener E, Oishi S, Lucas RE. Personality, culture, and subjective well-being: emotional and cognitive evaluations of life. Annu Rev Psychol., 2003; 54:403–425.
33. Eklund M, Backstrom M, Hansson L. Personality and self-variables: important determinants of subjective quality of life in schizophrenia out-patients. Acta Psychiatrica Scandinavica, 2003; 108: 134–143.
34. Yanos PT, Moos RH. Determinants of functioning and well-being among individuals with schizophrenia: An integrated model. Clin Psychol Rev. 2006; 27:58–77.
35. Ritsner M, Kurs R. Impact of antipsychotic agents and their side effects on the quality of life in schizophrenia. Expert Review and Pharmacoeconomic Outcomes Research, 2002; 2: 89–98.
36. Ritsner M, Kurs R. Quality of life outcomes in mental illness: schizophrenia, mood and anxiety disorders. Expert Review and Pharmacoeconomic Outcomes Research, 2003; 3: 189–199.
37. Awad AG, Voruganti LN. New antipsychotics, compliance, quality of life, and subjective tolerability–are patients better off? Canadian Journal of Psychiatry, 2004; 49: 297–302.
38. Naber D, Karow A, Lambert M. Subjective well-being under the neuroleptic treatment and its relevance for compliance. Acta Psychiatrica Scandinavica Supplementum, 2005; 427: 29–34.
39. Hansson L. Determinants of quality of life in people with severe mental illness. Acta Psychiatr Scand Suppl., 2006; 429:46–50.
40. Heinrichs DW, Hanlon TE, Carpenter WT. The Quality of Life Scale: an instrument for rating the schizophrenic deficit scale. Schizophr. Bull., 1984; 10:388–398.
41. Stone WS, Faraone SV, Seidman LJ, et al. Concurrent validation of schizotaxia: a pilot study. Biol Psychiatry, 2001; 50:434–440.
42. Tsuang MT, Stone WS, Gamma F, Faraone SV. Schizotaxia: current status and future directions. Curr Psychiatry Rep., 2003; 5:128–134.
43. Erickson DH, Beiser M, Iacono WG, et al. The role of social relationships in the course of first-episode schizophrenia and affective psychosis. Am J Psychiatry., 1989; 146:1456–1461.
44. Bechdolf A, Pukrop R, Kohn D, et al. Subjective quality of life in subjects at risk for a first episode of psychosis: a comparison with first episode schizophrenia patients and healthy controls. Schizophr Res., 2005; 79:137–143.
45. Browne S, Clarke M, Gervin M, et al. Determinants of quality of life at first presentation with schizophrenia. Br J Psychiatry, 2000; 176:173–176.
46. Melle I, Haahr U, Friis S, et al. Reducing the duration of untreated first-episode psychosis – effects on baseline social functioning and quality of life. Acta Psychiatr Scand., 2005; 112:469–473.

47. Czernikiewicz A, Gorecka J, Kozak-Sykala A. [Premorbid adjustment and quality of life in schizophrenia]. Pol Merkuriusz Lek., 2005; 19:659–662.
48. Malla A, Payne J. First-episode psychosis: psychopathology, quality of life, and functional outcome. Schizophr Bull., 2005; 31:650–671.
49. Rapaport MH, Clary C, Fayyad R, Endicott J. Quality-of-life impairment in depressive and anxiety disorders. American Journal of Psychiatry, 2005; 162:1171–1178.
50. Priebe S, Roeder-Wanner UU, Kaiser W. Quality of life in first-admitted schizophrenia patients: a follow-up study. Psychological Medicine, 2000; 30: 225–230.
51. Henkel H, Schmitz M, Berghofer G. [Quality of life of the mentally ill] Wiener medizinische Wochenschrift, 2000; 150:32–36.
52. Huppert JD, Smith TE. Longitudinal analysis of subjective quality of life in schizophrenia: anxiety as the best symptom predictor. Journal of Nervous and Mental Disease, 2001; 189:669–675.
53. Tempier R, Mercier C, Leouffre P, Caron J. Quality of life and social integration of severely mentally ill patients: a longitudinal study. Journal of Psychiatry and Neuroscience, 1997; 22: 249–255.
54. Skantze K. Subjective quality of life and standard of living: a 10-year follow-up of out-patients with schizophrenia. Acta Psychiatrica Scandinavica, 1998; 98:390–399.
55. Kentros MK, Terkelsen K, Hull J, et al.The relationship between personality and quality of life in persons with schizoaffective disorder and schizophrenia. Quality of Life Research, 1997; 6:118–122.
56. Hansson L, Eklund M, Bengtsson-Tops A. The relationship of personality dimensions as measured by the temperament and character inventory and quality of life in individuals with schizophrenia or schizoaffective disorder living in the community. Quality of Life Research, 2001;10:133–139.
57. Eklund M, Backstrom M. A model of subjective quality of life for outpatients with schizophrenia and other psychoses. Quality of Life Research, 2005; 14: 1157–1168.
58. Zubin, J., Spring, S. Vulnerability - a new model of schizophrenia. Journal of Abnormal Psychology, 1977; 88:103–128.
59. Green MF, Nuechterlein KH. Backward masking performance as an indicator of vulnerability to schizophrenia. Acta Psychiatr Scand Suppl., 1999; 395:34–40.
60. Berner P. Conceptualization of vulnerability models for schizophrenia: historical aspects. Am J Med Genet., 2002;114:938–942.
61. Szoke, A., Schurhoff, F., Ferhadian, N., et al. Temperament in schizophrenia: a study of the tridimensional personality questionnaire (TPQ). European Psychiatry, 2002; 17: 379–383.
62. Ormel J, Oldehinkel AJ, Vollebergh W. Vulnerability before, during, and after a major depressive episode: a 3-wave population-based study. Arch Gen Psychiatry, 2004; 61:990–996.
63. Ritsner, M, Susser, E. Temperament types are associated with weak self-construct, elevated distress, and emotion-oriented coping in schizophrenia: evidence for a complex vulnerability marker? Psychiatry Research, 2004; 128:219–228.
64. Jansen LM, Gispen-de Wied CC, Kahn RS. Selective impairments in the stress response in schizophrenic patients. Psychopharmacology (Berl)., 2000; 149:319–325.
65. Myin-Germeys, I, Jim van, OS, Schwartz, JE, Stone, AA, Delespaul, PA. Emotional reactivity to daily life stress in psychosis. Archives of General Psychiatry, 2001; 58:1093–1208.
66. Walker EF, Diforio D. Schizophrenia: a neural diathesis-stress model. Psychological Reviews, 1997; 104: 667–685.
67. Majewska, MD. Neurosteroids: endogenous bimodal modulators of the GABAa receptor. Mechanisms of action and physiological significance. Prog. Neurobiol., 1992; 38: 379–395.
68. Debonnel G, Bergeron R, de Montigny C. Potentiation by dehydroepiandrosterone of the neuronal response to N-methyl-D-aspartate in the CA3 region of the rat dorsal hippocampus: an effect mediated via sigma receptors. J Endocrinol, 1996;150 (Suppl):33–42.
69. Rupprecht R. The neuropsychopharmacological potential of neuroactive steroids. J Psychiatr Res, 1997;31:297–314.
70. Wen S, Dong K, Onolfo JP, et al., Treatment with dehydroepiandrosterone sulfate increases NMDA receptors in hippocampus and cortex. Eur J Pharmacol 2001; 430:373–374.

71. Migues, PV, Johnston, AN, Rose, SP. Dehydroepiandosterone and its sulphate enhance memory retention in day-old chicks. Neuroscience, 2002; 109: 243–251.
72. Wolf OT, Kirschbaum C. Actions of dehydroepiandrosterone and its sulfate in the central nervous system: effects on cognition and emotion in animals and humans. Brain Res Brain Res Rev 1999; 30:264–288
73. Hu Y, Cardounel A, Gursoy E, et al., Anti-stress effects of dehydroepiandrosterone: protection of rats against repeated immobilization stress-induced weight loss, glucocorticoid receptor production, and lipid peroxidation. Biochem Pharmacol., 2000; 59:753–62
74. Boudarene, M, Legros, JJ. Study of the stress response: role of anxiety, cortisol and DHEA-S. Encephale, 2002; 28:139–146.
75. Lapchak, PA, Araujo, DM. Preclinical development of neurosteroids as neuroprotective agents for the treatment of neurodegenerative diseases. Int. Rev. Neurobiol., 2001; 46:379–397.
76. Ritsner M, Maayan R, Gibel A, et al. Elevation of the cortisol/dehydroepiandrosterone ratio in schizophrenia patients. European Neuropsychopharmacology, 2004; 14: 267–273.
77. Ritsner M, Gibel A, Ram E, et al. Alterations in DHEA metabolism in schizophrenia: two-month case-control study. Eur Neuropsychopharmacol., 2006;16:137–146.
78. Johnson, MD, Bebb, RA, Sirrs, SM. Uses of DHEA in aging and other disease states. Ageing Res. Rev. , 2002; 1: 29–41.
79. Ritsner, MS, Gibel, A, Ratner, Y, et al. Improvement of sustained attention, visual and movement skills, but not clinical symptomatology, following dehydroepiandrosterone augmentation in schizophrenia: a randomized double-blind, placebo-controlled, crossover trial. Journal of Clinical Psychopharmacology 2006; 26:495–499.
80. Griebel G, Stemmelin J, Gal CS, Soubrie P. Non-peptide vasopressin V1b receptor antagonists as potential drugs for the treatment of stress-related disorders. Curr Pharm Des, 2005; 11:1549–1559.
81. Insel TR, Young LJ. The neurobiology of attachment. Nat Rev Neurosci, 2001; 2:129–136.
82. Phillips ML, Drevets WC, Rauch SL, Lane R. Neurobiology of emotion perception I: The neural basis of normal emotion perception. Biol Psychiatry, 2003; 54:504–514.
83. Wang J, Rao H, Wetmore GS, Perfusion functional MRI reveals cerebral blood flow pattern under psychological stress. Proc Natl Acad Sci U S A, 2005; 102:17804–17809.
84. Taylor SF, Phan KL, Britton JC, Liberzon I. Neural response to emotional salience in schizophrenia. Neuropsychopharmacology, 2005; 30:984–995.
85. Davidson RJ. Well-being and affective style: neural substrates and biobehavioural correlates. Philos Trans R Soc Lond B Biol Sci., 2004; 359:1395–1411.
86. Sala M, Perez J, Soloff P, et al. Stress and hippocampal abnormalities in psychiatric disorders. Eur Neuropsychopharmacol., 2004; 14:393–405.
87. Herman JP, Ostrander MM, Mueller NK, Figueiredo H. Contributions of the amygdala to emotion processing: from animal models to human behavior. Neuron, 2005; 48:175–187.
88. Phelps EA, LeDoux JE. Dopaminergic contribution to the regulation of emotional perception. Clin Neuropharmacol., 2005; 28:228–237.
89. Schneider F, Weiss U, Kessler C, et al. Differential amygdala activation in schizophrenia during sadness. Schizophr Res., 1998; 34:133–142.
90. Rauch SL, Shin LM, Wright CI. Neuroimaging studies of amygdala function in anxiety disorders. Ann N Y Acad Sci., 2003; 985:389–410.
91. Shayegan DK, Stahl SM. Emotion processing, the amygdala, and outcome in schizophrenia. Prog Neuropsychopharmacol Biol Psychiatry, 2005; 29:840–845.
92. van Rijn S, Aleman A, Swaab H, Kahn RS. Neurobiology of emotion and high risk for schizophrenia: role of the amygdala and the X-chromosome. Neurosci Biobehav Rev., 2005; 29:385–397.
93. Whittle S, Allen NB, Lubman DI, Yucel M. The neurobiological basis of temperament: Towards a better understanding of psychopathology. Neurosci Biobehav Rev., 2005; Nov 8; [Epub ahead of print]
94. Wright CI, Williams D, Feczko E, Neuroanatomical Correlates of Extraversion and Neuroticism. Cereb Cortex, 2006; 16:1809–1819.

CHAPTER 2

ROLE OF DOPAMINE IN PLEASURE, REWARD AND SUBJECTIVE RESPONSES TO DRUGS

The neuropsychopharmacology of quality of life in schizophrenia

LAKSHMI N.P. VORUGANTI[1] AND A. GEORGE AWAD[2]

[1]*McMaster University, Hamilton, ON, Canada*

[2]*Department of Psychiatry and the Institute of Medical Science University of Toronto, Chief of Psychiatry, Humber River Regional Hospital, Canada*

Abstract: This chapter will attempt to accomplish a seemingly impossible task of characterizing the concept of quality of life in terms of pleasure centres, neuronal circuits and chemical mechanisms in the brain. The supporting explanations will be presented in three parts: first, identifying three key characteristics of quality of life – first, the central doctrine of subjectivity, the time frame of appraisal, and the relevance of immediate affective tone in determining quality of life ratings; second, a review of the cumulative knowledge on the neuroanatomical and neurochemical mechanisms underlying subjective responses to pleasurable stimuli such as food and drugs. Thirdly, evidence is presented to support that interference with these brain reward mechanisms leads to feelings of lack of pleasure and poor quality of life, citing antipsychotic drug therapy in schizophrenia as an example. Laboratory experiments in animals and neurochemical imaging studies in humans suggest that dopaminergic mechanisms in nucleus accumbens, amygdala, hippocampus and prefrontal cortex are key players in determining the qualitative and quantitative aspects of subjective responses to pleasurable stimuli. Antipsychotic drug induced dopaminergic blockade in these neuronal circuits leads to persistent feelings of dysphoria and pervasive lack of pleasure, leading to subjective distress and compromised quality of life. The nuances of subjective responses to antipsychotic drugs thus have enormous implications for long term care of the mentally ill people

Keywords: Subjective responses, Dysphoria, Dopamine, Antipsychotic drugs, Quality of life

INTRODUCTION

There are three key characteristics of quality of life concept that are relevant to the discussion on its biological aspects – its subjective nature, the affective dimension and the time frame of reference of quality of life appraisal, which are described below.

M.S. Ritsner and A.G. Awad (eds.), Quality of Life Impairment in Schizophrenia, Mood and Anxiety Disorders, 21–31.

Quality of Life is an Idiosyncratic Subjective Phenomenon

Quality of life has remained an attractive theme of research and discussion among psychologists, sociologists, economists and politicians for over half century[1]. It is often quipped that the concept of quality of life is vague enough that people have been able to mould it to suit their needs and exploited its appeal[2]. An important aspect of this debate relevant to the present chapter is the distinction between the uses of quality of life at a "macro" level as opposed to its connotation at a "micro" level. Examples of the former are found in "social indicators" research, such as standard of living of nation states, crime statistics in cities, or the availability of parks and recreational facilities in communities[3]. Quality of life, in these circumstances, is used interchangeably with terms such as "human development index (HDI)" and "gross domestic product (GDP)". These indices of *quality of collective human living and wellbeing* fall within the realm of sociologists, economists, politicians and policy makers[4].

Quality of life at a *micro* level deals with overlapping concepts such as health, wellbeing, psycho-social adjustment, level of functioning and subjective satisfaction of individual human beings, and has been a field of study by psychometrists and clinicians. For the rest of this chapter quality of life is considered a subjective phenomenon, i.e. quality of life could only be perceived and appraised by an individual, but can not be witnessed and accurately judged by independent observers.

Affect Plays a Crucial Role in Quality of Life Appraisal

"How is your quality of life today?" is a question used frequently to assess global quality of life in clinical settings and epidemiological surveys. While it may not take more than few moments to appraise one's own life, assign it a score, or express it as "good", "bad" or "lousy", there has been considerable speculation, if not an abundance of original research, with regard to what goes through a person's mind, in response to such a question. There is a general consensus that quality of life appraisal includes two elements, "cognitive" and "affective", which are explained below.

It is intuitively believed that quality of life appraisal involves a comprehensive examination of all the relevant aspects of one's life, such as health, wealth, productivity, interpersonal relations, leisure pursuits etc. These individual dimensions are often elaborated as factors or sub-scales in "domain-specific" quality of life evaluations. Whether the appraisal is formal or informal, global or domain-specific, there is an element of mental arithmetic involving an assessment of one's performance in each of the individual aspects of one's life, assigning relative weights to the accomplishments and failures, and arriving at a composite index or a summated expression at the end. This series of psychological steps involving introspection, reflection, appraisal and adjudication constitutes the cognitive element of quality of life formulation[5].

Self appraisal of quality of life, however, is not a mere accounting exercise. The cumulated result of the systematic, rational, deductive "checklist" process is further modified by one's prevailing mood, which is referred to as the "affective" element of the quality of life appraisal. At this stage of the formulation, the logical conclusion from the original review is further embellished, minimized or distorted by the more immediate life circumstances and the emotional reactions consequent to them. Achievements considered as significant by most standards, could be misconstrued, and quality of life is reported as "poor" if the individual's mental status is coloured by depression, hopelessness and worthlessness. Modest accomplishments and suboptimal circumstances of living could be portrayed as glorious living, if the psychological perspective is magnified by elevated mood, optimistic outlook or an inflated self esteem.

There is credible evidence from general as well as clinical populations that both cognitive and affective components contribute significantly towards the ultimate quality of life appraisal process. Self-rated quality of life appraisals are known to be inconsistent and incorrect among people with impaired cognitive functions, such as those affected by mental retardation, head injuries or severe forms of psychosis[6]. These studies indicate that an optimal level of cognitive potential is essential to reliably appraise and report subjective well-being and satisfaction. The relative role of affective tone towards quality of life formulation has been supported by numerous studies in clinical populations involving people with mood disorders[7]. People who are depressed have been shown to consistently under-rate their quality of life; and changes in the validity of their ratings seem to correspond with the improvement in the severity of depression[8,9]. Euphoric mood, resulting from an innate disorder such as mania or an induced state such as drug intoxication, is known to inflate quality of life ratings irrespective of an individual's real life circumstances.

Do the cognitive and affective components have differential impact on quality of life appraisal process, in terms of their relative importance? Though the hierarchical order of factors determining the quality of life formulation has not been formally established, the abundance of literature points towards the value of prevailing mood state as an overwhelmingly important determinant of quality of life appraisal[10]. Considering the importance of the affective modulation of quality of life appraisals, it is essential to explore the neurobiological aspects of pleasure and altered mood states and discuss its relevance to quality of life appraisal in clinical settings.

Quality of Life Appraisals are Influenced by Immediate Life Circumstances

Feeling states are often distinguished as "affect" and "mood" with reference to the state of the self in relation to its environment. Affect is defined as transient, specific feeling directed towards objects, while mood is a more prolonged, prevailing state or disposition. Individual's responses to administered drugs, similarly, are categorized as subjective responses and subjective tolerability[11]. Subjective responses refer to the changes in the feeling state of an individual soon after the administration of a single dose of a drug, and the effects are directly attributable to the drug's pharmacokinetic and pharmacodynamic actions. Subjective tolerability, on the other hand, is the global appraisal of a drug, based on the pattern of cumulative subjective responses after ingesting several doses of a drug over a period of time. Besides the pharmacological effect of a drug, other factors such as personality attributes, cultural attitudes and health beliefs also play a role in its evolution. Subjective tolerability has a lasting impact on individual's behaviour with regard to the treatment adherence, and quality of life.

What is the time frame of quality of life appraisal while a subject is undertaking the cognitive task of evaluating one's own performance and rating the ensuing satisfaction – a week, a month, a year or an entire life span? Though the customary question aimed at eliciting quality of life appraisal is initiated with a preamble such as "Taking everything into account, ...", it has been shown that quality of life appraisals are often based on a shorter time frame and are significantly influenced by immediate life circumstances [12].

In summary, quality of life is conceptualized as an idiosyncratic subjective phenomenon, largely determined by the events occurring within a short time frame and the prevailing mood at the time of appraisal. Such a premise is essential for exploring the neurobiological aspects of subjective psychological states and their relevance to quality of life appraisal.

NEUROBIOLOGICAL BASIS OF SUBJECTIVE RESPONSES

Pleasure responsivity, also known as salutogenesis, refers to the experience of joy in response to hedonic stimuli. The following is a summary of neurobiological aspects of pleasure responsivity that is relevant to the discussion on psychotropic drugs and quality of life. The summary includes an overview of the neurochemical, neuro-anatomical and molecular events involved in the mediation of subjective responses, and their alteration in health and disease.

Dopamine in the Mediation of Pleasure

The putative role of dopamine in the mediation of pleasurable experiences has been studied for over half century, and data supporting this speculation is derived from laboratory research as well as clinical observations [13]. The basic science studies on the role of dopamine used strategies such as electrical stimulation of dopamine-rich areas in the brain with micro-electrodes, examining the consequences of chemically or mechanically induced brain lesions, measuring extra-cellular dopamine concentration using in-vivo micro-dialysis techniques during drug self-administration experiments, and studying the behavioural alterations in genetically altered mice that are lacking in dopamine receptor (DR) or dopamine transporter (DAT) genes. Cumulated data from these studies confirmed that dopamine has a central, if not an exclusive, role in the mediation of pleasure responsivity [14]. The original research in laboratory animals is now confirmed by studies involving human subjects in clinical settings as well as in the laboratory. Human studies on dopamine function included studies on natural alterations in pleasure responsivity in clinical conditions involving increased (e.g. cocaine intoxication and levodopa therapy) or lowered (e.g. Parkinson's disease and dementia) dopamine levels in the brain, drug self-administration experiments, controlled chemical probing trials (with amphetamine or alphamethylparatyrosine) and in-vivo brain imaging studies [11]. These studies not only confirmed the crucial role of dopamine release in the mediation of pleasure, but also identified the specific anatomical regions in the brain and the neural circuits.

Dopaminergic Circuits Underlying Subjective Responses and Tolerability

It is now widely accepted that the mesolimbic and the mesocortical dopamine systems are the neuroanatomical substrates associated with the varied subjective responses to food and drugs, and much of the relevant knowledge and the subsequent speculations has been largely derived from studies in drug addiction. There is evidence to indicate that interfering with these mechanisms with drugs or disease could impair pleasure responsivity, induce negative subjective responses and leads to drug refusal in clinical settings.

The mesocortical dopaminergic circuit includes the prefrontal cortex, orbito-frontal cortex and anterior cingulate, and is implicated in the conscious experience of drug induced "high", drug incentive salience, drug expectation/craving, and repetitive drug administration. The mesolimbic dopamine circuit, which includes the nucleus accumbens, amygdala and hippocampus, is involved in remembering the acute reinforcing effects of a drug, crucial for establishing the drug-related cues and maintaining craving [15].

The mesolimbic and mesocortical circuits operate in tandem, and this interaction influences the response to a given drug and subsequent drug taking behaviour [16]. Typically, drug related cues activate the memory circuits (the hippocampus and amygdala) which in turn activates the orbitofrontal cortex and anterior cingulate creating an expectant situation, which in turn activates the dopamine cells, leading to a further increase in the craving sensation and a possible decrease in inhibitory control.

The activation of these circuits in succession perpetuates the cycle of addiction [17,18]. The attribution of salience to a given stimulus, which is a function of the orbitofrontal cortex, depends on the relative value of a reinforcer compared to simultaneously available reinforcers, which require knowledge of the strength of the stimulus as a reinforcer, a function of the hippocampus and amygdala. Consumption of the drug in turn will further activate cortical circuits (the orbitofrontal cortex and anterior cingulate) in proportion to the dopamine stimulation by favoring the target response and decreasing non-target-related background activity. The activation of these interacting circuits may be indispensable for maintaining the compulsive drug administration observed during bingeing and to the vicious circle of addiction.

Synaptic and Cellular Events Underlying the Mediation of Pleasure

It is now widely accepted that increased dopaminergic transmission in the mesolimbic system underlies the experience of pleasurable feelings, and impaired dopaminergic activity gives rise to lack of pleasure, anhedonia and dysphoria. Pleasurable events (e.g. food) and chemical stimuli (e.g. amphetamine or cocaine) lead to an increased release or a decreased reuptake of intrasynaptic dopamine, making dopamine available in larger amounts or for an increased amount of time within the synaptic cleft[19]. The binding of synaptic dopamine with the post-synaptic dopamine receptors triggers a series of intra-cellular changes; and the cascade of events eventually manifest as various subjective and behavioural changes. Drugs that exert their actions through binding with either G protein–coupled (e.g. cannabinoids, opiates and caffeine) or ligand-gated ion channel receptors (e.g. alcohol, phencyclidine and nicotine) also indirectly enhance dopaminergic activity through the interconnections between their respective receptors and the dopaminergic neurons.

How does increased dopaminergic activity manifest itself as pleasure and satisfaction? There has been considerable speculation and ongoing research with regard to deciphering the phenomenological aspects of dopamine-led "highs" and "lows". Earlier "reward" literature that was based exclusively on animal experiments theorized the mesolimbic system as the "pleasure pathway" and the dopamine as the critical mediator of "reward". Recent revisions of this hypothesis speculate that the nature of subjective experience is far more complex and still elusive. Some have suggested that the mesolimbic system does not necessarily mediate the pleasurable experience of the reward itself but determines *incentive salience*, or the state of *wanting* of a prospective reward. Increased dopaminergic activity in this circuit creates a motivational state of wanting (or an *expectation* of pleasure) but does not mediate the hedonic effects, or the affective state ("liking") associated with the reward. An alternative perspective is that increased mesolimbic dopaminergic transmission mediates *motivational learning* but not salience[20]. An integrated hypothesis proposes that mesolimbic system in general, and nucleus accumbens (NAc) in particular, acts as a sensorimotor integrator that is involved in higher order motor and sensorimotor processes underlying behavioural components such as *motivation, response allocation, and responsiveness to conditioned stimuli*[21].

Role of Dopamine in the Mediation of Subjective Responses to Drugs

Manipulating dopaminergic activity in clinical settings leads to corresponding changes in subjective mental status, which in turn has an impact on individuals' quality of life. It is now generally agreed that alterations in dopamine function can not be reduced to a mere quantitative "up and down" paradigm, but need to take into consideration the qualitative aspects such as the degree of affinity between dopamine and its receptors, site of its action (cortical or subcortical), and its modulation by other neurotransmitters such as serotonin and acetyl choline[22]. Consequently, the nature and magnitude of subjective responses to drugs may vary widely. Two types of clinical subjective responses, *euphoria* and *dysphoria*, are chosen to illustrate the effects of altered dopaminergic activity. It has been clarified recently that the commonly used expressions "euphoria" and "dysphoria" do not necessarily represent alterations in affect alone but encompass impairments in multiple psychological dimensions such as arousal, attention, concentration, motivation and mood, to varying degrees[23].

Much of the knowledge on the positive (euphoric) subjective responses to drugs has been derived from the drug addiction field. Two classical examples of research in this area include experimental administration of intravenous amphetamine or cocaine in controlled laboratory settings, and measuring the consequent changes in dopaminergic activity through real time in-vivo receptor binding studies using positron emission tomography (PET) or single photon emission computed tomography (SPECT) imaging techniques[11]. Simultaneous recording of subjective responses using self-rated standardized rating scales revealed a correlation between euphoric responses and increased availability of synaptic dopamine[24]. There is also some indication that the quantity of dopamine and quality of its binding determine the nature of subjective responses. While moderate increases in dopamine transmission leads to a corresponding elevation in mood, excess elevations could trigger positive psychotic symptoms[25].

Knowledge on negative subjective (dysphoric) responses has been largely derived from the study of disease states, especially schizophrenia and Parkinson's disease. While studies on increased dopaminergic transmission and the corresponding changes in subjective mental status in schizophrenia remain inconsistent due to a number of practical, ethical and methodological limitations, more has been learnt from the study of the effects of antipsychotic drugs used in its treatment. Antipsychotic drug therapy over the past 50 years has been based on the principle of blocking post-synaptic dopamine receptors, without the ability to sensitively titrate the site of action or the degree of blockade. Such an indiscriminate and widespread blockade of dopaminergic activity has been known to cause a variety of side effects including negative subjective responses (dysphoria). Recent imaging techniques have made it possible to establish a correlation between the extent of dopaminergic blockade and the severity of negative subjective responses[26]. The introduction of second generation antipsychotic drugs with modified dopamine blocking action has thus minimized the risk of dysphoria[27]. The association between dysphoric responses and the relative lack of dopamine is also demonstrated by alpamethyl paratyrosine (AMPT) induced dopamine depletion studies, and the occurrence of dysphoric mood states in Parkinson's disease[28].

In summary, clinical observations and neuroimaging studies support the speculation that enhanced dopaminergic transmission is associated with euphoria, and dopamine deficient states induced by drugs or disease states result in dysphoric responses. Based on the cumulated evidence from animal and human studies, it is also possible to conclude that dopamine plays a critical, if not exclusive, role in the experience of pleasure in health and disease.

SUBJECTIVE RESPONSES TO DRUGS AND QUALITY OF LIFE IN SCHIZOPHRENIA

The implications of these neurochemical mechanisms are indeed directly relevant individuals' sense of satisfaction and quality of life. The following is an overview of two common clinical scenarios, the vicissitudes of antipsychotic drug therapy in schizophrenia, and the emergence of comorbid substance abuse, which illustrate the link between altered dopaminergic function, the subjective responses, and their impact on the sense of well-being and quality of life.

Antipsychotic Drug Therapy and Quality of Life

In a majority of individuals afflicted by schizophrenia, the illness runs a chronic course, and the treatment requires a long term commitment to antipsychotic drug use. Antipsychotic drugs, by virtue of their dopaminergic blocking action, induce dysphoric responses in about 40% of the individuals; and the neuroleptic induced dysphoria has been known to impair motivation, concentration, arousal and a persistent inability to derive pleasure from routine life experiences over a prolonged period of time[22]. The resulting apathy and a motivational syndrome has a direct impact on daily routine and quality of life[29,30]. Besides, the dysphoric state also colours the subjective appraisal process and overshadows the modest benefits of antipsychotic drug therapy, resulting in a skewed perspective, and a perception of compromised quality of life[31]. The negative subjective responses to drug treatment, thus lead to a cascade of events including drug refusal, non-adherence to treatment, clinical instability, increased risk of suicidality, substance abuse and other psychosocial consequences which are discussed in greater detail in chapter 16. The problem of dysphoria and its sequelae are more often associated with the first generation antipsychotic drugs (neuroleptics), and the risk is considered significantly less with the second generation (atypical) antipsychotics. It has also been observed that a significant minority of people with schizophrenia could experience spontaneous dysphoric episodes, irrespective of their treatment status, presumably linked to the inherent instability of their dopaminergic system[32].

Schizophrenia, Substance Abuse and Quality of Life

One of the specific consequences of neuroleptic induced dysphoria is the development of comorbid substance abuse during the course of antipsychotic drug therapy of schizophrenia, which illustrates the imbalance and counter-balancing mechanisms of the dopaminergic system on one hand, and underscores the cumulative burden of the dual disorder on quality of life. There is an emerging picture from a series of studies that the subgroup of patients who experience neuroleptic induced dysphoria have an inherently vulnerable dopaminergic system with a low basal dopamine tone. Drug-induced dopaminergic blockade in such individuals leads to a quick impairment in dopaminergic activity and perpetuate the dysphoric feelings.

It has been further noted that it is the same subgroup of dysphoric responders that are also likely to resort to self medication strategy, using illicit drugs to relieve their dysphoric feelings. Not surprisingly, the drugs of choice among people with schizophrenia tend to be of the stimulant type whose effects are mediated through enhanced dopaminergic transmission[33,34]. The cumulative effects of this dual disorder are far more damaging, leading to a further impairment in functioning and quality of life[30].

CONCLUSION

There is increasing evidence that affective state not only contributes significantly towards the genesis of compromised quality of life in persons suffering from schizophrenia, but also may impact on the person's ability to correctly appraise their level of satisfaction and quality of life particularly among those who are on antipsychotic medications.

If enhancing quality of life of people with schizophrenia is a dream, neuroleptic dysphoria and its consequences represent a recurring nightmare attributable to antipsychotic drugs. There is convincing evidence to suggest that the molecule that can turn the dream into the nightmare is indeed dopamine, or specifically the lack of it! Antipsychotic drug development over the past 50 years has been centered on the principle of post-synaptic dopaminergic blockade, which is now recognized as responsible for perpetuating the problem of neuroleptic dysphoria. The arrival of a second generation of antipsychotic drugs with a dual dopamine and serotonin blocking action has significantly minimized the problem of drug-induced dysphoria. There is growing recognition now that dopaminergic system is inherently unstable in schizophrenia, and a proportion of the affected individuals seem to experience frequent dopamine "lows" that makes them susceptible to substance abuse. These insights highlight the need for developing novel pharmacological strategies such as dopaminergic stabilizers, which should help to overcome the limitations of current antipsychotic drugs and improve the quality of life of people affected by schizophrenia. Finally, it is imperative for researchers and clinicians to pay particular attention and note the affective state of the person at the time of their self-appraisal of their quality of life.

REFERENCES

1. Hornquist JO. The concept of quality of life. Scandinavian Journal of Social Medicine. 1982;10:57–61.
2. Cohen C. On the quality of life: some philosophical reflections. Circulation 1982;66(suppl.111): s29–s33.
3. Campbell A, Converse P, Rogers W. The quality of American life. New York: Russel Sage Foundation, 1976.
4. Diener E. Subjective Well-Being. Psychological Bulletin 1984;95(3):542–575.
5. George L, Bearon I. Quality of life in older persons: meaning and measurement. New York: Human Sciences Press, 1980.

6. Voruganti L, Heslegrave H, Awad AG & Seeman MV. Quality of life measurement in Schizophrenia – Reconciling the quest for subjectivity with the question of reliability. Psychological Medicine 1998;28:165–172.
7. Awad AG, Voruganti LN, Heslegrave RJ. A conceptual model of quality of life in schizophrenia: description and preliminary clinical validation. Qual Life Res. 1997; 6(1):21–6.
8. Corten P, Mercier C, Pelc I. "Subjective quality of life": clinical model for assessment of rehabilitation treatment in psychiatry. Social Psychiatry and Psychiatric Epidemiology 1994; 29:178-183.
9. Goethe JW & Fisher EH. Functional impairment in depressed patients. Journal of Affective Disorders, 1995;33: 23–29.
10. Wells KB, Stewart A, Hays RD et al. The functioning and well-being of depressed patients: results from the medical outcomes study. JAMA 1989;262:914–919.
11. Voruganti LP, Awad AG. Brain imaging research on subjective responses to psychotropic drugs. Acta Psychiatr Scand Suppl. 2005;(427):22–8.
12. Allison PJ, Locker D, Feine JS. Quality of life: a dynamic construct. Social science and medicine 1997;45:221–230.
13. Adinoff B. Neurobiologic processes in drug reward and addiction. Harv Rev Psychiatry 2004; 12(6):305–20.
14. Wise RA. Neurobiology of addiction. Curr Opin Neurobiol 1996;6:243–51.
15. James GA, Gold MS, Liu Y. Interaction of satiety and reward response to food stimulation. J Addict Dis. 2004;23(3):23–37.
16. Kalivas PW, Volkow ND. The neural basis of addiction: a pathology of motivation and choice. Am J Psychiatry. 2005;162(8):1403–13.
17. Koob GF, Ahmed SH, Boutrel B, Chen SA, Kenny PJ, Markou A, O'Dell LE, Parsons LH, Sanna PP. Neurobiological mechanisms in the transition from drug use to drug dependence. Neurosci Biobehav Rev. 2004;27(8):739–49.
18. Risinger RC, Salmeron BJ, Ross TJ, Amen SL, Sanfilipo M, Hoffmann RG, Bloom AS, Garavan H, Stein EA. Neural correlates of high and craving during cocaine self-administration using BOLD fMRI. Neuroimage. 2005;26(4):1097–108.
19. Volkow ND, Fowler JS, Wang GJ. The addicted human brain viewed in the light of imaging studies: brain circuits and treatment strategies. Neuropharmacology. 2004;47 Suppl 1:3–13.
20. Di Chiara G, Bassareo V, Fenu S, De Luca MA, Spina L, Cadoni C, Acquas E, Carboni E, Valentini V, Lecca D. Dopamine and drug addiction: the nucleus accumbens shell connection. Neuropharmacology. 2004;47 Suppl 1:227–41.
21. Salamone JD, Correa M, Mingote SM, Weber SM. Beyond the reward hypothesis: alternative functions of nucleus accumbens dopamine. Curr Opin Pharmacol. 2005;5(1):34–41.
22. Voruganti L, Awad AG. Neuroleptic dysphoria: towards a new synthesis. Psychopharmacology (Berl). 2004;171(2):121–32.
23. Voruganti LN, Awad AG. Subjective and behavioural consequences of striatal dopamine depletion in schizophrenia – Findings from an in vivo SPECT study. Schizophr Res. 2006 Sep 1; [Epub ahead of print].
24. Tsibulsky VL, Norman AB. Real time computation of in vivo drug levels during drug self-administration experiments. Brain Res Brain Res Protoc. 2005;15(1):38–45.
25. Laruelle M, Abi-Dargham A, van Dyck CH, Rosenblatt W, Zea-Ponce Y, Zoghbi SS, Baldwin RM, Charney DS, Hoffer PB, Kung HF, et al. SPECT imaging of striatal dopamine release after amphetamine challenge. J Nucl Med. 1995;36(7):1182–90.
26. de Haan L, van Bruggen M, Lavalaye J, Booij J, Dingemans PM, Linszen D. Subjective experience and D2 receptor occupancy in patients with recent-onset schizophrenia treated with low-dose olanzapine or haloperidol: a randomized, double-blind study. Am J Psychiatry. 2003;160(2):303–9.
27. Voruganti L, Cortese L, Owyeumi L, Kotteda V, Cernovsky Z, Zirul S, Awad A. Switching from conventional to novel antipsychotic drugs: results of a prospective naturalistic study. Schizophr Res. 2002;57(2–3):201–8.

28. Voruganti L, Slomka P, Zabel P, Costa G, So A, Mattar A, Awad AG. Subjective effects of AMPT-induced dopamine depletion in schizophrenia: correlation between dysphoric responses and striatal D(2) binding ratios on SPECT imaging.Neuropsychopharmacology. 2001;25:642–50.
29. Koivumaa-Honkanen H-T, Viinamaki H, Honkanen R, Tanskanen A, Antikainen R, Niskanen L, Jaaskelainen J, Lehtonen J, Correlates of Life satisfaction among psychiatric patients. Acta Psychiatrica Scandinavica. 1996;94:372–378.
30. Awad AG & Voruganti LP. Neuroleptic dysphoria, comorbid drug abuse in schizophrenia and the emerging science of subjective tolerability: towards a new synthesis. J Dual Diagnosis 2005;1(2):83–94.
31. Lehman A. The well-being of chronic mental patients: assessing their quality of life. Archives of General Psychiatry 1983;40:369–373.
32. Grace AA. The tonic/phasic model of dopamine system regulation and its implications for understanding alcohol and psychostimulant craving. Addiction. 2000;95 Suppl 2:S119–28.
33. Voruganti LN, Heslegrave RJ, Awad AG. Neuroleptic dysphoria may be the missing link between schizophrenia and substance abuse. J Nerv Ment Dis. 1997;185(7):463–5.
34. Voruganti LN, Slomka P, Zabel P, Mattar A, Awad AG. Cannabis induced dopamine release: an in-vivo SPECT study. Psychiatry Res. 2001;107(3):173–7.

CHAPTER 3

NEUROENDOCRINE FUNCTIONS, MOOD AND QUALITY OF LIFE

MARIANNA MAZZA[1] AND SALVATORE MAZZA[2]

[1]*Institute of Psychiatry, Catholic University of Sacred Heart, Rome, Italy*
[2]*Department of Neurosciences, Catholic University of Sacred Heart, Rome, Italy*

Abstract: The neuroendocrine system and the distribution of hormones through the brain and their modulatory role and influence in behaviour and mood have been studied for a long time. The most evident changes associated with mood disorders are in the hypothalamic-pituitary-adrenal (HPA), the hypothalamic-pituitary-thyroid (HPT), the hypothalamic-pituitary-GH (HPGH) and the hypothalamic-pituitary-gonadal (HPGn) axes. Hypotheses referring to the psychophysiological meaning and the development of these alterations are discussed. In order to improve the quality of life for patients suffering from mood disorders, it would be important to define which specific hormonal axes contribute to mood symptoms and which medications that normalize neuroendocrine function are conditioning the impact of mood symptoms. The identification and detailed characterization of these pathways will ultimately lead to the development of novel neuropharmacological intervention strategies. Future directions for research are described.

All data derived from studies focusing on central neuropeptidergic circuits and peripheral hormone systems add to the understanding of the pathophysiology of mood disorders and indicate the importance of investigating neuroendocrine dysfunctions in psychiatric patients both to ensure proper diagnosis and adequate pharmacotherapies

Keywords: Neuroendocrinology, Neuroendocrine System, Mood, Depression, Bipolar Disorder, Quality of life

MOOD DISORDERS AND QUALITY OF LIFE

Mood symptoms should be considered when assessing quality of life in severe mental disorder patients, since affective disturbances can seriously distort subjective appraisals of quality of life [1]. Although health-related quality of life is an amorphous and heterogeneous concept, it has become an important outcome measure in the treatment of psychiatric disorders. In particular, evaluation of quality of life is essential to the study of pharmacoeconomics of mental disorders [2]. One common element in definitions of quality of life is the individuals subjective sense of wellbeing and it is often postulated that quality of life and subjective wellbeing are determined by various circumstances or dimensions of physical, social and

M.S. Ritsner and A.G. Awad (eds.), Quality of Life Impairment in Schizophrenia, Mood and Anxiety Disorders, 33–56.

role functioning, mental and physical health perceptions. Quality of life should not only be defined by one's subjective sense of wellbeing but also by the ability to function in various life domains and by the ability to access resources and opportunities[2]. Awad[3] proposed a basic clinical quality of life model in which life satisfaction is viewed as the subject's perception of the outcome of an interaction between the severity of psychotic symptoms, side effects of medication and the level of psychosocial performance. In terms of the distress/protection model, resultant quality of life is an outcome of the interaction between distress and protective factors.

Despite similarities in levels of quality of life among patients belonging to different diagnostic categories, there are considerable between-group differences among the predictors of life satisfaction. In a naturalistic follow-up study, Ritsner et al.[4] observed that individual fluctuations in quality of life index scores among patients with schizoaffective/mood disorders were associated with changes in depression severity, sensitivity, task-oriented coping, and expressed emotion scores. Recently, Ritsner[5] demonstrated in a population of schizophrenia patients that changes in stress process-related factors are stronger predictors of fluctuations in quality of life domains than changes in illness symptoms and should be considered when evaluating life satisfaction. These findings have contrasted prior models that explained quality of life for mentally ill persons in terms of clinical factors alone and support those focusing on the complex interplay among symptoms, distress, and personal resources in quality of life appraisal[6,7].

Many studies have investigated quality of life in various psychiatric disorders. Patients with anxiety disorders and depressed patients generally reported greater dissatisfaction with quality of life than schizophrenia patients. Depression can be devastating to all aspects of everyday life, including family relationships, friendships and the ability to work or attend school. Subjective quality of life scores are inversely associated with severity of depressive symptoms among patients with major depression and dysthymia. Patients with bipolar disorder have substantial impairment in quality of life in comparison with the general population. Quality of life is compromised in bipolar patients even during periods of euthymia, when quality of life may be predicted by current subthreshold depressive symptoms.

NEUROENDOCRINE SYSTEM AND MOOD DISORDERS

The neuroendocrine system, which plays an important role in regulation of mood, is dysfunctional in patients suffering from mood disorders. In functional studies mood disorders are characterized by abnormalities of hypothalamic-pituitary-adrenal (HPA) function as reflected primarily in a blunted response to the dexamethasone suppression test (DST), blunted release of thyroid-stimulating hormone (TSH) in response to administration of exogenous thyrotropin-releasing hormone (TRH) and blunted prolactin (PRL) and growth hormone (GH) responses[8]. Some of these abnormalities may be attributed to central serotoninergic dysfunction. In order to improve the quality of life for patients, additional research is needed to define clinical implications of neuroendocrine dysfunction in mood disorders[9]. It would be important to define which specific hormonal responses that are blunted in affective disorders contribute to mood symptoms and which medications that normalize neuroendocrine function are conditioning the impact of mood symptoms[10].

The relationship between neuroendocrine dysfunction and affective disorders is best established for unipolar major depression but is becoming increasingly well defined for bipolar disorder.

ENDOCRINE SYSTEM AND MAJOR DEPRESSION

Many data suggest a high incidence of depression in hypothyroidism and an association between hyperthyroidism and both mania and depression. Correction of the primary metabolic defect often results in improvement of the associated affective syndromes. Some patients with primary depression may have an increased response of TSH or TRH suggesting possible thyroid hypofunction in a small segment of patients[11]. Blunting of the nocturnal increase of TSH and greater blunting of the TSH response to TRH in the evening compared with the morning have also been reported[12,107]. All these data, along with evidence of increased cerebrospinal fluid TRH[13] suggest a relative overactivity in the hypothalamic-pituitary-thyroid axis in some depressed patients. In many instances, affective disorders have been seen to improve following treatment with replacement doses of T3 or high dose of T4[14].

Hypothalamic-Pituitary-Adrenal Axis (HPA axis)

Several of the hypothalamic-pituitary-adrenal system dysregulation hypotheses of major depression suggest that: 1) basal output of a dysregulated system is more erratic than normal; 2) normal circadian rhythmicities are altered or disrupted; 3) peripheral manifestations of central dysregulation will normalize in remission of depression[15]. These criteria are consistent with the phenomenology and biological bases of major depression: neurovegetative symptoms, such as altered appetite, sleep architecture and activity patterns, and the characteristic diurnal variability in symptom severity may reflect disturbances in the normal rhythmic variation of neurotransmitters and neuroendocrine concentrations generated by underlying regulatory mechanisms. Partially disturbed relationships among these neurotransmitters, neuroendocrine factors and autonomic measures have been received more and more attention. In particular some studies suggest how, consistent with dysregulation hypotheses, certain central regulatory mechanisms may persist as vulnerabilities to relapse or recurrence of acute depression, while other systems that reflect more peripheral functions tend to normalize in remitted patients[16,17].

The temporal dynamics of HPA responses to stressors typically consist of three phases: 1) basal activity, which reflects unstimulated, non-stressed HPA activity; 2) a 'stress reactivity' phase in which cortisol increases from baseline levels following the onset of a stressor; 3) a 'stress recovery' phase in which cortisol levels return to baseline levels following the offset of the stressor. Each of these phases reflects different physiological processes, with mineralocorticoid receptors (MCRs) regulating cortisol levels during periods of low HPA activity (e.g. evening), and glucocorticoids receptors (GCRs) regulating cortisol responses to stress and cortisol levels during periods of high HPA activity (e.g. morning)[18].

The dexamethasone suppression test (DST), the TRH stimulation test and the stimulation tests of the GH have been widely applied in neuroendocrine studies of depression. When 1–2 mg of dexamethasone (DEX) are administered at 23.00 h to normal subjects, plasma cortisol level remains suppressed during the following day. However, a high percentage of patients with affective disorders do not exhibit the suppressive effect of DEX because of impaired glucocorticoid receptor signalling. This impairment can be primary (genetic) or secondary, due to an increased secretion of ACTH and cortisol which results in glucocorticoid receptor desensitization[19]. Nonsuppression in DST is found most frequently in endogenous depression, is highly state dependent, and is usually interpreted as a reduced efficacy of the feedback loop mediated by pituitary glucocorticoid receptors.

Although the pharmacological literature on HPA activity in depressed patients has yielded important findings for the field of psychoneuroendocrinology, it is not without limitations. Dexamethasone is a synthetic glucocorticoid, and the levels of dexamethasone used in pharmacological challenge studies are specifically designed to mimic the highest extreme of glucocorticoid functioning in order to suppress subsequent endogenous cortisol release. Such high levels of glucocorticoids may not accurately reflect the magnitude of endogenous HPA responses to psychosocial stressors. Further, dexamethasone specifically probes glucocorticoid but not mineralocorticoid receptor function, and poorly crosses the blood-brain-barrier. Finally, in contrast to psychological stressors, many pharmacological and neuroendocrine challenge tests ignore suprahypothalamic (e.g. limbic) input[18]. Not all patients affected by major depressive disorder are cortisol non-suppressors, suggesting that there may exist significant variability in HPA response patterns amongst depressed patients. Potential sources of variation include individual characteristics such as age and gender, depression characteristics such as subtype, severity, hospitalization status, early life stress or Post-Traumatic Stress Disorder (PTSD) comorbidity, stressor characteristics such as type of stressor and duration. Finally, the effect of depression on HPA responses to stress may depend on the time of day or HPA phase examined (e.g. basal activity, stress reactivity, stress recovery)[20]. For example, in a recent study Burke et al.[18] showed how studies conducted in the afternoon were more likely to find higher baseline cortisol levels and greater stress reactivity in clinically depressed patients than non-depressed individuals than studies conducted in the morning. Results also revealed that the blunted cortisol stress reactivity observed in depressed patients was most pronounced in older and/or more severely depressed patients. Similarly, poor stress recovery was most pronounced in more severely depressed and/or hospitalized depressed patients, as well as depressed patients with comorbid PTSD.

Recently, the combined dexamethasone/corticotrophin-releasing hormone (CRH) test or DEX/CRH test has been reported to be more sensitive than DST[21]. In this test, patients are pre-treated with 1.5 mg of DEX at 23.00 h and then given 100 μg of CRH i.v. at 15.00 h the following day.

CRF (corticotropin-releasing factor) is a 41-amino acid peptide that stimulates adenohypophysial production and release of adrenocorticotropic hormone (ACTH) and is the principal mediator in the activation of the HPA axis response to stress. Following a physical or psychological threat, CRF is released from nerve terminals in the parvicellular portion of the paraventicular nucleus of the hypothalamus and transported humorally to the hypothalamus-hypophysial portal vessels to activate CRF receptors on the anterior pituitary, resulting in release of ACTH into the general circulation. In response to systemic ACTH, the adrenal glands increase the synthesis and release of cortisol, which feeds back to the hypothalamus, hippocampus, and anterior pituitary to inhibit ACTH release and to suppress further release of CRF. Depression is associated with increased concentrations of CRF in cerebrospinal fluid, increased CRF immunoreactivity and CRF mRNA expression in the hypothalamic paraventicular nucleus, and down-regulation of CRF-R1 receptor in the cerebral cortex. Neural circuits containing CRF have been identified as an important mediator of the stress response. Early-life adversity, such as physical or sexual abuse during childhood, results in long-lasting changes in the CRF-mediated stress response and a greatly increased risk of depression in genetically predisposed persons[22].

Rates of impaired glucocorticoid responsiveness during major depression (as assessed by nonsuppression of cortisol on the DST or DEX/CRH test) vary from approximately 25% to 80%, depending on depressive symptoms (the highest rates are found for melancholic, or endogenous, subtypes), age (older subjects are more likely to exhibit nonsuppression), and the technique used for assessment (the DEX/CRH test is more sensitive than DST)[23]. Both the DES and the DEX/CRH test have been shown to powerfully predict clinical response. In the case of the DEX/CRH test, there is evidence that impaired glucocorticoid responsiveness represents a genetically based risk factor for the development of depression[24]. Complementing in vivo findings, results of in vitro studies have demonstrated that peripheral immune cells from patients with major depression exhibit decreased sensitivity to well-known immunosuppressive effects of glucocorticoids. However, we know of no data demonstrating that the structural integrity of the glucocorticoid receptor itself is altered in depressed patients[21]. Nevertheless, several lines of evidence suggest that alterations in neurotransmitter-linked signal transduction pathways that regulate glucocorticoid receptor function may contribute to diminished glucocorticoid receptor signaling in major depression. Indeed, altered glucocorticoid receptor signaling has been proposed as a major factor in the pathogenesis of the disorder[25].

Melancholic depression seems to represent an exaggerated and prolonged form of the hyperarousal seen with stress system activation, and patients with this condition show behavioural patterns that are similar in ways to rats treated with CRH[26]. In these patients, cortisol excretion is increased, and the ACTH response to endogenous CRH is decreased[27].

Successful antidepressant treatment is associated with normalization of altered glucocorticoid-mediated inhibitory feedback in patients with major depression, as assessed by either the DST or DEX/CRH test.

Similarly, long term antidepressant treatment for patients with major depression restores appropriate glucocorticoid inhibitory control of immune cell function, as assessed in vitro[25]. A number of animal studies have demonstrated that long-term antidepressant administration increases both the number and functional capacity of corticosteroid receptors in brain regions known to be of key importance in HPA axis regulation, including the hippocampus and hypothalamus. This effect has been observed most consistently with tricyclic antidepressants[21]. Antidepressants in fact are capable of translocating the glucocorticoid receptor from cytoplasm to nucleus, even in the absence of glucocorticoids. These drugs also enhance dexamethasone-induced gene transcription mediated by the glucocorticoid receptor and antidepressant effects may be related to the impact of these drugs on second messenger pathways involved in glucocorticoid receptor regulation[28].

On the basis of theories that excessive glucocorticoid activity plays an integral role in the pathophysiology of major depression, several clinical studies using glucocorticoid agonists and antagonists[29,30] and CRH receptors antagonists[31] are under way.

In a study by Yuuki et al.[32] seven depressed patients who failed trials of antidepressant medication and showed high cortisol response in the DEX/CRH test showed a remission after electroconvulsive therapy (ECT) that was accompanied by resolution of HPA dysregulation. Cortisol response in the DEX/CRH test was normalized after a successful course of ECT. However, measures of cerebral brain metabolism did not resolve.

Pfenning et al.[33] recently reported surprising results: in a retrospective analysis of 310 depressed patients receiving DEX/CRH challenge they observed various aspects of suicide (ideation, past attempt, recent attempt) and found a less activation of the HPA axis in suicidal patients. This finding contradicts the post-mortem studies of completed suicides, which have shown increased numbers of CRH neurons in the paraventicular nucleus of the hypothalamus, increased vasopressin expression in neurons of the paraventricular nucleus, increased CRH in the cerebrospinal fluid, decreased CRH receptors in the frontal cortex, increased mRNA levels of the adrenocorticotropic prohormone in the pituitary, and even adrenal hypertrophy reflective of increased adrenocorticotropic hormone secretion[34]. Thus the post-mortem studies point to excessive activation of the HPA axis. This paradox of a less active HPA axis in subjects with suicide attempts and a more active axis in brains of suicide victims has to be clarified.

There is no direct evidence indicating that cortisol levels influence the development and functioning of the brain in children, but several studies are suggestive[35]. Studies using single blood sampling have failed to show consistent differences in cortisol levels between children with depression and controls[36]. Subsequent study designs using multiple non-invasive salivary sampling have provided evidence that evening cortisol hypersecretion was found in about 24% of school-aged children and adolescents with current major depressive disorder[37]. This was specifically associated with comorbid dysthymia, indicating an association between chronic depressive syndromes and cortisol dynamics. Patients with a diagnosis of major depression without comorbid dysthymia presented low levels of dehydroepiandrosterone (DHEA), a steroid hormone derived like cortisol from pregnenolone. So the definition of 'endrocrine risk' would incorporate a common population of subjects with depression with two hormonal profiles: higher cortisol levels with normal DHEA, and normal cortisol levels with lower DHEA, leading to the same functional outcome[37].

A higher evening cortisol/DHEA ratio at entry in subjects with major depressive disorder has been shown to be a better predictor of persistent major depression in the short term. This suggests an integrative role for these two adrenal steroids in abnormal psychological processes that maintain or lead to further disturbed interpersonal relationships in already depressed patients.

A prospective investigation in well adolescents at high risk of psychopathology found significant associations between one or more daily morning peak levels for cortisol and subsequent major depressive disorder. There were also associations between evening DHEA peaks and subsequent major depressive disorder[38]. These endocrine indices predict major depressive disorder independently of recent life events, long-term difficulties or level of premorbid depressive symptoms. These antecedent features of circulating cortisol and DHEA may arise from more distal environmental and/or genetic origins. It remains to be determined whether or not these premorbid psychoendocrine characteristics are entirely specific for major depression.

Hypothalamic-Pituitary-Thyroid Axis (HPT axis)

TRH stimulation test reveals a lower or blunted TSH response to TRH in a substantial proportion of depressed patients[39,40].

GH response to clonidine or to dopaminergic agents such as apomorphine is also found to be blunted in depressive patients. GH response to clonidine is accepted to point to the abnormalities in the noradrenergic system, especially in the sensitivity of alpha 2 noradrenergic receptors, while GH response to apomorphine or L-DOPA to the dysfunction in the central dopaminergic system[41].

So hyperactivity in HPA axis, one of the most consistently documented endocrine abnormalities in depression, may be associated with the diminished TRH-induced TSH release in depression. It can be postulated that the hyperactivity in HPA axis and hypercortisolemia associated with depression may be a primary hormonal dysfunction, and produce changes in monoaminergic pathways that modulate hormonal responses[42]. In this way, it may cause both an increase in dopaminergic activity inducing a down-regulation in dopamine receptors, which can be measured by GH response to L-DOPA, and a decrease in pituitary TRH receptors, which can be evaluated by TSH and PRL response to THR challenge. A recent study by Esel et al.[43] suggests that the TSH response to TRH shows a tendency to be enhanced throughout antidepressant treatment while TRH-induced PRL and L-DOPA-induced GH responses do not change with treatment in depressive patients. Other authors have stressed the particular features and alterations of both HPT and HPA axes in affective disorders during sleep, by investigating nocturnal EEG and the secretion of thyrotropin, ACTH and cortisol in depressed patients[44]. Their data support the hypothesis that both hypophyseal hormones reflect a common dysregulation of both systems in depression probably due to impaired action of TRH-related corticotropin-release-inhibiting-factor (CRIF) and suggest the possibility to use the ratio TSH/ACTH as a tool to characterize alterations of both the HPT and HPA axis in depression during the first half of the night.

A review by Sullivan et al.[45] concluded that TSH response to TRH was not an impressive discriminator between depressed and control subjects, and the TSH variable was mainly associated with treatment response in depressed individuals. A study by Engum et al.[46] showed that earlier diagnosed thyroid disorder is associated with both anxiety and depression, independent of thyroid function, but this association is weak.

We have also to consider that the dilemma when diagnosing depression or anxiety disorders in concurrent illness is the distinction between symptoms due to psychiatric or physical diseases. For example, hyperthyroidism and syndromal depression-anxiety have overlapping features that can cause misdiagnosis during acute phase. For differential diagnosis, one should follow-up patients with hyperthyroidism with specific hormonal treatment and evaluate persisting symptoms thereafter. In addition to specific symptoms of hyperthyroidism, psychomotor retardation, guilt, muscle pain, energy loss and fatigue seem to appear more frequently in patients with comorbidity with depression; so the presence of these symptoms should be a warning sign to consider in differential diagnosis[47].

It is well known that patients with overt hypothyroidism can develop psychiatric symptoms, such as depressed mood and cognitive dysfunction, which are generally reversible with levothyroxine (T4) replacement therapy. However, some hypothyroid patients receiving levothyroxine replacement therapy complain of malaise and depressive symptoms despite adequate thyroid hormone replacement, as defined by a normal serum TSH concentration. As stressed before, it has been suggested that the addition of T3 (triiodothyronine) to T4 replacement therapy may improve mood, cognitive function, and general sense of well-being. Furthermore, T3 has been reported to augment the effect of antidepressant treatment in depressed patients who are already receiving levothyroxine replacement therapy[48]. However, the current data do not support the routine use of combined T3 and T4 therapy in hypothyroid patients with depressive symptoms[49].

Mood disorders often are associated with other thyroid dysregulations, particularly depression with autoimmune thyroiditis (Hashimoto's disease). Different studies have observed strong differences between different subtypes of depressive disorders in their association with autoimmune thyroiditis. In particular, most patients affected by Hashimoto's disease are suffering from a unipolar recurrent major depression[50]. There is no trend to a higher prevalence of autoimmune thyroiditis in affective disorders in comparison to the general population. Autoimmune thyroid dysfunctions warrant further testing in patients with acute mood disorders and probably offer new immunological considerations for psychopathology et nosology of affective disorders and for treatment of risk groups.

The simultaneous use of the TRH test and the DEX/CRH test seems to provide a more useful biological marker than the separate use of either test alone in patients with major depressive disorder. The TRH test and the DEX/CRH test are unlikely to provide a valid means of predicting the response of depression to treatment when one of these two tests is conducted separately. However, their combined use can serve as a useful marker for predicting response to treatment[51].

Hypothalamic-Pituitary-Gonadal Axis (HPGn axis)

There is increasing scientific attention to the modulation of the neuroendocrine system by fluctuating gonadal hormones. The lifetime prevalence of mood disorders in women is approximately twice than that of men. This higher incidence of depression in women is primarily seen from puberty on and is less marked in the years after menopause. The lifetime rate of major depression is two to three times higher in women than men. Women are prone to depression during times of reproductive hormone change such as puberty, the postpartum period, the premenstrual phase of the menstrual cycle, the perimenopause, and during oral contraceptive use. Thus, female susceptibility to mood disorders appears to be influenced by reproductive system function[52]. The underlying causality of this gender difference is not yet understood. The higher prevalence of mood disorders in women could be related to either an increased genetic predisposition, an increased vulnerability/exposure to stressful life events, modulation of the neuroendocrine system by fluctuating gonadal hormones, or a combination of any or all of these factors[53].

The sudden appearance of higher levels of estrogen in puberty alters the sensitivity of the neurotransmitter systems. Moreover, the constant flux of estrogen and progesterone levels throughout the reproductive years portends constant modification of the neurotransmitter systems. Premenstrual syndromes may be the result of an altered activity or sensitivity of certain neurotransmitter systems. The serotoninergic system is in close reciprocal relationship with gonadal hormones. In the hypothalamus, estrogen induces a diurnal fluctuation in serotonin, whereas progesterone increases the turnover rate of serotonin. Drugs facilitating serotoninergic transmission, such as selective serotonin reuptake inhibitors (SSRIs), are very effective in reducing premenstrual symptoms. These evidences imply, at least in part, a possible change in the serotonin receptor sensitivity in women with premenstrual dysphoria[54].

Pregnancy and delivery produce dramatic changes in estrogen and progesterone levels as well as significant suppression along the HPA axis, possibly increasing vulnerability to depression[55]. It is argued that postpartum withdrawal of gonadal hormones may cause changes along the serotoninergic cascade which may lead to a mood disorder in vulnerable or genetically predisposed women. In some women depressive symptoms are associated with positive thyroid antibody status during the postpartum period. It is believed that 1% of all postpartum women will show a mood disorder associated with transient thyroid dysfunction and treatment of the thyroid condition must be part of the management. The direct and/or indirect effect of the rate of the postpartum withdrawal of some of the other major hormones and neuromodulators (beta-endorphin withdrawal, fall in circulating estrogen concentrations) involved is nevertheless still intriguing. These changes are believed to be the triggers to a cascade of changes at central and peripheral monoamine centres (increased sensitivity of dopamine receptors, changes in sensitivities of serotoninergic receptors). At menopause, estrogen levels decline while pituitary LH and FSH levels increase. The loss of modulating effects of estrogen and progesterone may underlie the development of perimenopausal mood disorders in vulnerable women.

Since these hormonal changes occur in all women, it seems safe to speculate that the development of mood disorders requires more than just fluctuating levels of hormones, but also a genetic predisposition. These genetics as yet unidentified 'defects' probably relate to subtle alterations in number and function of various receptors and enzymes and to subtle structural and anatomical differences in the central nervous system. These differences caused by genetic polymorphism, combined with the flux in the hormonal milieu determine how the system reacts to multiple environmental stresses and predicts the development of mood disorders. Further research into this complex system is needed to be able to identify specific 'genetic markers' which might help us better understand how the balance between estrogen, progesterone, testosterone and other steroid hormones affect neurotransmitter function[53].

A recent study by Bao et al.[56] showed how estrogens may directly influence CRH neurons in the human hypothalamic paraventricular nucleus. The increased number of neurons expressing CRH in mood disorders is accompanied by increased estrogen receptors colocalization in the nucleus of these neurons. These changes seem to be trait- rather than state- related. The post-mortem investigation has been conducted on 13 subjects suffering from major depression/major depressive disorder (eight cases) or bipolar disorder (five cases) and 13 controls, studied with double-label immunocytochemistry. It was found that the total number of CRH-immunoreactive neurons in patients with mood disorders was nearly 1.7 times higher than in controls. Besides, this is the first study to find a significantly larger volume in the sub-region of the paravenrticular nucleus, which is delineated on the basis of CRH neurons, in patients with mood disorders than in controls, and this observation is consistent with the presence of a permanent hyperactivity of the HPA axis. Depressive illness is presumed to be the result of an interaction between the effects of environmental stress and genetic and developmental predisposition. The set point of the HPA axis activity is not only programmed by genotype, but can be changed to another level by early life events. In particular, stressful life events during development may predispose individuals to adult-onset depression by a permanent hyperactivity and hyperresponsiveness of the HPA axis[57]. Interestingly, prenatal estrogen administration may also increase the risk of affective disorders, as appeared from a study on individuals who, during foetal life, were exposed to diethylstilbestrol, a synthetic non-steroidal estrogen[58]. The results showed by Bao et al.[56] raise the possibility that estrogens are involved in prenatal programming of the level of activity of CRH neurons, thus influencing the increased risk of mood disorders. Besides, these results suggest that estrogens may contribute to CRH activation via estrogen receptors. However, such hypotheses should be further investigated.

Based on current knowledge, estrogen treatment for affective disorders may be efficacious in two situations: 1) to stabilise and restore disrupted homeostasis - as occurs in premenstrual, postpartum or perimenopausal conditions; 2) to act as a psychomodulator during periods of decreased estrogen levels and increased vulnerability to dysphoric mood, as occurs in postmenopausal women[59].

Studies examining the relationship between testosterone and depression in men have found conflicting results. It is likely that the relationship is quite complex, with genetic, environmental and personality factors playing a role. It is unclear whether hypogonadism causes major depression or depressive states, increases stress vulnerability or leads to resistance to standard treatments. Due to the cross-sectional nature of most studies, one cannot also rule out the hypothesis that depression leads to low testosterone levels in some individuals due to complex interactions between hormones and emotion. However, it does seem that mildly reduced testosterone levels are not sensitive or specific in predicting any type of depression[60]. Testosterone replacement has demonstrated short-term tolerability and efficacy in augmenting antidepressants to alleviate treatment-refractory depression in adult males. In conclusion, the exact role of testosterone in depression has yet to be clarified. Evidence supports a relationship, although perhaps weak, between low levels of testosterone and depressive symptoms in elderly males. Randomised studies of testosterone for the prevention of depression in males with hypogonadism have not yet been conducted, but may provide important information. Future studies of testosterone replacement therapy in depression should evaluate androgen receptor polymorphisms and their importance to response[60].

In depressed women, plasma levels of estrogen are usually lower and plasma levels of androgens are increased, while testosterone levels are decreased in depressed men. This is explained by the fact that both in depressed males and females the HPA-axis is increased in activity, parallel to a diminished HPGn-axis, while the major source of androgens in women is the adrenal, whereas in men it is the testes. It is speculated, however, that in the aetiology of depression the relative levels of sex hormones play a more important role than their absolute levels[61]. So far the exact mechanism of the sex hormones' involvement in mood or affective diseases remains unclear. Apart from the effects on the HPA-axis, an interaction between sex hormones and the serotoninergic system has been proposed.

Other Neuroendocrine Systems

Alterations in other neuroendocrine systems may also play a role in the pathogenesis of depression[62,63].

Vasopressin (AVP) is known to synergize with CRH in regulating ACTH release from the anterior pituitary[62]. After prolonged stress, AVP is increasingly expressed and released from the hypothalamic neurons in both humans and rodents[64].

Similarly, a marked increased synthetic activity of hypothalamic AVP neurones has been described in depressed patients, which is reflected by increased AVP plasma concentrations[65]. Recently, administration of a AVP receptor antagonist was shown to display anxiolytic and antidepressant-like effects in rodents[66]. Very recently, evidence was provided that paroxetine, a clinically well established antidepressant, normalises aberrant behavioural and neuroendocrine patterns in the psychopathological rat model for high anxiety behaviour. This was accompanied by a down-regulation of hypothalamic AVP overexpression[67]. The suprachiasmatic nucleus (SCN), the pace-maker of the circadian rhythm in mammals and an additional source of AVP, may also be involved in the pathophysiology of depression. Early morning awaking is a common symptom in depressed patients. In rats it was shown that AVP neurones of the SCN inhibit CRH neurones in the paraventricular nucleus[68]. Furthermore, the functional ability of the SCN in maintaining normal biological rhythms seems to be diminished in depression. The increase of AVP plasma levels have been related to an enhanced suicide risk[69].

Leptins or hypocretins are peptide hormones secreted from white adipose tissue and involved in the regulation of food intake. Plasma concentrations of leptins positively correlate with the body mass index (BMI)[70,71]. Loss of appetite, decreased food intake and weight loss are frequent symptoms of major depression. Leptins were found to inhibit CRH release from the hypothalamus in vitro and to reduce plasma ACTH and corticosterone responses to restraint stress in vivo. Clinical studies have shown contrasting leptins levels in depressed patients[62]. Some studies reported that leptins plasma levels did not differ between depressed and control subjects[72], while others demonstrated elevated nocturnal serum leptins levels, despite a reported weight loss[73]. On the other hand, despite the fact that antidepressants clearly result in weight gain, they have no effect or a very subtle one on leptins concentrations.

So the relationship between leptins and the HPA axis in stress response and mood disorders requires further investigation.

ENDOCRINE SYSTEM AND BIPOLAR DISORDER

Abnormalities of hypothalamic-pituitary-adrenal function have frequently been reported in patients with bipolar disorder. These patients generally have a blunted response to the dexamethasone suppression test[39]. Administration of the cortisol analogue dexamethasone fails to provoke the decrease in peripheral cortisol levels normally arising as a result of negative feedback to the hypothalamus and anterior pituitary[74]. The DEX/CRH test is abnormal in both remitted and non-remitted patients with bipolar disorder. This measure of HPA axis dysfunction is a potential trait marker in bipolar disorder and thus possibly indicative of the core pathophysiological process in this illness[75].

Patients with bipolar disorder also have a blunted release of thyrotropin-stimulating hormone in response to administration of exogenous thyrotropin-releasing hormone[76]. Moreover, some studies showed how in a sample of female patients whose psychiatric symptoms fluctuated in association with the menstrual cycle were observed abnormalities in hypothalamic-pituitary-gonadal hormones including elevated basal luteinizing hormone, reduced basal follicle-stimulating hormone and elevated serum testosterone and androstenedione[77]. These findings are consistent with researches showing a high incidence of menstrual abnormalities in women affected by bipolar disorder.

A dysfunction in the central serotoninergic system, among other neurotransmitters, may contribute to neuroendocrine abnormalities observed in bipolar disorder. Serotonin influences the release of many of the hypothalamic and pituitary hormones shown to be dysregulated in this psychiatric disorder. In studies of the effects of administration of serotonin-releasing or serotonin-stimulating agents on neurohormonal responses patients affected by bipolar disorder exhibit abnormal responses compared with individuals without a mood disorder[78].

There are also interesting studies using neuroimaging techniques, in particular using magnetic resonance imaging (MRI), demonstrating that patients with bipolar disorder have significantly smaller pituitary volumes compared with healthy controls, whereas the pituitary volume of patients with unipolar depression did not differ from that of healthy controls[79]. The authors suggested that the reduced pituitary volume may reflect a dysfunctional hypothalamic-pituitary-adrenal (HPA) axis in bipolar disorder.

Medications, such as valproate, may contribute to neuroendocrine abnormalities in patients bipolar disorder[80]. Possibly, weight gain associated with use of valproate mediates this effect; however, the evidence regarding the possible impact of weight on neuroendocrine dysfunction in bipolar disorder is inconclusive. Gilmor et al.[81] chronically administered valproic acid and lithium, two clinically effective mood stabilizers, in nonstressed rats. Chronic valproic acid administration decreased CRF mRNA expression in the paraventricular nucleus of the hypothalamus; lithium administration increased CRF mRNA expression in the central nucleus of the amygdala. These and other results suggest that the therapeutic actions of these mood stabilizers may, in part, result from their actions on central CRF neuronal systems. The distinct actions of each drug on CRF systems may underlie their synergistic clinical effects.

Some medications, such as mood stabilizers, may also affect concentrations of biologically active hormones by displacing them from hormone-binding proteins or by altering hormone metabolism. Mood stabilizers, including carbamazepine, valproate, vigabatrin and gabapentin, cause weight gain that can be substantial in some patients. This weight gain, particularly when associated with or leading to obesity, reduces insulin sensitivity and may lead to the development of polycistic ovary syndrome-like condition[82]. Weight gain and obesity play a fundamental role in causing endocrine abnormalities in women suffering by bipolar disorder. In fact weight gain may contribute to the development of hyperinsulinemia that leads to insulin resistance, which may stimulate the ovaries in overproducing androgens and lead to lipid abnormalities. In addition, it may contribute to hyperandrogenism by stimulating steroid production by means of adipose tissue[83].

Notwithstanding the progress in documenting neuroendocrine abnormalities in bipolar disorder, much work remains before the clinical significance of these abnormalities is understood.

According to most available data, no mood stabilizer should be excluded from the range of treatment options available for patients with bipolar disorder. However, the possibility that mood stabilizers can cause or exacerbate neuroendocrine dysfunctions should not be discounted. In particular, the obese patient or the patient who gains significant weight after initiating a pharmacological therapy with a mood stabilizer should be monitored closely in view of the relationship of weight gain and obesity with neuroendocrine abnormalities. The possible benefits of a patient's change in mood stabilizer therapy should be considered carefully against the risks of adverse events and efficacy of alternative therapeutic options.

There is still debate whether rapid-cycling bipolar patients are predisposed to thyroid axis abnormalities and whether such dysfunction may contribute to the development of rapid mood shifts. Several studies have found an association among indices of low thyroid function or clinical hypothyroidism or even both situations and rapid-cycling bipolar disorder[84,85], others suggest that this commonly-cited relation has to be reevaluated[86].

It is important to stress the concept that most of these studies evaluated samples of patients with rapid-cycling bipolar disorder who were receiving prophylactic long-term lithium treatment, an agent which has been demonstrated to have "antithyroid" properties. On the other hand, some prospective studies indicate that short-term treatment with lithium leads to diminished thyroid function both in healthy control subjects and in unselected bipolar patients. Cross-sectional studies of unmedicated rapid-cycling bipolar patients, on the other hand, found no abnormalities in basal thyrotropin (TSH) and thyroxine levels in this patient population[87].

Bauer et al.[87] propose that individuals with rapid-cycling bipolar disorder are sensitive to goiterogenic stressors. A recent interesting study[88] demonstrates no baseline, medication-free differences between patients and control subjects on any measure of thyroid function including the TRH stimulation test. The finding of thyroid hypofunction unmasked by lithium treatment provides a pathophysiologic model that may explain the potential role of the hypothalamic-pituitary-thyroid axis in precipitating the rapid-cycling phenotype of bipolar disorder. Under this model, a dysfunction in the hypothalamic-pituitary-thyroid system is latent until the axis is challenged by the thyroprivic effect of lithium. So lithium challenge may offer a research paradigm valuable in uncovering latent hypothalamic-pituitay-thyroid (HPT) system abnormalities in affective disorders, including non-rapid-cycling bipolar disorder, and other psychiatric syndromes[10].

There is good evidence that triiodothyronine (T3) may accelerate the antidepressant response to tricyclic antidepressants, and some studies suggest that T3 may augment the therapeutic response to antidepressants in refractory depressed patients. Open studies have also indicated that adjunctive supraphysiological doses of thyroxine (T4) can ameliorate depressive symtomatology and help stabilize the long-term course of illness in bipolar and unipolar patients, especially women refractory to standard medications[89,90].

As we have seen for depressive disorder, there is robust evidence demonstrating abnormalities of the HPA axis in bipolar disorder[91]. Hypercortisolism may be central to the pathogenesis of depressive symptoms and cognitive deficits, which may in turn result from neurocytotoxic effects of raised cortisol levels, although we have no demonstration of this hypothesis[61]. Manic episodes may be preceded by increased ACTH and cortisol levels, leading to cognitive problems and functional impairments. Identification and effective treatment of mood and cognitive symptoms of mood disorders are clinical goals, but currently available treatments may fall short of this ideal. Manipulation of the HPA axis has been shown to have therapeutic effects in preclinical and clinical studies, and recent data suggest that direct antagonism of glucocorticoids receptors may be a future therapeutic strategy in the treatment of mood disorders[30,92].

MOOD, NEUROENDOCRINE FUNCTION AND QUALITY OF LIFE

Over the past few years, together with progress of research in psychoneuroendocrinology, there has been growing interest in the psychological aspects of clinical care in endocrine disease. In particular, some issues such as life events preceding disease onset, psychological distress associated with acute illness and convalescence, abnormal illness behaviour and several other aspects of quality of life in endocrinology and mood disorders have received more and more attention.

The World Health Organization defines quality of life as an individual's perception of where he/she lives within his or her cultural and value judgments, his/her goals, expectations, standards and concerns. Quality of life related to health is the totality of one's physical health, psychological condition, beliefs, social relationships and relationships within the environment.

The life expectancy of women is increasing throughout the world and in most countries, women who reach the age 50 can expect to live another 30-40 years. Adult women will, therefore, live almost as long after menopause as they do before. The perception of the menopausal transition in women is strongly influenced by socio-cultural and lifestyle factors[93]. Vasomotor problems like hot flashes and night sweats are menopausal symptoms that are experienced in the premenopausal, menopausal and firsts few years of the postmenopausal periods and can have a negative effect on a woman's daily activities. Nevertheless, a recent study by Ozkan et al.[93] did not find any significant difference in the quality of life of pre and postmenopausal women.

The results of several observational studies suggest that the use of estrogen replacement is associated with better mood, cognitive function and quality of life. Such findings are consistent with those of laboratory-based research showing that estrogen promotes neuronal sprouting, enhances cholinergic activity in the brain, decreases brain and plasma levels of β-amyloid, increases serotonin postsynaptic responsivity and the turnover of noradrenaline, and inhibits monoamine oxidase activity. A very recent randomized controlled trial of estradiol replacement therapy for women aged 70 years and older indicate that the use of a relatively high dosage of unopposed estrogen replacement for 20 weeks is not associated with significant changes in cognitive function, mood and quality of life[94].

Currently it is estimated that 3–8% of women of reproductive age meet strict criteria for premenstrual dysphoric disorder (PMDD). The impairment and lowered quality of life for PMDD is similar to that of dysthymic disorder and is not much lower than major depressive disorder[95]. PMDD shares many features with these other conditions commonly including symptoms of depression and anxiety, headache, and abdominal discomfort. Conversely, this disorder is cyclical, tending to be relatively stable over time, and may lead to both increased use of healthcare services and impaired worker productivity. An increased number of sick days and impaired work productivity in women with PMDD was reported. Premenstrual impairment may be more severe at home, influencing marital relationships and homemaking, as compared to social and out-of-home occupational impairment. Nevertheless, this disorder is still under-recognized.

Appropriate recognition of the disorder and its impact should lead to treatment of more women with PMDD, aiming to reduce individual suffering and impact on family, society and economy.

Growth hormone (GH) replacement therapy with duration of several years is known to be safe and beneficial in GH-deficient adult patients. A study showed that 10 yeas of GH therapy is beneficial in terms of well-being and cognitive functioning in childhood-onset GH-deficient men. It may be concluded that once the decision to start GH treatment has been taken, this may imply that GH therapy has to be continued for a long period to maintain the psychological improvements and to prevent a relapse[96]. Stouthart et al.[97] outlined that GH treatment mainly improves quality of life in the first six months of treatment, but more than six months are needed to improve mood. This study showed that mood impairment after GH discontinuation was counteracted by one year of GH retreatment. Authors suggested that mood impairments in GH-deficient subjects are related with a reversible GH-specific disturbance in neural cell metabolism.

In men, plasma total testosterone levels decline progressively over the lifespan, and this occurs to a greater extent in some men than others. The effect of this decline on cognitive function, mood and quality of life in aging men is presently unclear. This issue is of importance, since testosterone is considered by many to be beneficial in improving mood and health-related quality of life. A study by Haren et al.[98] showed that 12-month supplementation with oral testosterone does not affect scores on visuospatial tests or mood and quality of life scales in older men with low-normal gonadal status. On the other hand, Delhez et al.[99] reported that andropause, defined as hypogonadism or as hypergonadotrophic hypogonadism, is not characterised by specific psychological symptoms. In particular, andropause could be associated with minor depressive symptoms that are not considered pathological. Authors affirmed that the correlations between testosterone and depression should be interpreted with caution since they are weak.

Hyperthyroidism causes several physical complaints, such as exhaustion, excessive perspiration, loss of weight, as long as a number of neurologic, cognitive and emotional impairments. This disease may also compromise the social and professional performance of patients and thus endanger their social and professional status. A study by Fahrenfort et al.[100] investigated the issue of residual complaints after treatment for hyperthyroidism in current euthyroid patients and found that over one third of patients with a full-time job were unable to resume the same work after treatment. More frequently residual complaints included depressive symptoms, anxiety and lack of energy, confirming the importance of psychological support for many patients during and after recovery from hyperthyroidism. Hyperparathyroidism also causes subjective neurobehavioral symptoms ranging from subtle to severe: lethargy, drowsiness, depressed mood, neurasthenia, paranoia, hallucinations, disorientation, confusion and cognitive complaints, decreased ability to complete daily tasks at home or work and decreased social interaction. Justification of the studies about quality of life in patients suffering from hyperparathyroidism is based on the presumption that some vague symptoms, such as fatigue, mood swings, irritability and physical pain, would affect quality of life and could be self-reported. In particular, some authors have suggested that formal neuropsychological testing and evaluation of health-related quality of life are useful tools that may assist physicians in choosing whom to refer for surgical treatment[101].

Also asymptomatic patients with thyroid disorders may suffer a reduction in perceived health status due to distress related to physical appearance and awareness of disease. A study by Bianchi et al.[102] showed that in a sample of 368 patients with thyroid disorders health-related quality of life was impaired also in the absence of altered hormone levels. Mood disorders and behaviour disturbances were present in a large proportion of patients and were significantly associated with a poor quality of life. Besides, quality of life was significantly reduced in patients with thyroid diseases referred to a secondary level endocrinology unit. So perceived health status may be considered as an additional outcome of management and therapy of thyroid disorders.

Many studies showed how quality of life, as determined by illness intrusiveness, is compromised in subjects with bipolar disorder. Illness intrusiveness is defined as the extent to which an illness and/or its treatment interferes with important activities and interests across some specific life domains. It is hypothesized to compromise psychosocial well-being in several ways, by reducing personal control and opportunities to engage in valued activities to achieve positive outcomes and avoid negative ones[103].

Life events have also been determined to play a significant role in the exacerbation of mood disorders. Numerous authors have reported the significance of undesirable or unpleasant life events precipitating episodes of depression and mania. It is also suspected that life events cause more instability and episodes in the early stages of a mood disorder. Robb et al.[103] have suggested that perception of life quality and the ability to pursue valued activities and interests is compromised in subjects with bipolar disorder. Their findings have demonstrated impairment persisting in clinically euthymic patients and have provided ongoing support for the idea that there are residual complications that persist beyond clinical symptom resolution.

Moreover, resolution of the clinical symptoms of depression did not appear to be indicative of a return to premorbid functioning. In other words, a growing body of work indicates that episodes of mania and hypomania are not typically characterized by euphoric mood and sense of increased well-being, but rather by significant dysphoric symptoms. Moreover, depressive symptoms appear to be the primary determinant of quality of life in bipolar disorder, although other factors may be associated with both depression and reduced quality of life [104].

Bipolar disorder is at least as intrusive as several chronic medical conditions. Those with a type II Bipolar Disorder report greater impairment in all domains compared with type I [103]. Somatic and mood symptoms, stigmatization, comorbid conditions, associated cognitive deficits and the burden of treatment all can combine to severely undermine patient quality of life and functioning at all levels. The tolerability of treatment is an important consideration when choosing a therapeutic option as patient satisfaction with, and adherence to, treatment can influence health outcomes and quality of life [105].

In summary, quality of life is reportedly lowest for anxiety and mood disorder patients and somewhat higher for schizophrenia patients. Extreme satisfaction with quality of life reported by manic patients is symptomatic of the illness [2]. Through the relationship between quality of life and personality traits has been discussed in the literature there is a lack of data concerning this issue among mood and anxiety disorder patients. Personality-related factors, such as self-esteem and self-efficacy play a prominent role in the appraisal of subjective quality of life. Studies regarding relationships between mood stabilizers or antidepressants and quality of life outcomes of mood and anxiety disorders are scant and should be replicated and expanded upon [2].

As Ritsner & Kurs [2] have affirmed, in the next years there will be an increase in empirical data regarding the quality of life outcomes of various mood and anxiety disorders and the role of clinical and psychosocial factors on quality of life measures.

Clinical data on pharmacologic and non pharmacologic interventions that are effective at improving the quality of life for patients with mood disorders and endocrine disturbances is emerging. With the development of more effective treatment options many more patients with mood disorders and endocrine disturbances will achieve healthy levels of functioning and quality of life, which will alleviate the burden that the illness imposes on patients, their families, and caregivers.

CONCLUSIONS

For a long time, clinicians suspected a causal link between mood disorders and the endocrine system. In fact for many years, the widespread and selective distribution of hormones throughout the brain, and their neuromodulatory role and influence in regulating behaviour, including mood, have been studied.

The most frequently occurring endocrine abnormality in depressed subjects is hyperactivity of the HPA axis. CRH and AVP are likely to play a substantial role in the pathophysiology of mood disorders, and their receptors appear to be a specific target for future antidepressant drugs[62]. Mood disorders also affects the HPGH and HPT axes. Taken together, most of the studies suggest an association between mood disorders and both subclinical, mild and overt hypothyroidism. The literature suggests an association between various degrees of thyroid dysfunction and depressive disorder, although the mechanism for this is unclear. There is a variety of hypotheses considering biological causality, but we also have to accurately investigate how these common disorders can coexist and influence each other, and which causal or pathogenic effects they play on each other. It is important to add that even if future research confirms that there is not an association between thyroid hormones and mood disorders in the general population, there may well be individuals who are more sensitive to fluctuations of thyroid hormones, even within the normal range[106].

Alterations in the reproductive system may also be involved in the pathogenesis of mood disorders. In addition, there is increasing evidence that leptins and neurosteroids are implicated in mood disorders.

Studies of rats have shown that high levels of cumulative corticosteroid exposure and rather extreme chronic stress induce neuronal damage that selectively affects hippocampal structure. Studies performed under less extreme circumstances have so far provided conflicting data. The corticosteroid neurotoxicity hypothesis that evolved as a result of these initial observations is, however, not supported by clinical and experimental conditions. Although cortisol and CRH may well be causally involved in the signs and symptoms of depression, there is so far no evidence for any major irreversible damage in the human hippocampus in this disorder. In fact, in depressed subjects or in patients treated with synthetic corticosterois the hippocampus is intact and does not show any indication of neuropathological alterations or major structural damage in post-mortem material[61].

Consideration and evaluation of endocrine status remains important in psychiatric patients, both to ensure proper diagnosis and adequate treatment. At the same way, improved quality of life of patients suffering from mood disorders has become an important treatment goal. Changes in psychosocial or stress process-related factors, rather than psychopathology, predict change in perceived quality of life and should be considered when evaluating quality of life outcomes.

REFERENCES

1. Ritsner M, Modai I, Endicott J, et al. Differences in quality of life domains and psychopathologic and psychosocial factors in psychiatric patients. J Clin Psychiatry 2000; 61 (11): 880–889.
2. Ritsner M, Kurs R. Quality of life outcomes in mental illness: schizophrenia, mood and anxiety disorders. Expert Rev Pharmacoeconomics Outcomes Res 2003; 3 (2): 89–99.
3. Awad AG. Quality of life of schizophrenic patients on medications and implications for new drug trials. Hops Com Psychiatry 1992; 43: 262–265.
4. Ritsner M, Kurs R, Gibel A, et al. Predictors of quality of life in major psychoses: a naturalistic follow-up study. J Clin Psychiatry 2003; 64 (3): 308–315.
5. Ritsner M. Predicting changes in domain-specific quality of life of schizophrenia patients. J Nerv Ment Dis 2003; 191: 287–294.

6. Ritsner M, Ben-Avi I, Ponizovsky A, et al. Quality of life and coping with schizophrenic symptoms. Quality of Life Res 2003; 12: 1–9.
7. Ritsner M, Gibel A, Ratner Y. Determinants of changes in perceived quality of life in the course of schizophrenia. Quality of Life Res 2006; 16: 515–526.
8. Rasgon NL. Anatomic, functional, and clinical studies of neuroendocrine function in bipolar disorder. Adv. Stud. Med 2003; 3 (8A): S726–732.
9. Kaplan HI, Sadock BJ, Grebb JA. Psichiatria: manuale di scienze del comportamento e psichiatria clinica (VII Ed), CSI, Torino, 1996.
10. Mazza M, Satta MA, Bria P, Mazza S. Neuroendocrinology of mood disorders. Clin Ter 2004; 155 (11-12): 537–541.
11. Loosen PT. The TRH-induced TSH response in psychiatric patients: a possible neuroendocrine marker. Psychoneuroendocrinology 1985; 10:237–260.
12. Duval F, Macher JP, Mokrani MC. Difference between evening and morning thyrotropin responses to protirelin in major depressive episode. Arch of Gen Psychiatry 1990; 47: 443–448.
13. Kirkegaard C, Faber J, Hummer L, et al. Increased levels of TRH in cerebrospinal fluid from patients with endogenous depression. Psychoneuroendocrinology 1979; 4: 227–235.
14. Joffe RT, Singer W, Levitt AJ, et al. A placebo-controlled comparison of lithium and triiodothyronine augmentation of tricyclic antidepressants in unipolar refractory depression. Archives of General Psychiatry 1993, 50: 387–393.
15. Siever LJ, Davis KL. Overview: toward a dysregulation hypothesis of depression. Am J Psychiatry 1985; 142: 1017–1031.
16. Trestman RL, Yehuda R, Coccaro E, et al. Diurnal neuroendocrine and autonomic function in acute and remitted depressed male patients. Biological Psychiatry 1995; 37:448–456.
17. Ehlert U, Gaab J, Heinrichs M. Psychoneuroendocrinological contributions to the etiology of depression, posttraumtic stress disorder, and stress-related bodily disorders: the role of the hypothalamus-pituitary-adrenal axis. Biol Psychology 2001; 57: 141–152.
18. Burke HM, Davis MC, Otte C, et al. Depression and cortisol responses to psychological stress: a meta-analysis. Psychoneuroendocrinol 2005; 30: 846–856.
19. Holsboer F. Stress, hypercortisolism and cosrticosteroid receptors in depression: implications for therapy. J Affect Disord 2001; 62: 77–91.
20. Otte C, Hart S, Neylan TC, et al. A meta-analysis of glucocorticoid reactivity in human aging: implications for Alzheimer's disease and depression. Psychoneuroendocrinol 2005; 30: 80–91.
21. Raison CL, Miller AH. When not enough is too much: the role of insufficient glucocorticoid signaling in the pathophysiology of stress-related disorders. Am J Psychiatry 2003; 160: 1554–1565.
22. Nemeroff CB, Vale WW. The neurobiology of depression: inroads to treatment and new drug discovery. J Clin Psychiatry 2005; 66(7): 5–13.
23. Holsboer F. The corticosteroid hypothesis of depression. Neuropsychopharmacol 2000; 23: 477–501.
24. Zobel AW, Nickel T, Sonntag A, et al. Cortisol response in the combined dexamethasone/CRH test as predictor of relapse in patients with remitted depression: a prospective study. J Psychiatr Res 2001; 35: 83–94.
25. Pariante CM, Miller AH. Glucocorticoid receptors in major depression: relevance to pathophysiology and treatment. Biol Psychiatry 2001; 49: 391–404.
26. O'Connor TM, O'Halloran DJ, Shanahan F. The stress response and the hypothalamic-pituitary-adrenal axis: from molecule to melancholia. Q J Med 2000; 93: 323–333.
27. Gold PW, Goodwin FK, Chrousos GP. Clinical and biochemical manifestations of depression: relationship to the neurobiology of stress. N Engl J Med 1988; 319: 348–353.
28. Pariante CM, Pearce BD, Pisell TL, et al. Steroid-independent translocation of the glucocorticoid receptor by the antidepressant desipramine. Mol Pharmacol 1997; 52:571–581.
29. Belanoff JK, Flores BH, Kalezhan M, et al. Rapid reversal of psychotic depression using mifepristone. J Clin Psychopharmacol 2001; 21: 516–521.
30. Young AH. Cortisol in mood disorders. Stress 2004; 7(4): 205–208.

31. Holsboer F. Corticotropin-releasing hormone modulators and depression. Curr Opin Investig Drugs 2003; 4(1): 46–50.
32. Yuuki M, Ida I, Oshima A, et al. HPA axis normalization, estimated by DEX/CRH test, but less alteration on cerebral glucose metabolism in depressed patients receiving ECT after medication treatment failures. Acta Psychiatr Scand 2005; 112: 257–265.
33. Pfennig A, Kunzel HE, Kern N. Hypothalamus-pituitary-adrenal system regulation and suicidal behaviour in depression. Biol Psychiatry 2005; 57: 336–342.
34. Nemeroff CB, Owens MJ, Bissette G, et al. Reduced corticotropin releasing factor binding sites in the frontal cortex of suicide victims. Arch Gen Psychiatry 1988; 45: 577–579.
35. Goodyer IM, Park RJ, Netherton CM, et al. Possible role of cortisol and dehydroepiandrosterone in human development and psychopathology. Br J Psychiatry 2001; 179: 243–249.
36. Birmaher B, Dahl RE, Perel J. Corticotropin-releasing hormone challenge in prepuberal major depression. Biol Psychiatry 1996; 39: 267–277.
37. Goodyer IM, Altham PME. Short term outcome of major depression. A high cortisol/DHEA ratio and subsequent disappointing life events predict persistent depression. Psychol Med 1997; 28: 265–273.
38. Goodyer IM, Tamplin A. Recent life events, cortisol, dehydroepiandrosterone and the onset of major depression high-risk adolescents. Br J Psychiatry 2000; 177: 499–504.
39. Rush AJ, Giles DE, Schlesser MA, et al. Dexamethasone response, thyrotropin releasing hormone stimulation, rapid eye movement latency, and subtypes of depression. Biol Psychiatry 1997; 41(9):915–928.
40. Mokrani MC, Bailey P. Thyroid axis activity and serotonin function in major depressive episode. Psychoneuroendocrinol 1999; 24: 695–712.
41. Mokrani MC, Duval F, Crocq MA, et al. HPA axis dysfunction in depression: correlation with monoamine system abnormalities. Psychoneuroendocrinol 1997; 22 (Suppl 1), S63–S68.
42. Dinan TG. Glucocorticoids and the genesis of depressive illness. A psychobiological model. Br J Psychiatry 1994; 164: 365–371.
43. Esel E, Kartalci S, Tutus A, et al. Effects of antidepressant treatment on thyrotropin-releasing hormone stimulation, growth hormone response to L-DOPA, and dexamethasone suppression tests in major depressive patients. Progr in Neuropsychopharmacol and Biol Psychiatry 2004; 28: 303–309.
44. Peteranderl C, Antonijevic IA, Steiger A, et al. Nocturnal secretion of TSH and ACTH in male patients with depression and healthy controls. J Psychiatric Research 2002; 36: 189–196.
45. Sullivan PF, Wilson DA, Mulder RT, et al. The hypothalamic-pituitary-thyroid axis in major depression. Acta Psychiatrica Scandinavica 1997; 95: 370–378.
46. Engum A, Bjoro T, Mykletun A, et al. An association between depression, anxiety and thyroid function: a clinical fact or an artefact? Acta Psychiatrica Scandinavica 2002; 106:27–34.
47. Demet MM, Ozmen B, Deveci A, et al. Depression and anxiety in hyperthyroidism. Archives of Medical Research 2002; 33: 552–556.
48. Cooke RG, Joffe RT, Levitt AJ. T3 augmentation of antidepressant treatment in T4-replaced thyroid patients. J Clin Psychiatry 1992; 53: 16–18.
49. Sawka AM, Gerstein HC, Marriott MJ. Does a combination regimen of Thyroxine (T4) and 3,5,3'-Triiodothyronine improve depressive symptoms better tha T4 alone in patients with Hypothyroidism? Results of a double-blind, randomised, controlled trial. J Clin Endocrinol Metab 2003; 88(10): 4551–4555.
50. Degner D, Meller J, Bleich S, et al. Affective disorders associated with autoimmune thyroiditis. J Neuropsychiatry Clin Neurosci 2001; 13: 4, 532–533.
51. Isogawa K, Nagayama H, Tsutsumi T, et al. Simultaneous use of thyrotropin-releasing hormone test and combined dexamethasone/corticotropine-releasing hormone test for severity evaluation and outcome prediction in patients with major depressive disorder. J Psy Research 2005; 39: 467–473.
52. Shively CA, Bethea CL. Cognition, mood disorders and sex hormones. ILAR Journal 2004; 45(2): 189–199.

53. Steiner M, Dunn E, Born L. Hormones and mood: from menarche to menopause and beyond. J Affective Disorders 2003; 74: 67–83.
54. Steiner M, Born L. Advances in the diagnosis and treatment of premenstrual dysphoria. CNS Drugs 2000; 13: 286–304.
55. Bloch M, Daly RC, Rubinow DR. Endocrine factors in the etiology of postpartum depression. Compr Psychiatry 2003; 44 (3): 234–246.
56. Bao AM, Hestiantoro A, Van Someren EJW, et al. Colocalization of corticotropin-releasing hormone and oestrogen receptor-α in the paraventricular nucleus of the hypothalamus in mood disorders. Brain 2005; 128: 1301–1313.
57. Swaab DF, Fliers E, Hoogendijk WJ, et al. Interaction of prefrontal cortical and hypothalamic systems in the pathogenesis of depression. Progr Brain Res 2000; 126: 369–396.
58. Meyer-Bahlburg HFL, Ehrhardt AA. A prenatal-hormone hypothesis for depression in adults with a history of fetal DES exposure. In: Halbreich U, editor. Hormones and depression. New York: Raven Press; 1987: 325–338.
59. Halbreich U, Kahn LS. Role of estrogen in the aetiology and treatment of mood disorders. CNS Drugs 2001; 15(10): 797–817.
60. Carnahan RM, Perry PJ. Depression in aging men. Drugs Aging 2004: 21(6): 361–376.
61. Swaab DF, Bao AM, Lucassen PJ. The stress system in the human brain in depression and neurodegeneration. Ageing Res Rev 2005; 4: 141–194.
62. Tichomirowa MA, Keck ME, Schneider HJ, et al. Endocrine disturbances in depression. J Endocrinol Invest 2005; 28: 89–99.
63. Spinelli MG. Neuroendocrine effects on mood. Rev Endocr Metab Disord 2005; 6(2): 109–115.
64. Antoni FA. Vasopressinergic control of pituitary adrenocorticotropin secetion comes of age. Front Neuroendocrinol 1993; 14: 76–122.
65. van Londer L, Goekoop JG, van Kempen GM. Palsma levels of arginine vasopressin elevated in patients with major depression. Neuropsychopharmacol 1997; 17: 284–292.
66. Griebel G, Sirniand J, Serradeil-Le Gal C. Anxiolotic and antidepressant-like effects of the non-peptide vasopressin V1b receptor antagonist, SSR149415, suggest an innovative approach for the treatment of stress-related disorders. Proc Natl Acad Sci USA 2002; 99: 6370–6375.
67. Keck ME, Welt T, Muller MB. Reduction of hypothalamic vasopressinergic hyperdrive contributes to clinically relevant behavioral and neuroendocrine effects of chronic paroxetine treatment in a psychopathological rat model. Neuropsychopharmacol 2003; 28: 235–243.
68. Zhou JN, Riernersma RF, Unmehopa UA. Alterations in arginine vasopressin neurons in the suprachiasmatic nucleus in depression. Arch Gen Psychiatry 2001; 58: 655–662.
69. Inder WJ, Donald RA, Prickett TCR, et al. Arginine vasopressin is associated with hypercortisolemia and suicide attempts in depression. Biol Psychiatry 1997; 42: 744–747.
70. Mazza M, Della Marca G, Paciello N, et al. Orexin, sleep and appetite regulation: a review. Clin Ter 2005; 156(3): 93–96.
71. Della Marca G, Farina B, Mennuni GF, et al. Microstructure of Sleep in Eating Disorders: Preliminary Results, J Eating and Weight Disorders 2004; 9: 77–80.
72. Deuschle M, Blum WF, Englaro P. Palsma leptin in depressed patients and healthy controls. Horm Metab Res 1996; 29: 714–717.
73. Antonijevic IA, Murck H, Frieboes RM, et al. Elevated nocturnal profiles of serum leptin in patients with depression. J Psychiatr Res 1998;32: 403–410.
74. Schmider J, Lammers CH, Gotthardt U, et al. Combined dexamethasone/corticotropin-releasing hormone test in acute and remitted manic patients, in acute depression and in normal controls. Biol Psychiatry, 1995; 38:797–802.
75. Watson S, Gallagher P, Ritchie JC, et al. Hypothalamic-pituitary-adrenal axis function in patients with bipolar disorder. Br J Psychiatry 2004; 184: 496–502.
76. Linkowski P, Brauman H, Mendlewicz J. Thyrotropin response to thyrotropin-releasing hormone in unipolar and bipolar affective illness. J affect Disorders 1981; 3: 9–16.

77. Matsunaga H, Sarai M. Elevated serum LH and androgens in affective disorder related to the menstrual cycle with referrence to polycystic ovary syndrome. Jpn J Psychiatry Neurol 1993; 47: 825–842.
78. Sobczak S, Honig A, van Duinen MA, et al. Serotonergic dysregulation in bipolar disorders: a literature review of serotonergic challenge studies. Bipolar Disorders 2002; 4: 347–356.
79. Sassi RB, Nicoletti M, Brambilla P, et al. Decreased Pituitary volume in patients with bipolar disorder. Biol Psychiatry 2001; 50: 271–280.
80. O'Donovan C, Kusumakar V, Graves GR, et al. Menstrual abnormalities in polycystic ovary syndrome in women taking valproate for bipolar mood disorder. J Clin Psychiatry 2002 ; 63: 322–330.
81. Gilmor ML, Skelton KH, Nemeroff CB, et al. The effects of chronic treatment with the mood stabilizers valproic acid and lithium on corticotropin-releasing factor neuronal systems. JPET 2003; 305: 434–439.
82. Rasgon NL, Harden CL. Neuroendocrine dysfunction in women with epilepsy or bipolar disorder: implications for patient management. Adv Stud Med 2003; 3(8A): S740-S744.
83. Isojarvi JI, Laatikainen TJ, Knip M, et al. Obesity and endocrine disorders in women taking valproate for epilepsy. Ann Neurol 1996; 39: 597–584.
84. Bartalena L, Pellegrini L, Meschi M, et al. Evaluation of thyroid function in patients with rapid-cycling and non-rapid-cycling bipolar disorder. Psychiatry Research 1990; 34: 13–17.
85. Kusalic M. Grade II and grade III hypothyroidism in rapid-cycling bipolar patients. Neuropsychobiology 1992; 25: 177–181.
86. Post RM, Kramlinger KG, Joffe RT, et al. Rapid cycling bipolar affective disorder: lack of relation to hypothyroidism. Psychiatry Research 1997; 72: 1–7.
87. Bauer MS, Whybrow PC. Rapid cycling bipolar affective disorder II. Treatment of refractory rapid cycling with high-dose levothyroxine: a preliminary study. Arch Gen Psychiatry 1990; 47: 435–440.
88. Gyulai L, Bauer M, Bauer MS, et al. Thyroid hypofunction in patients with rapid-cycling bipolar disorder after lithium challenge. Biological Psychiatry 2003; 53: 899–905.
89. Bauer M, Whybrow PC. Thyroid hormone, neural tissue and mood modulation. World J Biol Psychiatry 2001; 2(2): 59–69.
90. Cole DP, Thase ME, Mallinger AG, et al. Slower treatment response in bipolar depression predicted by lower pretreatment thyroid function. Am J Psychiatry 2002; 159: 116–121.
91. Cervantes P, Gelber S, Kin F, et al. Circadian secretion of cortisol in bipolar disorder. J Psychiatry Neurosci 2001; 26 (5): 411–416.
92. Daban C, Vieta E, Mackin P, et al. Hypothalamic-pituitary-adrenal axis and bipolar disorder. Psychiatr Clin North Am 2005; 28(2): 469–480.
93. Ozkan S, Alatas E, Zencir M. Women's quality of life in the premenopausal and post menopausal periods. Qual Life Res 2005; 14: 1795–1801.
94. Almeida OP, Lautenschlager NT, Vasikaran S, et al. A 20-week randomized controlled trial of estradiol replacement therapy for women aged 70 years and older: effect on mood, cognition and quality of life. Neurobiol Aging 2006; 27: 141–149.
95. Halbreich U, Borenstein J, Pearlstein T, et al. The prevalence, impairment, impact and burden of premenstrual dysphoric disorder. Psychoneuroendocrinol 2003; 28: 1–23.
96. Arwert LI, Deijen JB, Muller M, et al. Long-term growth hormone treatment preserves GH-induced memory and mood improvements: a 10-year follow-up study in GH-deficient adult men. Hormones Behav 2005; 47: 343–349.
97. Stouthart PJHM, Deijen JB, Roffel M, et al. Quality of life of growth hormone deficient young adults during discontinuation and restart of GH therapy. Psychoneuroendocrinol 2003; 28: 612–626.
98. Haren MT, Wittert GA, Chapman IM, et al. Effect of oral testosterone undecanoate on visuospatial cognition, mood and quality of life in elderly men with low-normal gonadal status. Maturitas 2005; 50: 124–133.

99. Delhez M, Hansenne M, Legros JJ. Andropause and psychopathology: minor symptoms rather than pathological ones. Psychoneuroendocrinol 2003; 28: 863–874.
100. Fahrenfort JJ, Wilterdink AML, van der Veen EA. Long-term residual complaints and psychosocial sequelae after remission of hyperthyrioidism. Psychoneuroendocrinol 200; 25: 201–211.
101. Coker LH, Kashemi R, Cantley L, et al. Primary hyperparathyroidism, cognition, and health-related quality of life. Annals of Surgery 2005; 242 (5): 642–650.
102. Bianchi GP, Zaccheroni V, Solaroli E, et al. Health-related quality of life in patients with thyroid disorders. Qual Life Res 2004; 13 (1): 45–54.
103. Robb JC, Cooke RG, Devins GM, et al. Quality of life and lifestyle disruption in euthymic bipolar disorder. J Psychiatr Res 1997; 31 (5): 509–517.
104. Vojta C, Kinosian B, Glick H, et al. Self-reported quality of life across mood states in bipolar disorder. Compr Psychiatry 2001; 42 (3): 190–195.
105. Kasper SF. Living with bipolar disorder. Expert Rev Neurotherapeutics 2004; 4 (6 Suppl 2): 9–15.
106. Marangell LB. Thyroid hormones and mood: are population data applicable to clinical cohorts? Acta Psychiatrica Scandinavica 2002; 106: 1–2.
107. Duval F, Mokrani MC, Bailey P, et al. Thyroid axis activity and serotonin function in major depressive episode. Psychoneuroendocrinology 1999; 24: 695–712.
108. Mazza M, Della Marca G, Mennuni GF, et al. Sleep disturbances and depression: a review. Minerva Psichiatr 2005; 46: 175–188.

CHAPTER 4

IN THE MIND OF THE BEHOLDER NEURONAL MEDIATORS FOR THE EFFECT OF EMOTIONAL EXPERIENCE ON QUALITY OF LIFE

TALMA HENDLER[1,2,3], ROEE ADMON[1,2] AND DAVID PAPO[1,2]

[1]*Functional Brain Imaging Unit, Wohl Institute for Advanced Imaging Tel Aviv Sourasky Medical Center, Israel*
[2]*Medical science, Faculty of medicine, Tel Aviv University, Israel*
[3]*Psychology Department, Tel Aviv University, Israel*

Abstract: It has been suggested that Quality Of Life (QOL) is greatly affected by the individual way in which one emotionally experience the world. The nature of the emotional experience, however, is particularly divergent among people. It seems that this individual uniqueness depends more on mental representations than physical attributes of a stimulus[1]. In line with this idea, it was suggested that subjective emotional experiences are determined by the individual tendency to either focus attention on internal self-oriented or external world-grounded signals[2]. This personal characteristic depends by and large on the unique operating system of attention and awareness, as driven mainly by vigilance or cognition. Accordingly, an individual bias for enhanced focus on negative signals can be attributed to modified attention operations through fast engagement, slow disengagement, or poor signal differentiation[3]. In other words, a resilient affective style may be associated with weak reaction and fast recovery from negative stressful events, while affective vulnerability may result in excessive response to and long standing distress from the same stressful event. Thus, the ability to assign appropriate emotional significance to incoming information and to form suitable associations between stimuli and emotional state are probably essential for QOL. The present chapter aims to present possible brain mechanisms that subserve the individual emotional experience, and through that mediate QOL

Keywords: Neuroticism, fMRI, Amygdala, Hippocampus, Pre Frontal Cortex, PTSD

INTRODUCTION

For most people, stressful events that may lead to mental tension are a common daily experience. Epidemiological studies show that about 40% of the general population will experience one or more significantly stressful events in their life[4]. In order to continue functioning despite such stressors, people rely on protective mechanisms that are either innate, or acquired through experience. Following such an event there may be

M.S. Ritsner and A.G. Awad (eds.), Quality of Life Impairment in Schizophrenia, Mood and Anxiety Disorders, 57–66.

a change in the person's QOL, but eventually, with individually suited stress handling abilities (i.e. protective mechanisms) QOL recovers. However, about 8% of the people will develop a full syndrome of Post Traumatic Stress Disorder (PTSD) following what is experienced as an extreme stressful event[4], while around 60% will report a severe impairment in their QOL[5]. Based on this, around 5% (8%-the probability of developing PTSD following a stressful event * 60%-the chance for QOL impairment following the development of PTSD) of the general population will suffer from a severe QOL impairment following a significantly stressful event. What distinguishes those 5% of the population, and makes them more vulnerable to impaired QOL?

There is growing recognition that the profile of personality traits may influence an individual's pattern of reaction to a stressful event, and affect their proneness to a pathological reaction. In a survey study that included 6104 people with one or more reported significantly stressful event (i.e. a traumatic event which included a threat to the physical integrity of self or others), the subjective component of experiencing powerful emotions (i.e. intense fear, helplessness or horror) was found in 97% of the people that later on suffered from traumatic memories, a mediating factor in the development of PTSD[6]. The conclusion was that the onset of PTSD was related not only to the experiencing of the traumatic event itself (i.e. objective component), but also depends on the degree of subjective emotional reaction to the event (i.e. emotional experience). The degree of severity of such a reaction seems to depend on the individual's cluster of traits or personality profile. Prior studies suggest that elevated levels of neuroticism are significantly correlated with occurrence of PTSD, as found for fire fighters[7], Vietnam veterans[8], burn survivors[9], victims of road accidents[10], and even in a general population survey at the city of Detroit[11].

Neuroticism is a relatively broad spectrum of personality traits that includes characteristics such as restlessness, tendency for mood fluctuations, and high tempers[12]. Elevated levels of neuroticism increase the intensity of response to negative stimuli, contribute to emotional instability, and interfere with the ability to handle daily social interactions[13]. Neuroticism was also found to increase the risk for a number of mental disorders that are related to stress[14]. It is often assumed that personality is an inborn factor of behavior. However, lately it appears that stress in early life contributes to the development of neuroticism at adulthood[15]. Interestingly, Bolger and Zuckerman found that levels of neuroticism in healthy adult subjects were directly related to increased likelihood for stress exposure, greater magnitude of the reaction to stress and decreased effectiveness of coping[16]. It thus seems that neuroticism and stress are reciprocally affected[17]. Therefore, neuroticism is a relevant behavioral parameter for characterizing the individual's style of handling stressful life events, and for discerning the biomarker of vulnerability to stress and impaired QOL.

BRAIN REGIONS MEDIATING STRESS RELATED RESPONSE

In order to identify a brain biomarker for stress vulnerability, it is important to first characterize the brain's functional state following a pathological reaction to trauma. Many studies have dealt with that by comparing groups that are similar in their objective stress component, and differ in the subjective one (i.e. comparing the brain's functional response between groups of PTSD patients and healthy controls that experienced similar traumatic events). Three regions seem to appear repeatedly in those studies as differentiators of PTSD and healthy controls; *amygdala, hippocampus/parahippocampal gyrus* and parts of the *prefrontal cortex.* Interestingly, the same three main brain regions were also proposed as key brain regions in mediating the greater tendency for expression of negative affect, a core personality feature in neuroticism [18,19].

Hippocampus and Parahippocampal Gyrus

Higher activity was found for PTSD patients in comparison with healthy subjects at the *hippocampus* and *parahippocampus*, in response to individually-tailored stress related content presented visually [20]. This result may be related to the *hippocampal* area association with episodic and autobiographical memory. Furthermore, it has become increasingly apparent that the *hippocampus* also plays a more general role in information processing and behavioral regulation. PTSD patients suffer from deficits in identifying safe contexts, which may be a result of an abnormal *hippocampal* structure and function [21]. Gray [22] proposed that the *hippocampus* is involved in an inhibitory mechanism that underlies the regulation of negative affective information. He further suggested that anxiety stems from a *hippocampal* processing deficit.

Amygdala

In a comparison between combat veterans, with or without PTSD, the *amygdala* was found to be hyperactive in PTSD patients irrespective of content [23] (see Figure 1), in response to general negative stimuli such as fearful faces [24,25], and trauma relevant negative words [26]. This is not surprising since the *amygdala* is thought to participate in the perception and production of negative affects and associative aversive learning [27]. The *amygdala* may also play a role in the symptom of hyperarousal that accompanies PTSD development, since its response is known to be modulated by arousal level [28]. The amygdala's hyperresponsivity can also explain the permanency of emotional memory attached to the traumatic event [21]. In a recent neuroimaging study with healthy subjects, Canli [29] gave further support for the role of the amygdala in personality profile, by finding a correlation between activity in the *amygdala* and the individual's level of neuroticism while subjects were viewing negative stimuli.

Pre Frontal Cortex

The *prefrontal cortex* has a set of connections suggesting that its distinct domains have different roles in cognition, memory and emotion. The *orbitofrontal* and *medial prefrontal cortices* (*OFC* and *mPFC*, respectively) receive robust projections from the *amygdala*, associated with emotional memory, and project to hypothalamic visceromotor centers for the expression of emotions [30]. The same two studies that

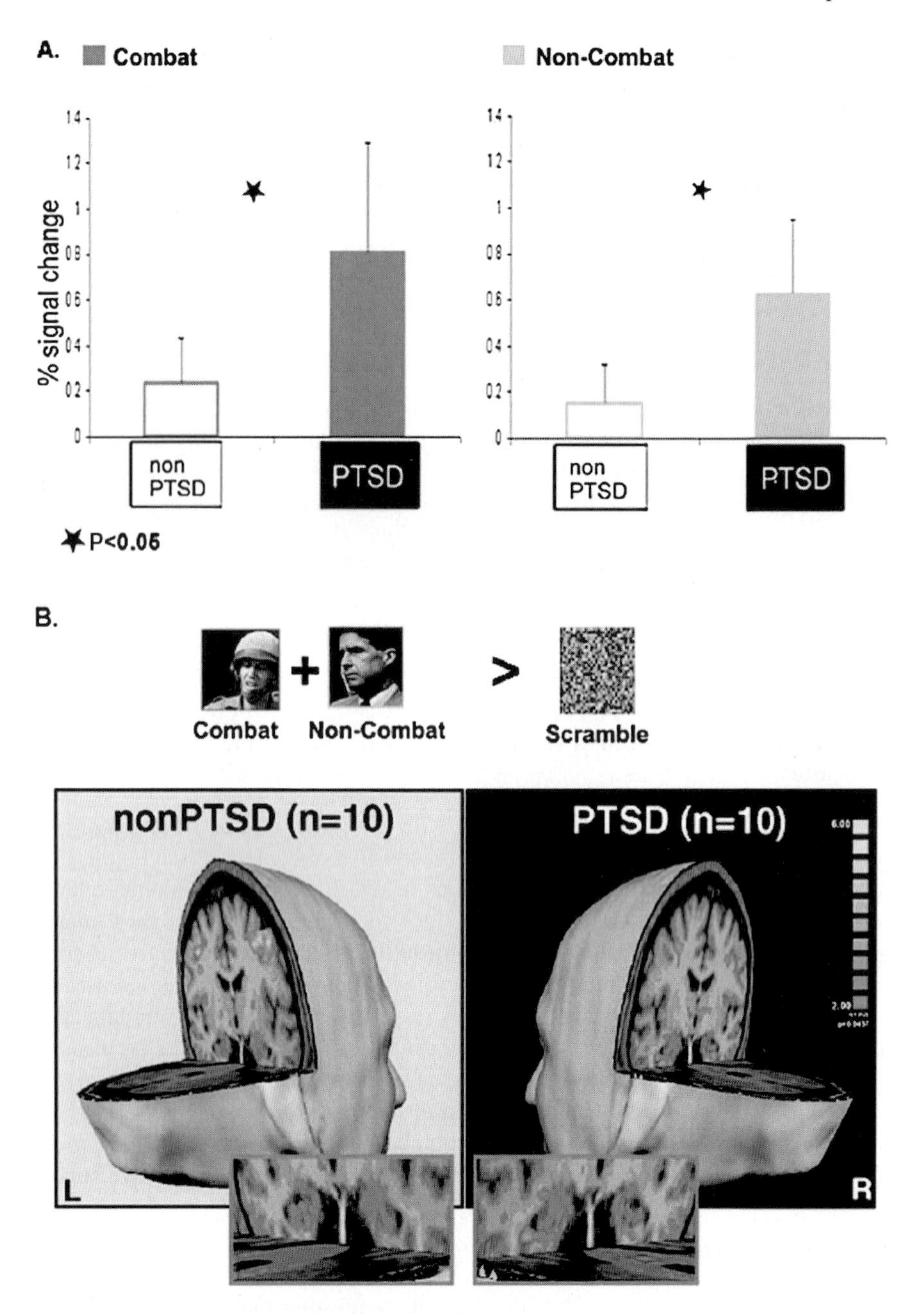

Figure 1. Amygdala hyper activation in PTSD. Activation in the amygdala complex. (A) Averaged percent signal change for combat (red) and noncombat (green) stimuli content. Activation for non-PTSD ($n = 10$) is marked in open bars, and activation for PTSD ($n = 10$) is marked in on colored bars. Error bars stand for standard error of the mean (SEM). Asterisks indicate significant simple effects. (B) Parametric maps of amygdala activation for PTSD (right) and non-PTSD (left) subjects. *(Contd.)*

found an increase in *amygdala* activity, also found a negatively correlated decrease in activity of the *mPFC*, thus suggesting that frontal disinhibitation might contribute to the generalized hyperarousability in PTSD[24,25]. It seems like the *mPFC* has the capacity to restrain the emotional (amygdalar) response to the stressful event-related content[21]. Zubieta[31] found a significant increase in blood flow to the *mPFC* in PTSD patients compared with a control group, during combat-sound stimuli. According to the authors, this pattern of *mPFC* activation is anxiety associated, which is another PTSD symptom. Davidson[32] suggested that certain parts of the *prefrontal cortex* are important components of Gray's inhibitory system, which further supports a link to its role in shaping personality profile.

Inter-regional Connections Modulating Stress Related Response

Considerable evidence implicates a balanced activation in several brain regions, rather than a focal regional activation, as the main factor upon which appropriate emotional association is greatly dependant (see Figure 2 for suggested model in PTSD). One relevant interregional interaction is suggested between *mPFC* and *amygdala* nuclei. The mammalian *mPFC* and basolateral division of the amygdala (*BLA*) are anatomically and functionally interconnected[33], and are critical for the processing and integration of emotionally salient sensory information and learning. The *mPFC* is also involved in the mediation of extinction (unlearning) of conditioned associations between environmental stimuli and fear-inducing events, such as footshock[34,35] In contrast, the *BLA* appears to be primarily involved in the processing and initial acquisition of fear associations[36,37,38]. It was thus suggested that long term appropriate emotional experience depends on the regulation of *BLA* activation via descending inputs from *mPFC*[39]. Recent evidence points to cannabinoid signalling through CB1 receptor transmission in the *mPFC* as involved in the neural encoding of a new set of associations[40]. This is in accordance with the long standing observation that cannabis consumption in humans can modify the experience of sensory stimuli via potently modulating their emotional salience. Thus, modulation of the *BLA-mPFC* circuit activation by the CB1 system may contribute to the acquisition and expression of emotional learning[41]. Indeed, stress-induced activation of the *amygdala* is potentiated by blockade of CB1 receptors in rodents[42]. Interestingly, abnormalities in CB1 receptor expression, and in the levels of endocannabinoids, have been reported in subjects suffering from schizophrenia [43]. This might explain the association found between development of schizophrenia and cannabinoid consumption[44]. Thus, cannabinoid signalling might be a potential mediator of aberrant emotional processing observed in schizophrenia[40].

Figure 1. The activation maps were obtained by multistudy GLM with object conditions (combat and noncombat) as positive predictors and scrambeld conditions as negative predictors, across all durations. Rectangles are enlarged inlay of the amygdala region to demonstrate difference in activation between the groups. Adapted from Hendler[23]

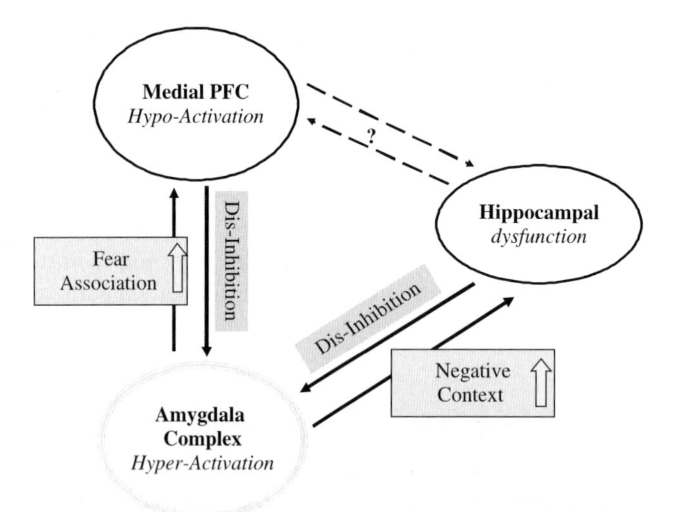

Figure 2. PTSD brain connectivity model. A proposed model for brain connectivity in PTSD. Hyper activation in the Amygdala (resulting in hyper-arousal and fear response). A deficient inhibitory activation in the medial PFC (lower capacity to suppress response to trauma related content), as well as a deficient Hippocampal activation (leading to a decrease in safe context association)

Another circuit suggested as relevant for regulating emotional experience, comprises the septohippocampal system and the *amygdala*. The septohippocampal system seems to serve a goal-conflict resolution without arousal, while the activation of the *amygdala* represents fear states, without the cognitive aspect of approach-avoidance conflict. Therefore, in order to resolve a distressed response to negative occurrence, a co-activation of these regions is essential[45]. This inter-regional relation can also contribute to the known bias for greater recollection of emotional events[46,47,48]. However, the *amygdala* might also improve the retrieval of emotional memories indirectly, by interacting with the arousal system in the brainstem[49,50,51,52].

The described relation between the emergence of emotional experience and focus of attention suggests that its establishment involves 'binding'-like neuronal phenomena, much like the perceptual and conscious experiences[53,54]. Similarly, its occurrence is expected to facilitate long-range transient connections between different brain regions[55]. Accumulating data from in vitro and in vivo studies suggests the importance of medium- to fast-oscillations (i.e. β and γ EEG bands, respectively) in coding of sensory information, as well as in promoting cellular mechanisms of synaptic plasticity[56]. Fast oscillations appear to be specifically associated with mechanisms used by the brain to establish communication between spatially separate brain regions[57,58], thus allowing for high-level integration of information as required for binding-like phenomena. The cellular mechanisms of fast oscillations and their synchronization over large cortical distances are not completely understood. In the cortex, GABA-mediated inhibitory interneurons and synchronizing principle pyramidal cells were postulated as both necessary and sufficient conditions for the generation of synchronized high frequency activity in neuron assembly[59]. Recent data demonstrating alterations in fast oscillation in diffused brain pathologies, such as dementia and schizophrenia, support their essential role in integrated behavior[60,61]. However, several studies presented contradictory findings to the claim that high-frequency synchronization is specifically involved in such integration[62]. It is therefore plausible to assume 'binding'-like phenomena occur through more than one frequency band[63].

CONCLUSION

Prior brain imaging studies point to the *amygdala, hippocampus* and *prefrontal cortex* as core brain regions that mediate the expression of personality traits of negative affect (i.e. neuroticism), as well as differentiate between PTSD patients and healthy matched controls. Considering the significant correlation between elevated levels of neuroticism, and the occurrence of PTSD, we propose here that a reciprocal connection exists between brain function, personal traits and pathological reactions to stress. Furthermore, we suggest that some of the differences in brain reactivity between PTSD patients and healthy control groups are related to a personality profile that determines the individual's vulnerability to stress and his emotional experience style. It is therefore assumed that the activity pattern of core brain areas such as the *amygdale, the prefrontal cortex* and the *hippocampus* might present a predisposing biomarker for developing PTSD following an extreme stressful event. So far, studies concerned with the human emotional experience have focused on estimating regional brain activity. However, it is proposed here that measuring inter-regional co-activations at different oscillation frequencies by electrophysiological methods might reveal the neural mechanism that underlies the effect of emotional experience on individual QOL.

ACKNOWLEDGEMENT

We would like to thank Michal Zarithky for assistance with the manuscript. We would also like to thank Israeli Ministry of Science, Adams Super Center for Brain Sciences and the Israeli Science Foundation "Bikora" for financial support.

REFERENCES

1. Frijda, N. H. (2005). "Emotion experience." *Cognition and Emotion*, **19**: 473–498.
2. Lambie, J. A. and A. J. Marcel (2002). "Consciousness and the varieties of emotion experience: a theoretical framework." *Psychol Rev* **109**(2): 219–59.
3. Lang, P. J., M. Davis, et al. (2000). "Fear and anxiety: animal models and human cognitive psychophysiology." *J Affect Disord* **61**(3): 137–59.
4. Breslau, N. (2001). "The epidemiology of posttraumatic stress disorder: what is the extent of the problem?" *J Clin Psychiatry* **62 Suppl 17**: 16–22.
5. Rapaport, M. H., C. Clary, et al. (2005). "Quality-of-life impairment in depressive and anxiety disorders." *Am J Psychiatry* **162**(6): 1171–8.

6. Creamer, M., A. C. McFarlane, et al. (2005). "Psychopathology following trauma: the role of subjective experience." *J Affect Disord* **86**(2-3): 175–82.
7. McFarlane, A. C. (1988). "The aetiology of post-traumatic stress disorders following a natural disaster." *Br J Psychiatry* **152**: 116–21.
8. Talbert, F. S., L. C. Braswell, et al. (1993). "NEO-PI profiles in PTSD as a function of trauma level." *J Clin Psychol* **49**(5): 663–9.
9. Fauerbach, J. A., J. W. Lawrence, et al. (2000). "Personality predictors of injury-related posttraumatic stress disorder." *J Nerv Ment Dis* **188**(8): 510–7.
10. Holeva, V. and N. Tarrier (2001). "Personality and peritraumatic dissociation in the prediction of PTSD in victims of road traffic accidents." *J Psychosom Res* **51**(5): 687–92.
11. Breslau, N., G. C. Davis, et al. (1991). "Traumatic events and posttraumatic stress disorder in an urban population of young adults." *Arch Gen Psychiatry* **48**(3): 216–22.
12. Costa, P. T., Jr. and R. R. McCrae (1997). "Stability and change in personality assessment: the revised NEO Personality Inventory in the year 2000." *J Pers Assess* **68**(1): 86–94.
13. Goldberg, L. R.,. The development of markers for the big-five factors structure Psychological Assessment, 1992.
14. McCrae, R. R. and O. P. John (1992). "An introduction to the five-factor model and its applications." *J Pers* **60**(2): 175–215.
15. McFarlane, A., C. R. Clark, et al. (2005). "The impact of early life stress on psychophysiological, personality and behavioral measures in 740 non-clinical subjects." *J Integr Neurosci* **4**(1): 27–40.
16. Bolger, N. and A. Zuckerman (1995). "A framework for studying personality in the stress process." *J Pers Soc Psychol* **69**(5): 890–902.
17. Aarstad, H. J., A. K. Aarstad, et al. (2003). "The personality and quality of life in HNSCC patients following treatment." *Eur J Cancer* **39**(13): 1852–60.
18. Whittle, S., N. B. Allen, et al. (2006). "The neurobiological basis of temperament: towards a better understanding of psychopathology." *Neurosci Biobehav Rev* **30**(4): 511–25.
19. Canli, T. (2004). "Functional brain mapping of extraversion and neuroticism: learning from individual differences in emotion processing." *J Pers* **72**(6): 1105–32.
20. Sakamoto, H., R. Fukuda, et al. (2005). "Parahippocampal activation evoked by masked traumatic images in posttraumatic stress disorder: a functional MRI study." *Neuroimage* **26**(3): 813–21.
21. Rauch, S. L., L. M. Shin, et al. (2006). "Neurocircuitry models of posttraumatic stress disorder and extinction: human neuroimaging research–past, present, and future." *Biol Psychiatry* **60**(4): 376–82.
22. Gray, J. A. (1983). "A theory of anxiety: the role of the limbic system." *Encephale* **9**(4 Suppl 2): 161B-166B.
23. Hendler, T., P. Rotshtein, et al. (2003). "Sensing the invisible: differential sensitivity of visual cortex and amygdala to traumatic context." *Neuroimage* **19**(3): 587–600.
24. Rauch, S. L., P. J. Whalen, et al. (2000). "Exaggerated amygdala response to masked facial stimuli in posttraumatic stress disorder: a functional MRI study." *Biol Psychiatry* **47**(9): 769–76.
25. Shin, L. M., C. I. Wright, et al. (2005). "A functional magnetic resonance imaging study of amygdala and medial prefrontal cortex responses to overtly presented fearful faces in posttraumatic stress disorder." *Arch Gen Psychiatry* **62**(3): 273–81.
26. Protopopescu, X., H. Pan, et al. (2005). "Differential time courses and specificity of amygdala activity in posttraumatic stress disorder subjects and normal control subjects." *Biol Psychiatry* **57**(5): 464–73.
27. Adolphs, R. and A.R. Damasio.. Neurobiology of emotion at a systems level In: Borod, J.C. (Ed.), The neuropsychology of emotion. Oxford University Press, 2000. p. 194–213.
28. Zald, D. H. (2003). "The human amygdala and the emotional evaluation of sensory stimuli." *Brain Res Brain Res Rev* **41**(1): 88–123.
29. Canli, T., et al., Behav Neurosci, 2001. 115(1): p. 33–42.
30. Barbas, H. (2000). "Connections underlying the synthesis of cognition, memory, and emotion in primate prefrontal cortices." *Brain Res Bull* **52**(5): 319–30.

31. Zubieta, J. K., J. A. Chinitz, et al. (1999). "Medial frontal cortex involvement in PTSD symptoms: a SPECT study." *J Psychiatr Res* **33**(3): 259–64.
32. Davidson, R. J., D. Pizzagalli, et al. (2002). "Depression: perspectives from affective neuroscience." *Annu Rev Psychol* **53**: 545–74.
33. Aggleton JP (2000) The amygdala. A functional analysis. Oxford University Press, Oxford
34. Milad, M. R., I. Vidal-Gonzalez, et al. (2004). "Electrical stimulation of medial prefrontal cortex reduces conditioned fear in a temporally specific manner." *Behav Neurosci* **118**(2): 389–94.
35. Santini, E., H. Ge, et al. (2004). "Consolidation of fear extinction requires protein synthesis in the medial prefrontal cortex." *J Neurosci* **24**(25): 5704–10.
36. LeDoux, J. E. (2000). "Emotion circuits in the brain." *Annu Rev Neurosci* **23**: 155–84.
37. Nader, K., G. E. Schafe, et al. (2000). "Fear memories require protein synthesis in the amygdala for reconsolidation after retrieval." *Nature* **406**(6797): 722–6.
38. Grace, A. A. and J. A. Rosenkranz (2002). "Regulation of conditioned responses of basolateral amygdala neurons." *Physiol Behav* **77**(4-5): 489–93.
39. Sotres-Bayon, F., C. K. Cain, et al. (2006). "Brain mechanisms of fear extinction: historical perspectives on the contribution of prefrontal cortex." *Biol Psychiatry* **60**(4): 329–36.
40. Laviolette, S. R. and A. A. Grace (2006). "Cannabinoids Potentiate Emotional Learning Plasticity in Neurons of the Medial Prefrontal Cortex through Basolateral Amygdala Inputs." *J Neurosci* **26**(24): 6458–68.
41. Marsicano, G., B. Moosmann, et al. (2002). "Neuroprotective properties of cannabinoids against oxidative stress: role of the cannabinoid receptor CB1." *J Neurochem* **80**(3): 448–56.
42. Patel, S., C. T. Roelke, et al. (2005). "Inhibition of restraint stress-induced neural and behavioural activation by endogenous cannabinoid signalling." *Eur J Neurosci* **21**(4): 1057–69.
43. Giuffrida, A., F. M. Leweke, et al. (2004). "Cerebrospinal anandamide levels are elevated in acute schizophrenia and are inversely correlated with psychotic symptoms." *Neuropsychopharmacology* **29**(11): 2108–14.
44. Semple, D. M., A. M. McIntosh, et al. (2005). "Cannabis as a risk factor for psychosis: systematic review." *J Psychopharmacol* **19**(2): 187–94.
45. McNaughton, N. and J. A. Gray (2000). "Anxiolytic action on the behavioural inhibition system implies multiple types of arousal contribute to anxiety." *J Affect Disord* **61**(3): 161–76.
46. Kilpatrick, L. and L. Cahill (2003). "Amygdala modulation of parahippocampal and frontal regions during emotionally influenced memory storage." *Neuroimage* **20**(4): 2091–9.
47. Dolcos, F., K. S. LaBar, et al. (2004). "Interaction between the amygdala and the medial temporal lobe memory system predicts better memory for emotional events." *Neuron* **42**(5): 855–63.
48. Richardson, M. P., B. A. Strange, et al. (2004). "Encoding of emotional memories depends on amygdala and hippocampus and their interactions." *Nat Neurosci* **7**(3): 278–85.
49. Davis, M. and P. J. Whalen (2001). "The amygdala: vigilance and emotion." *Mol Psychiatry* **6**(1): 13–34.
50. Berntson, G. G., M. Sarter, et al. (2003). "Ascending visceral regulation of cortical affective information processing." *Eur J Neurosci* **18**(8): 2103–9.
51. Bouret, S., A. Duvel, et al. (2003). "Phasic activation of locus ceruleus neurons by the central nucleus of the amygdala." *J Neurosci* **23**(8): 3491–7.
52. Strange, B. A. and R. J. Dolan (2004). "Beta-adrenergic modulation of emotional memory-evoked human amygdala and hippocampal responses." *Proc Natl Acad Sci U S A* **101**(31): 11454–8.
53. Metzinger T (1995) Conscious Experience. Imprint Academic
54. Engel, A. K., P. Fries, et al. (1999). "Does time help to understand consciousness?" *Conscious Cogn* **8**(2): 260–8.
55. Lamme, V. A. (2004). "Local versus global recurrency commentary on: Cortex, countercurrent context, and dimensional integration of lifetime memory by Bjorn Merker." *Cortex* **40**(3): 580–1; discussion 582–3.
56. Engel, A. K. and W. Singer (2001). "Temporal binding and the neural correlates of sensory awareness." *Trends Cogn Sci* **5**(1): 16–25.

57. Whittington, M. A., H. J. Faulkner, et al. (2000). "Neuronal fast oscillations as a target site for psychoactive drugs." *Pharmacol Ther* **86**(2): 171–90.
58. Munk MH, Roelfsema PR, Konig P, Engel AK, Singer W (1996) "Role of reticular activation in the modulation of intracortical synchronization." *Science* 272: 271–274.
59. Traub, R. D., M. A. Whittington, et al. (1996). "Analysis of gamma rhythms in the rat hippocampus in vitro and in vivo." *J Physiol* **493 (Pt 2)**: 471–84.
60. Lee KH, Williams LM, Breakspear M, Gordon E. (2003) "Synchronous gamma activity: a review and contribution to an integrative neuroscience model of schizophrenia." *Brain Research review* 41:57–78.
61. Spencer KM, Nestor PG, Niznikiewicz MA, Salisbury DF, Shenton ME, McCarley RW (2003) "Abnormal neural synchrony in schizophrenia." *J Neurosci* 23: 7407–7411.
62. Tovee, M. J. and E. T. Rolls (1992). "Oscillatory activity is not evident in the primate temporal visual cortex with static stimuli." *Neuroreport* **3**(4): 369–72.
63. Newman, J. and A. A. Grace (1999). "Binding across time: the selective gating of frontal and hippocampal systems modulating working memory and attentional states." *Conscious Cogn* **8**(2): 196–212.

CHAPTER 5

CROSS-CULTURAL QUALITY OF LIFE RESEARCH IN MENTAL HEALTH

Conceptual approaches, assessment strategies, empirical results and potential impact

MONIKA BULLINGER[1,*], SILKE SCHMIDT[1] AND DIETER NABER[2]

[1] *University Medical Centre of Hamburg Eppendorf Institute and Policlinic for Medical Psychology, Hamburg, Germany*

[2] *University Medical Centre of Hamburg Eppendorf, Department of Psychiatry, Center of Psychosocial Medicine, Hamburg, Germany*

INTRODUCTION

Quality of Life (QoL) is one of the concepts reflecting multiple scientific perspectives, including psychological, political, economic and philosophical approaches. As a research topic in sociology, QoL emerged already within the first half of the last century, with early studies in mostly western nations or societies. In the meantime, the topic is well established in the social sciences, and current knowledge is accumulated in scientific Societies (eg International Society for Quality of Life Studies, ISQOLS or International Society for Quality of Life research, ISOQOL) and represented in respective Journals (e.g. in Social Indicator Research, Quality of Life Research). Recently the term 'Quality of Life' has also been adopted by health scientists, and has been taken to reflect subjective perception of health in terms of along the lines of the WHO definition of health. To date, the Quality of Life has a history both in social sciences as well as health sciences, and a communication between the two has just started to develop and expand.

Health related quality of life (HrQoL) denotes the patients' perception of well-being and function in physical, emotional, mental, social and every day life domains. In medicine, HrQoL is used in clinical studies to evaluate patient reported treatment

* University Medical Centre of Hamburg Eppendorf, Institute and Policlinic for Medical Psychology, University of Hamburg, Martinistr. 52, 20246 Hamburg / Germany, Tel.: ++49/40/ 42803-6430, Fax: ++49/40/ 42803-4940

M.S. Ritsner and A.G. Awad (eds.), Quality of Life Impairment in Schizophrenia, Mood and Anxiety Disorders, 67–98.

outcomes, in epidemiological studies to describe population health, in quality assurance to document the quality of care and in health economic research to assess health care benefits. Over the past decades, several psychometrically sound patient-based instruments to assess generic and disease specific HrQoL have been developed and applied, also in Psychiatry.

Most current papers in Quality of Life research refer to industrialised nations or so called western world. By comparison, information about the Quality of Life of countries in development is still lacking. This is less true for structural indicators of Quality of Life of a population, such as infant mortality or gross national product, which are well documented the data banks of respective organizations (for example the UN or the WHO, e.g. The World Health Report[1]). However, beyond the assessment of Quality of Life from aggregate levels as reflected by structural aspects of a given societies current state, Quality of Life as viewed from the individual perspective is not well understood.

Anthropological research as well ethnological studies have for a long time attended to issues related to Quality of Life, however methods of study were not primarily focused on understanding the subjective representation of a persons world, but rather on documenting and understanding behaviours, cultural rites or specific beliefs of the population studied. This view of other cultures places them more into the position of an interesting subject to study rather than attempting to understand well-being from the individual perspective. Within the scientific tradition of descriptive research, the aim to understand how people view themselves, their lives, their immediate surroundings as well as their larger social situation, is a relatively recent focus. The question is how to approach, measure, understand peoples quality of life and how to implement this knowledge – by acting together and making use of the Quality of Life information given by individual members or groups within the given nation or culture.

Reflections on cultural issues in quality of life and mental health include a thorough discussion about the meaning of the term culture, a review of current knowledge in cultural differences in mental health as well as a thorough examination of cross-cultural quality of life research in mental health. The three main pillars of the current chapter, namely culture, mental health and quality of life, will be explored in order to better understand how quality of life in mental health can be measured across cultures and how these results can be interpreted and used to improve research and practice.

To address this topic, a definition of the meaning of Quality of Life, its methods of assessment and the review of the current state of knowledge is necessary. On the basis of available research on Quality of Life both from the social sciences as well as the health sciences, the aim is to develop a frame work, with in which Quality of Life can be studied internationally and to define a purpose for which such research might be useful. A general topic underlying this exercise is the question to which extent Quality of Life can be considered a human universal and how concepts, methods and results of Quality of Life assessment might be comparable across cultures.

One of the major challenges lies within the international and cross-cultural development and use of quality of life measures. The role of culture is a conceptual issue; empirically, however, results regarding cultural differences or similarities strongly depend on the type of instruments used. According to recent guidelines of instrument development, necessary steps of cross-cultural instrument development include item formulation (using focus groups and item writing), response scale choice (in terms of format and time frame), translation (with forward and backward translations as well as international harmonization), pilot-testing (with cognitive debriefing and revision) and, after field-testing, the statistical test-theoretically based analysis (on the level of single countries or the combined data set), using classical or modern psychometric methods). The concept of differential item functioning (DIF) is essential for understanding the implication of cross-cultural differences, demonstrating the implications of minimizing or enhancing cross-cultural DIF.

Two strategies have been followed to approach the issue of cross-cultural differences in constructing quality of life instruments. One is to minimize cross-cultural differences by identifying and eliminating culture-specific responses to items in order to achieve as much as possible a "culture-free" assessment. The other is to explicitly attend to and even enhance such differences in order to produce as much as possible "culturally-sensitive" instruments. Depending on the measurement purpose, both approaches are useful, but have diverging implications.

CONCEPTUAL APPROACHES TO HEALTH AND QUALITY OF LIFE

Definitions

It has been proposed that quality of life is identical with the subjective representation of health. The WHOQoL Group and other groups have suggested that health-related quality of life can best be understood as the operationalization of the WHO definition of health, through the inclusion of self-report data on well-being and functioning in physical, emotional and social domains of life[2]. There is also a general recognition that other factors, e.g. cultural, spiritual and environmental aspects, are associated with health-related quality of life[3]. These have been most clearly stated in the QoL definition of the WHOQoL Group[4]:

"Quality of life is an individuals' perceptions of their position in life in the context of the culture and value systems in which they live, and in relation to their goals, expectations, standards and concerns. It is a broad ranging concept affected in a complex way by the persons' physical health, psychological state, level of independence, social relationships and their relationship to salient features of their environment".

In a review of definitions of quality of life used in the literature, 27 were identified and of these, 85% include emotional well-being, 70% include physical health, 70% social and family connections, 59% material wealth or well-being and 56% work or other form of productive activity[5]. In a more recent review, Garrett et al.[6] have shown that 55% of instruments assessing quality of life (total $N = 690$)

were related to physical functioning, pain or symptoms, 38 % to psychological well-being, and 8 % to social well-being, however, there was also a range of other issues (13 % to personal constructs). Garret and co-workers[5] interpreted this as a shift towards physical components of health in Qol instrument developments, particularly in health-related quality of life research. Recently published compendia on quality of life instruments for health conditions underscore this tendency to assess functioning in health-related quality of life assessment[7,8].

The discussion on the types of components operationalised in QoL research (physical vs. emotional) is often associated with the gap between a functionalist approach towards health-related and subjective well-being research, with most approaches demanding that a clear distinction has to be made. In health related quality of life research, functional impairments, such as being not able to walk, have been treated separately from the subjective consequences of these functional limitations.

Another controversial conceptual debate in quality of life research pertains to the issue whether it should be conceived as a functional state or as the subjective evaluation of health status. This is related to the distinction between more subjective and more objective, tangible indicators of quality of life[9,10,11], with external or environmental indicators serving as objective indicators of the people's life situation, and the personal evaluation and subjective well-being as subjective indicators[12,13]. In general, the subjective refers to the personal evaluation of symptoms or life contexts, while the objective relates to measures independent of the subjective evaluation.

Current notions of quality of life emphasise the role of perceptions as well as cognitive and emotional evaluations of health-related information, rather than simply self-reported behaviour and functioning[7,8]. The examination of relationships between objective and subjective health indicators shows that persons are informed about their health status and can account for symptoms and diagnosis in accordance with clinical information. There is, however, evidence that the evaluation of symptoms, experiences and behaviours neither directly nor proportionally relates to clinical indicators, such as the severity of symptoms[5]. Thus, in order to fully capture the concept, health status measurements should be operationalized both via clinical information and self-reports, which is in line with the classical approach to health indicator research[14]. Indicators of perceived quality of life also should complement it, which is a new component of health interview surveys.

Subjective Versus Objective Perspective on Quality of Life

Of major interest is the question whether quality of life is of a subjective or objective nature. In the international 'life quality literature', which is related to the interrelationship of standard of living or wealth and subjective well being, external or environmental factors were treated as the objective point of reference whereas the perspective of the individual is considered to be subjective. The distinction between the subjective and the objective depends on the theoretical framework and does not fully reflect health-psychological theories on the subjective.

There has been evidence that the subjective and the objective in measuring health or wealth are only marginally correlated. For example, a recent study of the relationship between life satisfaction and standard of living as measured by Gross National Product and Gross Domestic Product in 41 different countries showed that while there are strong correlations between economic wealth, measured by GNP and quality of life, only 17% of the variance of subjective well-being can be explained by GNP[11], though there were differences between rich and poor nations in the study. It is this discrepancy that has led researchers to reject subjective indicators because of their discordance with objective economic indicators.

Several studies have investigated the interrelationship between objective and subjective indicators of health and quality of life. Cummins reviewed some of the evidence on the association between different types of objective indicators and different types of subjective indicators as well as the interrelationship between subjective and objective indicators[5]. The results were summarized according to the population the data came from, the general population versus specific populations that have to face putative threats to their life conditions, such as people who are very old, very poor or suffering from chronic and severe diseases or disabilities. The intercorrelation of QoL measures within either the objective (O vs. O) or subjective (S vs. S) dimension has been shown to be higher than the intercorrelation of QoL measures between dimensions (O vs. S). However, the magnitude of the correlation between objective and subjective measures of quality of life also seems to be part of a population effect.

Cummins[5] concluded on the basis of a review of studies that the intercorrelation of QoL measures between the objective and subjective dimensions increases, if objective life quality decreases to low levels because putative threats to objective life conditions may defeat the homeostasis of subjective well-being. The correlation between objective and subjective measures was $r = .26(+/-.12)$ in vulnerable populations, while the correlation in studies referring to the general population was $r = .12(+/-.08)$*. The results of this meta-analytic revie w have to be interpreted carefully because of the considerable heterogeneity of the indicators included and the way they have been operationalized.

While research generally is aimed at having a high concordance between subjective and objective indicators, the discordance can also be an important source of information. The capacity for adaptive or homeostatic processes indicates that the correlation between subjective and objective indicators is notable and straightforward, but that this relationship changes dynamically across time and life situations[5,15].

The crucial distinction between the objective and the subjective was not primarily made in the health sector in respect to different dimensions of the health status of people, but evolved from the economic debate of the state[16].

---

* The sample size of the respective studies varied between $n = 156$ and $n = 1980$

The work of Amartya Sen[17,16] on the relationship between well-being and socioeconomic wealth, e.g. income or the cross-national product has shown that populations with the same per capita incomes can differ significantly in health status. This can best be demonstrated in non-westernized countries. Costa Rica, Sri Lanka and Kerala state in India, for example, have about the same average level of income as Pakistan, Afghanistan and Morocco. Yet the infant mortality rates for the former group average 64 per 1000 life births, compared with 173 in the latter. Average life expectancy is 61 in the first groups, 45 in the second group. Sen[16] discussed a variety of causes for this discrepancy, however, attributed it specifically to the fact that countries with a better health status have placed a greater emphasis on the importance of children and women in their culture and social environment and in their social policies.

The relationship between health and income distribution has attracted intense research interest when differentiating between low-income countries, moderate and high-income countries. Developed countries show rather strong correlations between the degree to which cross-national income is distributed and health status[18]. Comparing health status measures for a group of countries with measures of the inequality of income distribution, Wilkinson[18] found a positive correlation between inequality and mortality suggesting that high unequal societies are less healthy overall. Marmot's studies of UK civil servants[19], for example, found health differences that were unambiguous and large (three-to-one differences in mortality rates) and that showed a gradient across groups – linked to hierarchy per se and not to deprivation or poverty. One of the most prominent findings of the studies of Marmot was that these differences could not be interpreted simply by life style choices. An important question is whether this health inequality gap diminishes over time or not, e.g. triggered by improved health care systems. Mackenbach[20] concluded in a review referring to data from different countries and his own studies that the evidence for a correlation between income inequality and the health of a population is slowly dissipating, however, the supporting evidence to date is almost entirely restricted to analyses in the United States.

The Cross-cultural Focus in Quality of Life Research

The expanding field of quality of life research is the increase in the demand for international quality of life research in terms of conceptual clarifications, methodological approaches available and practical applications in the international context[21,22].

The term "international" has different meanings: politically it refers to a nation, geographically it refers to country, anthropologically it refers to culture, sociologically it refers to society and psychologically it refers to the identity of its members. All of these meanings are reflected in the notion of language which makes it one of the key issues in working with quality of life assessments internationally.

For the following discussion it might be helpful to distinguish between the terms "international" and "cross-cultural". Usually the term "international" is used to refer to phenomena concerning more than one nation or culture with a possible extension to cultural groups within one nation. In its focus on quality of life research, the term "international" denotes primarily activities of different countries in the quality of life field (i.e. studies from different countries concerning specific research questions). Cross-cultural" quality of life research, by contrast, denotes an additional collaborative and comparative effort in the quality of life field (i.e. using the same instrument to assess quality of life eg. with the aim of comparison across cultures). Thus, the use of quality of life tools in different cultures as well as their application across cultures is a challenge for researchers.

While anthropology has focussed on quality of life indicators across cultures and nations, this research mainly concerns the so-called objective or structural indicators of quality of life (such as gros national product, infant mortality, life years). Subjective indicators of quality of life have more recently been included in sociological surveys on well-being of citizens of e.g. the United States[23] or Germany[24]. These surveys focus on quality of life in terms of satisfaction with different life domains including the material, financial as well as political aspects of well-being. The interest in quality of life research has also reached the medical field concerns the description of function and well-being of populations with and without medical conditions (epidemiological perspective), its use as an outcome criterion for interventions (clinical perspective) and its contribution to decision making in the health care field (political perspective). These objectives are not only of relevance at a national level, also international efforts are directed at these goals.

Health related quality of life in contrast focuses directly on dimensions of function and well-being that are relevant for a person's judgement of his or her health status. Especially as concerns chronically ill patients, this perspective is directly linked to concept of disease.

From a cross-cultural perspective it is necessary to realize that illness is a patient's perspective of and response to disease, the meaning of which is largely determined by cultural schemata. As Hutchinson[25] points out these meanings of disease are most clearly noticeable in so called folk-illnesses. These include conditions like susto in Latin Americans (depressive anxiety), koro in Asia (the fear of penis withdrawal into the body), windigo in native Americans (cannibalistic obsession) or as heart distress in Arabia (a condition occurring under specific distressing life conditions). Other examples for the cultural bases of illnesses as referred to by Hutchinson[25] include the tribe of the Mano in Liberia who do not consider Malaria a disease because so many suffer from it or the perception of measles, mumps and whooping cough as inherent conditions of normal growing up in rural Greece. Well known are conceptions of disease as function of a balance of different forces (e.g. between hot and cold in Mexico or between Ying and Yang in China) or as an activity of supernatural forces (as viewed by the Abron of the Ivory Cost) or of enemies (as viewed by the Doubans of Melanesia).

If disease, as anthropological research suggests, is so very much culture-bound, how could quality of life be culture free? The basic scepticism, especially of colleagues from Anthropology, is captured in the following citation: "Although some researchers may desire a scale or similar instruments for global assessments of cultures, permitting comparison of the "nature" of one culture with that of another, no such scale exists. In fact, given the multiplicity of variables or domains comprising a culture, that goal is unrealistic, both theoretically and methodologically"[26].

In reviewing attempts at measuring health-related quality of life cross-nationally, Guarnaccia[27] points out that "researchers start with an underdeveloped notion of culture and its impact on quality of life assessment. In focusing on particular ethnic populations, there is a lack of attention to inter- and intracultural diversity among study populations. Inadequate approaches are applied to the adaptation and translation of quality of life instruments"[27]. Thus it seems that the call for construction of measures of quality of life that should be sensitive to language and dialect, customs, beliefs and traditions as well as education and socio-economic status of respondents has remained unheard.

ASSESSMENT STRATEGIES

The Cross Cultural Quality of Life Measures

Given the growing interest in international quality of life research and in cross-culturally applicable measures, both for research in a given country as well as in terms of comparisons across countries, the following questions are essential:

1. Is quality of life a relevant concept in a given nature/culture?
2. Do nations/cultural groups share an identical set of concepts about quality of life?
3. Can quality of life concepts be assessed with quality of life instruments?
4. Is quality of life measurable across nations/cultures with the same instrument?
5. Can quality of life data be compared across nations/cultures?
6. Do cross-cultural quality of life results provide a sound basis for decision making in the health care field?

Brislin et al.[28] have pointed out early that in translating measures from one culture to another the aspect of semantic equivalence (i.e. comparable meaning), content equivalence (i.e. the relevance of questions across cultures), technical equivalence (the types of question used), and criterion equivalence (the functioning of the questionnaire in the respective culture) are of importance. In order to approach these key issues, a set of basic criteria to judge the equivalence of instrument versions across cultures, as provided by Hui and Triandis[29] is necessary. These criteria include *functional equivalence*, which concerns the adequacy of translations, *scale equivalence*, which concerns the comparability of response scales, *operational equivalence*, which concerns the standardization of psychometric testing procedures and *metric equivalence* concerning the order of scale values across a continuum.

In developing a cross-nationally usable measure of quality of life, three goals can be distinguished[22]. The first goal would be to develop an instrument which is universally applicable across all cultures. The more modest second goal would include the development of a core instrument, which might be universally applicable but which contains specific add-on national modules. The third option pertains to the development of a series of national instruments, which are specific to each culture.

So far, the efforts in existing research on cross-cultural instruments focuses on the first aim with the question whether instruments are universally applicable across cultures. The oldest example for Quality of Life instrument used in both developed and developing countries is the adaptation of the Cantrill Self-Anchoring Striving Scale[30], which is a very easy and internationally usable ladder scale with rungs, the top rung representing the best possible and the bottom rung representing the worst possible description of a feeling state. The scale is flexible, can be offered with different types of questions at different time points and has been used in general population studies in over 40 western and non-western countries involving over 20.000 interviews. In spite of its frequent use for the purpose, the scale does not in the essential sense represent a quality of life instrument, because it is not multi-dimensional in nature, but rather representing a method of questioning, which uses person-defined endpoints. Similarly instruments used in psychiatric research are available in different languages, but are not truly QOL instruments.

Subsequently, from the 1980's, different working groups have been active in cross-cultural development mostly related to a specific instrument. One of the first groups to join efforts in the endeavour was the European Organization for Research and Treatment of Cancer (EORTC), which began to develop the EORTC quality of life Questionnaire in 1986[31,32]. Also active in the field is the European group for quality of life and health measurement group (EGQLHM) working with the Nottingham Health Profile (NHP). Along the guidelines of the instrument's authors, the group provided translation and psychometric testing of the Nottingham Health Profile in several languages[33]. In 1991, the International Quality of Life Assessment Project Group (IQOLA) was founded, which works with the SF-36 Health Survey[34]. In parallel, the European Quality of Life Project Group developed, which contributed to the development of the EUROQOL/EQ5D Questionnaire[35]. The World Health Organization quality of life (WHOQOL) Group followed around the same time with the simultaneous effort at developing a quality of life instrument from different cultures[36]. Although the Sickness Impact Profile[37] (SIP) has been a widely used instrument to assess health-related quality of life, international efforts to work with the SIP were only begun in 1994[38]. In parallel the Functional Assessment of Cancer Treatment (FACT) Group began its work in translating and testing the FACT Questionnaire[39,40,41] as did the group Functional Living Index Cancer[42] (FLIC).

Three approaches can be distinguished in cross-culturally developing an instrument[43]. The first concerns the *sequential* approach, which refers to transferring an existing questionnaire from one culture to another. This approach was used with the SF-36 Health Survey (Ware et al.[44] 1996), the Functional Assessment of Cancer Treatment[39] (FACT) questionnaire and with the Nottingham Health Profile[45] (NHP). The second constitutes the *parallel* approach, which includes assembling an instrument based on existing scales from different countries, which e.g. was used by the European Organization for Research and Treatment of Cancer quality of life Working Group[31] (EORTC-QLQ-C30). The last is the *simultaneous* approach, which involves the cooperative cross-cultural development of a questionnaire, which so far was only used by the World Health Organization quality of life Working Group[36].

Each of these approaches includes as basic steps in the developing process 1. the *identification* of relevant items 2. the *translation* of the questionnaire, 3. its *psychometric testing* and 4. the *norming* process[22]. Also comparisons between instruments are to be made on the ground of these steps[46].

Steps in Cross-cultural Instrument Development

Identifying relevant questions

In case of already existing and sequentially derived questionnaires, item development is not the issue, however when exploring quality of life under certain conditions such as specific health conditions or living situations it is important to sample the items that are relevant for the topic. One way to do this, which has been used in different studies, is to conduct focus groups in which persons concerned by the conditions under study are asked individually (interview version) or in groups (with a mediator) to discuss and expand on issues surrounding the condition. In both cases – focus interview or focus group- it is vital to collate questions (mostly open questions) that could stimulate the reflection of the respondents. In the interview, semi-standardized versions with a defined topic selection are preferable to highly structured interviews. In focus groups, it is very helpful to have a manual or study protocol describing the sequence of the procedure. Careful protocoling of the discussions is important in order to be able to derive adequate items, and a special challenge is then to record, select, sort, screen and modify the statements.

Item development as a conducted via focus groups and interviews is a complicated issue because sentences or statements have to be recorded adequately, selected appropriately and adapted or modified in order to avoid redundancies. Nevertheless it is important to capture the meaning of each individual statement as closely as possible, taking to account language considerations. Once item identification, which is helped by a process of two persons reviewing the statement, has been completed, item writing, i.e. the actual formulation of items follows.

Either as a separate step of as part of a pilot study, a second process- the so-called cognitive debriefing - should be conducted. Within this process, formulated items are presented to potential respondents in order to find out whether the items are clear understandable and capture the experiential frame of the issue concerned. To do this an interview can be conducted along with the presentation of the items or space for written comments can be provided, in which in addition to rating the performance of the items alternative formulations can be noted. After item generation, item writing and cognitive debriefing a preliminary questionnaire as well as the decision on answer scales, an questionnaire can be put together for the next step which after pilot testing subsequently is psychometric testing.

Translating instruments

The aspect of translation has been most intensively been dealt with in recent as well as older literature[47,48]. From cross-cultural and comparative sociological research as well as from cross-cultural psychiatry and educational psychology, theoretical foundations and methodological approaches to translating instruments from one culture to another have been suggested. Here, each working group has developed its own procedures for translation, which are essentially based upon a forward translation. However, the number of translators necessary as well as the use of back translations is debated. While e.g. in the Nottingham Health Profile Group strong emphasis is placed upon discussing forward translations in a focus group of health care professionals and patients suffering from the condition under question, the use of back translators was emphasized in the translation of the SF-36 Health Survey. In the FACT Group, the issue of translation was strongly emphasized by including in addition to several translators from each country a group of experts in the field, which were asked to review the translations, and a linguist who revised the translations. In the WHOQOL Group, translations were even more complicated by the fact that they were to be performed from a wide variety of original languages back into English, a process, which also was surveyed by quality ratings of translations, as was the case in the IQOLA SF-36 Group. In the EORTC Group the translations issue was pragmatically solved by obtaining different translations, which were then to be reviewed by the national coordinators in each country. The FACT relies on double translation methodology, the use of an expert advisory committee pilot testing and thorough linguist revision of the translations.

In reviewing different approaches to translation, Acquadro et al.[49] stress the need to include at least two forward translations with a comparative discussion and is sceptical about the use of back translations, which often are hampered by potential inferior quality of translations, which then unruly affects judgement of the forward translations. Most important is the international harmonization of translations into different countries by getting together a group of bilingual persons from different countries, which are able to interact and critically review each other's translations. While the basic philosophy of most guidelines for translation focussed upon the adequacy of the translation from the original into the target language, Guyatt[50] questions the attempt to transpose the measure from one country to another as closely as possible, arguing that during translation inconsistencies and illogical formulations as well as culturally unique expressions can occur, which should be the basis for reformulation of the question (also for the original) rather than adaption in the target language. In spite of the differences of the translation approaches, most authors agree that the use of two forward translators is absolutely mandatory, the use of a back translator is discussable and the use of focus groups to evaluate the applicability of the translated questionnaire in a specific country is recommendable.

As concerns the answer categories it is not only important to identify appropriate answer categories: answer categories also have to be translated in different languages and their meaning should be comparable between cultures. To do this, several international working groups have already compiled such item lists, which for example represent an interval scale characterized by the explicit distance between item words. Using a Thurstone scaling method, the equidistance of the translation in different languages were tested so that for answers on the intensity or frequency as well as the satisfaction domain, appropriate answer cases are available. The choice of an adequate answer scale is so especially important because it affects the scalability and scorability of the instrument.

Testing for psychometric properties

Basically, psychometric testing relies on methods and procedures from psychometric theory. This includes item descriptive statistics, measures of reliability, validity and sensitivity. Basically, reliability refers to the measurement accuracy of an instrument, validity to its representation of the relevant construct and sensitivity to its ability to identify change over time (or differences between groups). The international working groups, however, differ in the procedures employed for psychometric testing. The SF-36 IQOLA Group e.g. gives specific importance on the item discriminant validity and on the item response theory as a means to distinguish patterns of item responses across cultures. In addition, emphasis is placed on the performance of the questionnaire in terms of known group differences that is testing whether the SF-36 is able to differentiate between patients differing in the degree of disease severity. In psychometric testing, the FACT Group uses item analysis on the basis of the Rasch model, structural equation models and multivariate statistics to replicate the factor structure of the measure across countries. The WHOQOL Project and also the SF-36 Group mainly employ structural equation models (SEM) to test the measurement model of the questionnaire across countries. In the WHOQOL study involving data of over 4500 persons from 15 countries the SEM model is first fitted for the global data set and then replicated in each country[3]. In addition, item scale correlations as well as item descriptive statistics are used to test whether items are applicable across cultures. In the EORTC Group, item and scale statistics were used to decide, whether specific items followed the measurement model in one country as compared to another.

Norming

Of all the international working groups on quality of life assessment only the IQOLA Group had the opportunity to rely on population based data to assess the quality of life of the general population. So far, data from seven countries are available, these include the U.S., Great Britain, Germany, the Netherlands, Sweden, Denmark and Italy. More IQOLA member countries are in the process of collecting national norms (for instance in Denmark, or France). A comparison of the measurement model of the SF-36 dimensional structure across countries showed that western countries are highly similar in these models. In addition, comparisons of scale values of SF-36 sub-scores across countries shows a similarity in rating with only slight differences in country profiles. This, however, only applies to industrialized western countries. The similarity of the SF-36 structure as well as convergent scale values across cultures suggests that identical weighing systems can be used. The normative data of the SF-36 can be employed in each country to obtain age and gender specific reference groups for clinical quality of life data, which can be expressed as deviation from the respective age and gender specific norm.

Other working groups such as the NHP Group were able to collect a convenience sample of the general population, which could be traced back using available census data. Thus e.g. the NHP in Germany was used within a sample of over 500 inhabitants of a north German city, which can now be used as reference data for clinical groups[51]. Similar data sets on a norm study basis recently emerged also for the WHOQOL (e.g. Germany), however the number of countries is still small.

The Whoqol Experience

There have been approaches towards preserving culture-specificity in the development of subjective well-being measures. To ensure cross-cultural validity in developing countries during instrument construction and validation multinational collaboration is required at three levels. The conceptual latent model underpinning an instrument's manifest model must reflect different cultural nuances[52,53]; the manifest model – an instrument's descriptive system – must measure, in a representative sense, the universe of interest as defined cross-culturally in order to take account of emic (those effects that are specific to a culture) and etic effects (those effects that are invariant across cultures); and the observed model of QoL elicited by the instrument must demonstrate validity in different cultural settings[54].

Several research groups have tested the cross-cultural existence of structural components of health. The factor structure of the SF-36 health survey in 10 countries has been tested within the IQoLA Project[55]. It was not only possible to show the equivalence of SF-36 summary health scores estimated using standard and country-specific algorithms in 10 countries[56], but also the equivalence of the factor structure. However, despite its tendency to universality, several cultural differences were demonstrated. Japan, for instance, showed major differences in how the mental and physical health component is made up by its subfacets of which general health, role emotional functioning and bodily pain showing differential loadings on the principal components[57,55,56].

The development of The World Health Organization Quality of Life Assessment (WHOQoL) is an example for testing the universality of quality of life. It provides an example of how the approach towards cross-cultural testing is already implemented in the development of the measure. While in the IQoLA project the scale structure was analysed a posteriori by testing it in data from one country to the next, the WHOQoL study was unique by already bringing in culture specific items in the conceptual phase and piloting of the instrument. For instance, there were specifically marked items from Israel on security issues, and items from China and Japan on aspects of eating and its impact on QoL. However, it has been shown that the impact of culture-specific items does not contribute significantly to the scale structure within a country. This either suggests universality or that methods involved in cross-cultural testing are still not sensitive enough towards culture specific elements.

The universal approach towards developing quality of life and subjective well-being measures has sometimes been criticized discussed because of its emphasis on cross-cultural equivalent measures and its lack of integrating an anthropological perspective and the perspective of ethnic minorities[58,53]. Several studies have shown that existing health-related QoL items do not necessarily convey the same meaning for different ethnic groups.

To date the WHOQOL Group has used a common international protocol to develop two generic QoL profile measures, the WHOQOL-100 and the WHOQOL-BREF[59,60], and one age-specific module, the WHOQOL-Old for older adults. The group made use of a method of the simultanous approach to instrument development, described by Bullinger et al.[43]. All WHOQOL centres contributed to the definition of the domains and facets that were agreed to characterise Quality of Life, and, subsequently, all centres contributed items to the pilot version of the WHOQOL measure. Questions were drafted by focus groups, which generated ideas within each centre as to how and in which form to ask questions relating to quality of life. A second step of retaining cross-cultural equivalence was to rank order the importance of questions in each centre. This research showed that despite a high equivalence of the importance of domains and items across countries, a few culture-specific national items (these were culture specific items from focus groups entered into the global item pool as national items) showed high importance in some countries, such as items on energy and food in Asian countries, and items on security in Israel[61]. As a consequence, these items were retained as national items in the specific language version of the WHOQOL. Similar care was taken to generate cross-culturally comparable response scales by using three types of Likert scales which were worded in each culture according to the spacing of words falling at the 25%, 50% and 75% distances between the two anchors (low and high) of each scale.

Research with the different WHOQOL versions has shown that although country populations show different levels of Qol across domains, the overall structures has a high cross-cultural validity[53] suggesting a high degree of universality in this simultaneously developed measure. The WHOQOL Bref has been tested for its cross-cultural applicability in more than 50 countries[53]. The work of the WHOQOL group is thus an excellent example for the aim of developing a universal model but nevertheless preserving the possibility to retain culture-specific items and modules. In recent own research on simultaneously developed measures, many indices for a better cross-cultural performance of simultaneously developed measures were found. Examples are the cross-culturally developed DISABKIDS measure for children with chronic conditions[62], and healthy children (KIDSCREEN)[63] as well as in the EUROHIS measure[64,65,66] and the WHOQOL-Old measure[3]. In these developments, structural equation modelling was used to analyse cross-cultural differences or equalities in subjective health.

RESEARCH RESULTS

Cross-cultural Assessment in Population Studies

In studying heterogeneities in population health on a descriptive level, the most traditional strand is concerned with mapping differences in health outcomes in geographical space. This strategy is, in general, followed by National Health Surveys as well as international health surveys, the European Survey and the World Health Survey [1,67]. A typical example for cross-cultural comparisons on a descriptive level is data form the World Health Report [1]. The Global Burden of Disease study has further elaborated the traditional descriptive approach by generating estimates of the levels of major disease categories by region using a consistent methodology of health economics [68,69].

The second strand in analysing heterogeneities in international population health elaborates the descriptive approach and relates to differences between aggregates of individuals, defined in terms of shared socioeconomic characteristics including occupation, education, and income, respective of the particular target of health outcome [70,71]. Studies provided evidence for the fact that health status may be accounted by underlying differences in education, income and other socioeconomic indicators. Although the studies were set up on a cross-national basis, their primary objective was to study health inequalities in social groups, blue-printed by the health inequality approach of the Whitehall study [72].

According to the findings of the literature search, there were only a few studies based on individualised data [14,73,74]. These studies compared the impact on quality of life of socioeconomic subgroups across countries. A general finding was that the inequality gap did not differ significantly across health status measures between countries, however, with some tendencies on specific countries showing a higher health gradient, such as that northern countries had a higher inequality in health indicators [14,73].

The strategy of these studies was to apply the inequality approach used in specific countries to the cross-national setting in order to test whether the same gradients can be observed in cross-national comparisons. Authors who are concerned with socio-economic differences have interpreted their data as showing that differentials in mortality between ethnic groups within one country are essentially due to income or class inequalities, while other studies suggest that this explanation is at best partial [75]. Ethnic studies observed differences in both health status and health services use across ethnic groups. These have been variously attributed to cultural, socio-economic and genetic differences, as well as the impact of individual and institutional factors.

It was considered that the potential of cross-cultural comparative studies is to initiate a better understanding of the limits of explanations of inequalities in health obtained in particular settings, and thus caution against uncritical extrapolation of results to all countries or cultures. The observations that health or ill-health follows a social gradient was mainly made within countries. However, it has been shown that the scope of the gradient has varied over time, and it is likely that socioeconomic and political transition affect this change. The question has been raised whether inequality is a cross-culturally universal phenomenon. Most studies included provide evidence that health inequality between socioeconomic groups is a major issue in each country.

The descriptive studies were primarily oriented at analysing cross-national differences in objective mortality indicators, but also included some kind of measure of subjective health[14,73,74]. Subjective health measures did not play a pronounced role in these studies. Concerning the type of health indicators employed in these descriptive studies, socio-demographic variables, diseases, disabilities, health care utilisation and health behaviour were predominantly included. More rare is the representation of subjective well-being, social support, sexuality or behavioural aspects. Current field tests highlight that there seems to be a lack of internationally useable instruments in general studies on health including quality of life as well as physical, mental, social health or personal resources.

Indicators Quality of Life in Health Surveys

In order to classify existing studies on health indicators according to the type of empirical approach employed, the evidence on existing multinational studies on health indicators was reviewed. A literature search was conducted to identify the existing empirical studies in the databanks MEDLINE and EMBASE (1990-2005) using the key words "health indicator", "subjective health" or "quality of life" in combination with one of the terms "international" or "cross-cultural" or "cross-national".

As a result of the literature research, most articles were identified in the area of "quality of life" and "international"/"cross-cultural" (N=510), and several pertained to cross-cultural adaptations of a given instrument (N =168). There were only a few published abstracts using the term "health indicator" in any combination with the other research terms (N = 12). Forty-six abstracts were found using the combination "subjective health" and "international" while only six abstracts were identified using the term cross-cultural or cross-national and subjective health. Specific literature searches on subpopulations such as children or chronic diseases highlighted some more cross-national investigations when using the term "multi-centre" instead of international.

For the purpose of the current investigations, all studies were reviewed that conformed with the following selection criteria:

(a) They employed any kind of self-report measure on health or health related quality of life in an international, empirical study, and being the most restrictive criterion.
(b) They pertained to cross-national field testing of health indicators (thus excluding studies testing merely the cross-cultural applicability of measures)
(c) Analyses were conducted on individualised micro data, although most frequently on a national basis.
(d) Studies conceptualised both health determinants and health outcomes.

Following these criteria, 37 studies of the literature research were retained for further review, of these only the ones including a standardized quality of subjective health measure were included. The majority of studies referred to the adult population (n=26) and only a few to older adults (6) and children/adolescents (5). The articles were classified according to the age groups of the respective sample. Tables 1 gives an overview of the seven international studies analysing subjective health and its determinants in adults.

In adults mostly indicators of general health or quality of life were used, and cross-cultural differences, if reported, were ambiguous.

During the review process, several aspects appeared:

(1) The term health indicator is mainly used in studies related to external exposures (such as ozone concentration) and mortality-related indicators.
(2) The term cross-cultural is predominantly used in research related to international instrument development, in particular in the area of quality of life research. Most frequently, a sequential translation approach was employed in a specific national sample – and rarely cross-national comparisons are made, e.g. translations of one existing instrument into another language.
(3) The majority of studies have been conducted in the US and Europe.
(4) Ethnic studies have been conducted in their own research tradition mainly related to the impact of migration. Ethnic studies predominantly employed qualitative methods.
(5) Subjective indicators have rarely been integrated into international studies.
(6) Only a few multinational studies on specific diseases were identified using the current literature research terms.

As a general conclusion of the literature research described above, country variation has not been systematically addressed in research. A great majority of studies have not reported country effects, several studies reported different odds ratios in respect to health indicators across countries, others compared the effects of regression analyses across countries, and very few analysed one model across countries. Multi-ethnic studies addressed cross-cultural issues within the process of acculturation to a new country, however, with data relying almost exclusively on one country[76].

Indicators of Quality of Life in Epidemiological Mental Health Studies

Most epidemiological mental health studies are based on psychiatric assessment. Activities related to an international standardised assessment of mental health have been particularly initiated by the WHO, which set up a focus on international clinical epidemiology on psychiatric conditions in the 1980‘s[77].

The World Health Organisation and clinically oriented research groups on mental health[78] undertook a range of studies to assess the feasibility of developing standardized instruments and procedures for psychiatric assessment that could be applied reliably in a variety of cross-cultural settings. Another aim was to explore whether treatment outcomes differ from one country to the other. While symptoms could be similarly diagnosed between countries, the course and outcome varied

Table 1. International European studies on subjective health and its determinants in the adult population

	Author	Year	Population	Title	Components	Psychometric*	Type of approach	Analytical technique	Most important results	Country comparison effect size†
1	Marmot et al.[1]	2000	Aggregate data from Eastern and Western Europe countries	International comparators and poverty and health in Europe	SES, **self-rated health**, living circumstances	X	East-west gap in health inequality	Odds ratios Regression	Strong effect of psychosocial variables in Eastern Europe, strong East West gap	Odds ratios .93-.99 ($p<.001$) of health in rel. to psych. aspects
2	The European Opinion Research Group[2]	2003	15 European countries N = 16000	The mental health status of the European population	**Mental health**, MHI5 Positive mental health, EVI (energy/ vitality), Oslo social support, sociodem.	X	Cross-cultural differences in mental health	ANOVA, T-Tests, Logistic regression	Large cross-country differences exist in descriptive as well as regression approach (in Great Britain, East Germany and Italy particularly females showed high scores)	Odds ratios .39 (Finland) -2.1 (UK)
3	Vitterso et al.[3]	2002	N = 6949; 41 nations (aged 16–69) world-wide	The concept of life satisfaction across cultures: Exploring its diverse meaning and relation to economic wealth	**Satisfaction** with life scale, GNP	X	Structure of subjective well-being; LS and GNP	Structural equation modelling (SEM): CFA, correlation statistics	A simple underlying construct can be assumed. After controlling for the psychometric properties of the SWS scale, the relation between national wealth and life satisfaction was reduced to .25 (zero order $r = .42$)	$r_{(GNP,LS)} = .42$, only 13% between-nation variance (ANOVA)

4	Diener et al.[4]	1995 a	55 nations (mean size 1406)	Factors predicting the subjective well-being of nations	Income, individualism, human rights, societal equality, **basic needs fulfilment**	X	Cross-cultural differences in well-being	Partial correlation; Regression	High correlations between income, individualism, wealth and well-being; cultural homogeneity had no consistent effects	High partial correlation (.14-.78) between predictors
5	Shmueli[5]	2003	N = 1999 Ethnic origin: Asia, Africa, Europe, Jewish Israelis (aged 45–75)	Socio-economic and demographic variation in health in its measures: the issue of reporting heterogeneity	**Self-reported health, SF-36**, Visual analogue rating scale	X	Universal, structural approach	SEM	Evidence of the age-related reporting heterogeneity in disease; income-related heterogeneity in the gen. health; age, sex, income, ethnic origin and religion-related heterogeneity SF-36, in particular in its mental component	R^2QoL = .18 R^2PCI = .24 R^2 MCS = .10 MIMIC/ RMSEA = .05
6	Gureje et al.[6]	2001	3197 primary care patients of 14 countries	A cross-national study of the course of persistent pain in primary care	Pain, physical symptoms, anxiety, depressive disorders **(CIDI)**, occupational role, disability	X	Psychological problems in general health care	Regression	The baseline characteristics predicting the onset of persistent pain disorder were psychological disorder, poor self-rated health, and occupational role disability	No spec. country effects reported

(Continued)

Table 1. (Continued)

	Author	Year	Population	Title	Components	Psychometric*	Type of approach	Analytical technique	Most important results	Country comparison effect size†
7	Vandiver[7]	1998	102 outpatient men and women with schizophrenia Canada, Cuba, US (aged 18–70)	Quality of life, gender and schizophrenia; a cross-national survey in Canada, Cuba, and US	Social relationship, health, living situation, leisure, finances, **general quality of life**	X	QoL interview	ANOVA T-test within each country	In Canada, women with schizophrenia reported only a higher QoL for social relationships. Findings from the three sites showed no differences for other QoL domains.	$t = 2.1$-2.8, $p < .05$, No effect size reported

[1]Marmot, M. & Bobak, M. (2000). International comparators and poverty and health in Europe. British Medical Journal, 321:1124–1128.

[2] The European Opinion Research Group (2003): The mental health status of the European population. Brussels, Report to Directory General 5, European Union.

[3] Vitterso, J., Roysamb, E. & Diener, E. (2002). The concept of life satisfaction across cultures: Exploring its diverse meaning and relation to economic wealth. In The universality of subjective wellbeing indicators, ed. E. Gullone & R. A. Cummins. Dordrecht/Boston/London: Kluwer Academic Publisher, 2002, pp. 81–103.

[4] Diener E, Diener M & Diener C. Factors predicting the subjective well-being of nations. Journal of Personality and Social Psychology, 1995; 69:851–864.

[5] Shmueli, A. (2003). Socio-economic and demographic variation in health and in its measures: the issue of reporting heterogeneity. Social Science and Medicine, 57:125–134.

[6] Gureje, -. O., Simon, G. E. & Von-Korff, M. (2001). A cross-national study of the course of persistent pain in primary care. Pain, 92:195–200.

[7] Vandiver, V. L. (1998). Quality of life, gender and schizophrenia: a cross-national survey in Canada, Cuba, and U.S.A. Community Mental Health Journal, 34:501–511.

**Characterizes whether a psychometric approach was taken in the indicators.*

†Only effect size were mentioned concerning the impact of country / culture.

significantly between different countries. Standard clinical and sociodemographic data did not account for differences between the centres[77]. A large part in the variance in the course and outcome of schizophrenia seemed to be due to factors that have not yet been identified[52].

Following this study, the WHO undertook a cross-cultural study designed to analyse the determinants of psychopathological symptoms across countries. These studies have shown that certain determinants, such as expressed emotions in carers of schizophrenic patients, show a controversial validity for some cultures[52,79]. For instance, the life event scale has been found to have no predictive potential for certain ethnic groups, e.g. the Indian. The question is whether this is a methodological limitation of the cross-cultural applicability of the employed scales (or training methodologies of interviews) or whether differences between the expression of emotion, family relations and health outcomes exist. In the meantime, considerable approaches have been made to study the impact of culture on depression, somatisation, schizophrenia, anxiety and dissociation[78], with a clear focus on clinical perspectives. Medical diagnoses have usually been used universally in these approaches. Another approach has been related to testing the cross-cultural universality of models on different concepts on mental well-being, and also quality of life.

In a review of data collected regarding mental health in Europe[80], 26 surveys conducted in 13 European countries containing information on mental health were identified. These studies – in general – used single items rather than standardized instruments to operationalize clinical measurement along the International Classification of Diseases (ICD) or the Diagnostic and Statistical Manual of the American Psychiatric Association (DSM). However, a few self-report instruments on mental health have been employed[81]. A recent summary of health surveys in OECD countries showed that over 50 % of the national health surveys included at least some measure of psychological and emotional well-being rather than single questions[82].

Only few research groups working on mental health specifically focussed on cross-cultural universality, in contrast to cultural specificity and variation in mental health and well-being[83,84,85,11,77]. These studies have not particularly been oriented at studying heterogeneities between different European areas, but analysed whether different cultures share concepts on mental health, whether universal relationships exist between different facets of mental health and how these can be assessed.

Interestingly, it was in the study of emotions that initially much interest was spent on the universal, biological aspects of emotions. However, recently research has focussed on the cultural differences of emotional responses[86]. When studying specific elements of the emotional reaction, such as antecedent events, the significance of events, the valence of the emotional reaction and the expression of emotions, it has been shown that cultural differences become much more obvious[86]. These findings as well as the evidence of cross-cultural variability seem to indicate that the identification of cultural variation may occur more pronounced on the level of distinct behavioural processes, while in constructs referring to internal states, such as quality of life, cross-cultural variation is less frequent. A universal concept of a concept is the basis for a universal approach to instrument development. Therefore, in the section on cross-cultural instrument development, this approach will be explained. Disease specific measures in quality of life and mental health have been described by Bullinger & Naber[87].

IMPLICATIONS

Cross-cultural Assessment of Quality of Life in Mental Health

An overview of information regarding the state of international work with generic and specific quality of life instruments shows that most widely-known instruments have undergone translation. Testing has been completed in some instruments but norming still has to be carried out. In general the languages in which instruments are available so far include mainly European languages (north, south, west, and more recently eastern European languages), South American (with exception of Spanish), Asian, Arabic or African languages are definitely underrepresented. However, the approach that has been taken towards cross-cultural instrument development strongly diverges between the measures. Most adaptations have been performed sequentially so that it is difficult to discern the cross-cultural validity of the measures[88]. As described below, only a few working groups that have used a simultaneous cross-cultural approach.

Furthermore, international adaptation is also ongoing outside the respective working groups, in particular those activities that have been performed in specific cultures. Reviewing the available literature on non-generic (targeted) measures, there is only little literature on country-specific efforts in instrument development involving not only items with wording, but as figures and symbols. This may be a result of the fact that some culturally specific working groups do not necessarily publish their research. A disadvantage of these measures is that they do not allow to conclude references outside a particular culture.

Quality of life instrument development has so far mainly evolved from developed countries and has only subsequently been applied to developing countries. As a consequence QoL instruments have been developed to reflect the values and concerns of clinicians, patients and the general public of the country of origin. Bowden and Fox-Rushby[89] have conducted a systematic and critical review of the process of translation and adaptation of generic health-related quality of life measures in Africa, Asia, Eastern Europe, the Middle East, South America. Specifically, the review evaluated how the nine most prominent generic HRQL instruments (15D, Dartmouth COOP/WONCA Charts, EuroQol, HUI, NHP, SIP, SF-36, QWB, WHOQOL) were translated and adapted for use in Africa, Asia, Eastern Europe, the Middle East, and South America (Table 2). The review adopted a universal model of equivalence[90] and evaluated conceptual, item, semantic, operational, measurement and functional equivalence of the approaches.

Fifty eight papers were reviewed that came from 23 countries. A majority of studies came from the East Asia and Pacific region, and the SF-36 dominated

Table 2. Types of equivalence reviewed by Bowden & Fox Rushby (2003)[89] across 9 measures (in %)

Category		15D	COOP	EQ5D	HUI	NHP	SF-36	SIP	WHOQOL
Conceptual equivalence	None	100	77.0	80.0	100	50.0	69.6	100	27.3
	Partial	0.0	25.0	20.0	0	50.0	30.40	0.0	45.5
	Extensive	0.0	0.0	0.0	0	0.0	0.0	0.0	27.3
Item equivalence	None	50.0	62.5	100	50.0	75.0	91.3	66.7	20.0
	Partial	50.0	37.5	0	50.0	25.0	30.4	33.3	60.0
	Extensive	0	0	0	0	0	8.7	0	20.0
Semantic equivalence	None	100	62.5	60.0	50.0	50.0	60.9	100	81.8
	Partial	0	37.5	20.0	50.0	50.0	26.1	0	18.2
	Extensive	0	0	20.0	0	0	13.0	0	0
Measurement equivalence	None	0	37.5	40.0	50.0	25.0	21.7	66.7	54.5
	Partial	100	37.5	40.0	0	50.0	30.4	33.3	33.0
	Extensive	0	25.0	20.0	50.0	25.0	47.8	0	27.3
Assessment of local conceptions of health and QOL		0	37.5	0	0	50.0	8.7	66.7	54.5

the research. Bowden and Fox-Rushby[89] showed that currently most adaptations focussed rather on measurement and scale equivalence than on conceptual equivalence with the exception of the WHOQOL approach. In the WHOQOL approach conceptual equivalence was prioritized which was presumably a result. For instance, it has been shown that items related to emotions can have very ambiguous meanings in Asian languages. Table 1 shows the extent to which the nine generic measures have been tested in terms of different types of equivalence and suggests that most measures focus on measurement equivalence.

Concerning the World Bank regions, conceptual equivalence has been more frequently tested in Southern American than in Asian and Pacific, African and Middle East regions. The conclusion of this review is that research practice and translation guidelines still need to change to facilitate more effective and less biased assessments of equivalence of HRQL measures across countries.

The Implications of Cross-cultural Quality of Life Assessment for the Health Care Field

As already indicated quality of life instruments are available in different languages[89] and have been psychometrically tested mostly in the languages of the so-called industrialized nations. One of the aims of quality of life instrument development on a cross-cultural scale is to design measures that might be used in multinational clinical or health outcome studies. Implications of quality of life assessment in the health care field thus concern on a methodological basis firstly the availability of language versions of generic as well as targeted instruments in respective countries of the developing world. To do this, many of the international working groups have devised a procedure according to which researchers from interested countries can contact the instrument developments steering group and are informed about the proceedings of translation and psychometric testing of a given instrument in their country (e.g. WHOQol). The problem with this procedure is not the lack of standardisation but the lack of funds to actually perform the work required for translation psychometric testing and even norming.

If an instrument is available in the given language, the next question is to what extent it can be used in the respective country and for which purposes. Going back to the original aims of quality of life research which concern the epidemiological perspective (understanding populations' subjective health or quality of life), the clinical perspective (evaluating the effects of interventions in terms quality of life) as well as the quality of health economic approach (valuing subjective health as an indicator of economic benefit) different possibilities arise.

In terms of the epidemiological perspective, a necessity would be to implement a quality of life measure in national health surveys if available or in national data collection regarding populations health (which does not necessarily have to be representative). In several countries this procedure may be discussed and transported further by association and organisation such as the WHO or the UN. This information would be helpful to identify per country and also in comparison to other countries the actual state of the quality of life or subjective health as perceived by the citizens of the country as well as to plan strategies for improvement.

The second aspect concerns the clinical perspective. An example would be a randomized clinical trial, conducted in a developing country in which quality of life outcomes could be used as a primary or additional endpoint[42]. However a clinical evaluative perspective also concerns the assessment of larger programs such as community interventions in which in addition to clinical or behavioural indicators also the quality of life of the respondent could be assessed. Thus, as regards the clinical perspective, study designs may range from single prospective studies over comparative cohort studies to randomized clinical trials. It is notable here, that recently, disease specific instruments have been developed for use in mental health clinical trials. This applies also to mental health and especially to schizophrenia. Due to methodological problems, prejudices and other reasons, the patients' perspective in the treatment of schizophrenia has been largely neglected for a long time. With the development of atypical antipsychotics, goals of antipsychotic treatment became finally more ambitious. The self-rating scale to measure "Subjective Well-Being under Neuroleptics" (SWN) has been developed and used in numerous double-blind and open trials[91,92,93]. Internal consistency and test-retest reliability are high, SWN score was found to be sensitive to treatment effects, several studies showed marked improvement after change of medication within 2 to 4 weeks. The SWN has been translated into 15 languages, experience in Central European, but also in Asian, American and East European countries did not reveal any major cross-cultural differences.

Finally the question of health economic assessment in developing countries is a largely unresearched area. In pharmaco-economic literature[34], quality of life is used as an indicator of patient benefit; for example in cost benefits studies. Measures such as the EQ5D as well as the SF-6D health economic version of the SF-36 might be implemented. These measures give an indication of quality of life in terms of a single indicator (between 0 - 100) which either in itself is constructed as an index or is derived from multi-variate information and condensed into a single number. This information is used to represent subjective health and can be combined with different indicators such as survival or symptom free time periods (e.g in terms of quality adjusted life year).

However, meaningful inclusion of quality of life indicators in pharmaco-economic research or health-economic research is yet a challenge to be addressed. In addition to this more research orientated reflection of the use of quality of life assessment in the health care field there is also a practical approach. Quality of life information can be used as a descriptor of well-being and function in the individual patient. If a data indicates that a patient shows impairments of quality of life with regard to her/ his reference group, this can be seen as a clinical valuable information and as a basis for an individually tailored intervention.

While this more individual use of quality of life information is slowly applied into medical practice in the so called developed world, in that quality of life assessment is seen as one indicator of variable routine documentation and monitoring, this is not yet the case in the developing world. One reason for this is certainly the lack of research mentioned above, especially regarding the availability of instruments and the meaning of scores. Information as derived from epidemiological, clinical or health economical studies are lacking and this makes an appropriate interpretation of individual scores difficult.

An underlining issue in the general discussion about the use of quality of life instruments in the health care field of developing countries is, to what extent the quality of life produces information that is valuable for decision-making in the health care field – either on a health political level or in the individual physician-patient interaction. The decision about the principal usefulness of quality of life assessment certainly should not be limited to be experienced from the so called developed world, but should necessarily take in account the perspectives and needs of the general public as well as the health care professionals in the respective countries. To implement these measures, not only their methodological quality or their clinical interpretability is of importance, but also the willingness to accept such assessment tools and outcome indicators in research and practice. The general ethical demand of development policies holds also true for quality of life assessment in the health care field: Provide possibilites, but leave decisions to the people concerned.

Testing the Universality of Models of Various Subjective Health Measures

The identification of a universal structure on the interrelationship between different health indicators has been led by theoretical assumptions[94], methodological requirements[62] and existing evidence[4,59,60]. Theoretically, the different components have been assumed to pertain to humans irrespective of their specific cultural and geographical background.

First results with instruments for which cross-cultural testing (and in part norming) was possible suggests cross-cultural applicability of the instruments. However, in developing countries this has not yet sufficiently been done[89,53].

Empirically, several studies have been conducted that were able to demonstrate universality of health and well-being indicators across cultures[4,11,59,60,83,84]. These studies have been based on vigorous efforts in international psychometric development and testing of cross-culturally applicable measures. A basic question is whether a model identified in a universal approach applies to specific populations (e.g. patients) and to specific subgroups or other partitions of the population. The core principle of this research strategy is that it first identifies universal relationships before analysing the differences caused by subgroups. Methodologically, these studies analysed variance in health measure across populations before looking at heterogeneities caused by subpopulations, e.g. in terms of geographical and cultural background, gender or socio-economic subgroups.

Quality of life seems a universal human concept as concerns its relevant dimensions, which is not to say that the intensity of endorsement of these dimensions is similar across countries. Different individual behaviours, societal conditions and cultural regulations may apply, but these concern the means rather than the results of pursuing well-being. Specific behaviours which are instrumental in obtaining positive quality of life may vary culturally, such as engaging in religious services, engaging in specific activities or engaging in specific social behaviours. Although cultures do differ in their basic conditions provided to strive at a favourable quality of life, the person's subjective perception is not a linear reflection of these conditions. This does not imply that improvement of societal conditions is irrelevant, but it draws attention to the fact, that human perception (which is also a function of access to societal goods, information and education) may be the most important common factor in quality of life research.

In conducting such research, transparency of underlying concept is mandatory as is modesty in using specific measurement approaches. Correctness in applying instruments and analyzing data is one of the methodological prerequisites for cross-cultural research and responsibility for the results also after their publication should be taken by the researchers.

Quality of life research has a descriptive as well as a prescriptive aspect, which suggests that independent of race, gender, age, social status, occupation and mental or physical health, well-being and function as perceived by the person as is a human concern. Efforts to scientifically study and to politically improve the quality of life of its members should be a moral imperative in each society. However, not all societies have the possibility to address the quality of life of their citizens in a similar manner, especially countries in which the living conditions for its people may need improvement. Provided, however, that the quality of life of citizens is a major concern in a society, quality of life data may give information about the respective status of populations, may thus suggest plans to improve the quality of life status of these populations by specific interventions, can be used to measure the effects of such interventions.

DISCUSSION

From the early writings of well known anthropologists, culture has encaptured researchers attention. The diversity of expressions of human living conditions and values, which are embedded in a cultural context, are repeatedly discussed with regard to the concept of cultural universality versus specificity. While in anthropology, the focus is on understanding cultural diversity, social indicator research is more interested in universal approaches to understanding quality of life. This chiasm between universalist and specialised approaches to the role of culture is present in many research areas, in the social, medical and behaviour sciences. The difficulty to conceptually define and understand culture has been reflected in the many approaches suggested. Commonly, culture is understood as value and behaviour system, that underlies and governs human behaviour in social contexts.

Consequently, the topic of comparisons across cultures have been described as cross-cultural, sometimes also international. Different meanings of the term international can be distinguished. Independent of how the term international, language seems to be one major determinant of cultural identity and the ethnological and anthropological literature is full of examples of cultural specificity in value systems and behaviours across the world.

The WHO has been active in describing cultural variations of mental health across many publications and research projects. Specifically the work of the mental health branch under the guidance of Prof. Norman Satorius has advanced this research substantially. Many measurement instruments have been developed to assess or diagnose mental health and been tested across languages with the intent to find a common language in diagnosis, treatment and evaluation of mental health. This is reflected by the large literature on distribution of mental health problems across the world from epidemiology (see also global world health report), multinational clinical trial on therapeutic interventions for mental health problems worldwide as well as the many descriptions of addressing cultural factors in treatment approaches, which is an innovative point and important for the topic of cultural diversity.

With regard to quality of life outcome assessment in mental health cross-culturally, quality of life has been included as a relevant endpoint as well as descriptor in studies recently. Epidemiological and clinical research is moving towards international levels increasingly and here, cross-cultural applicable health indicators are of importance. The fundamental assumption in assessing quality of life in self-reports cross-culturally is, whether or not there is a universal construct[11]. Consequently, conceptual approaches have examined the universality of health care components and quality of life[95]. In term of quality of life, this implies that quality of life definitions postulate that the same dimensions should be valid across different cultures and nations[95]. In trying to answer the essential questions posed above, the following set of problems has to be kept in mind.

One of the main problems concerns possible *ethnocentrism* in terms of instruments and approaches used. All of the measures developed so far depart from the notion of verbal expressions of inner feelings and experiences. The main focus of existing questionnaires has been critically perceived as a white Anglo-Saxon middle-class outlook on quality of life so far it is prevailing and is even noticeable in a WHOQOL-Questionnaire.

A second problem is the possible *normativity* of the quality of life concept. There is a concern that quality of life dimensions are not value-neutral but act as standards according to which the individual in the society is expected to live up. In addition, biases in assessment may occur in terms of the choice of convenient samples for quality of life assessment and in terms of the very mode of questioning employed.

Finally, *ethical consequences* in cross-cultural quality of life research have to be kept in mind, which concern the freedom of personal information as well as the abuse potential of quality of life information collected within specific cultures and are ready to be included in clinical studies[96,97,98].

To sum up, international efforts to assess quality of life cross-culturally do exist. The developed instruments have passed the translation phase and mostly are in a testing phase and need to be reviewed for their cross-cultural performance.

REFERENCES

1. WHO. World Health Report 2002. Reducing Risk, Promoting Healthy Life. Geneva: World Health Organization, 2002.
2. Bullinger M. Quality of life – definition, conceptualization and implications: A methodologists view. Theor Surg, 1991; 6:143–148.
3. Power M, Harper A, Bullinger M. The World Health Organization WHOQOL-100: tests of the universality of Quality of Life in 15 different cultural groups worldwide. Health Psychology 1999; 18:495–505.
4. WHOQoL-Group, The. The World Health Organization Quality of Life Assessment (WHOQoL): Position paper from the World Health Organization. Social Science and Medicine, 1995; 41:1403–1409.
5. Cummins RA. Objective and subjective quality of life: An interactive model. Social Indicators Research 2000; 52:55–72.
6. Garrett M, McElroy AM & Staines A. Locomotor milestones and babywalkers: cross sectional study. British Medical Journal, 2002; 324:1494.
7. Bowling A. Measuring health: A review of quality of ife measurement scales. Buckingham/ Philadelphia: Open University Press, 1997.
8. Bowling A. Measuring disease: A review of disease-specific quality of life measurement scales. Buckingham: Open University Press. 2001.
9. Diener E & Suh E M, Editors. Culture and subjective well-being. Cambridge, MA: MIT Press. 2000.
10. Diener E & Suh EM. Measuring subjective well-being to compare the quality of life of cultures. In Culture and Subjective Well-being, ed. E Diener & EM. Suh. London: The MIT Press, 2000, pp. 3–12.
11. Vitterso J, Roysamb E, Diener E. The concept of life satisfaction across cultures: Exploring its diverse meaning and relation to economic wealth. In The universality of subjective wellbeing indicators, ed. E Gullone, RA Cummins. Dordrecht/Boston/London: Kluwer Academic Publisher. 2002: pp. 81–103.
12. Zapf W. Social reporting in the 1970s and in the 1990s. Social Indicators Research, 2000; 51:1–15.
13. Zautra A & Goodhart, D. Quality of life indicators: A review of the literature. Community Mental Health Review, 1979; 4:1–10.
14. Cavelaars, AE, Kunst, AE, Geurts, JJ, Crialesi R, Grotvedt L, Helmert U, Lahelma E, Lundberg O, Matheson J, Mielck A, Mizrahi A, Mizrahi A, Rasmussen NK, Regidor E, Spuhler T. & Mackenbach JP. Differences in self reported morbidity by educational level: A comparison of 11 Western European countries. Journal of Epidemiolology and Community Health, 1998; 52:219–227.

15. Kahnemann D, Diener E & Schwarz N, Editors. Well-being: The foundations of hedonic psychology. New York: Russell Sage Foundation, 1999.
16. Sen A. Economic progress and health. In Poverty, inequality and health, ed. D. A. Leon & G. Walt. New York: Oxford University Press, 2001, pp. 333–345.
17. Sen A. Development as freedom. New York: Knopf, 1999.
18. Wilkinson RG. Unhealthy societies: The afflictions of inequality. London, England: Routledge, 1996.
19. Marmot MG, Davey Smith G, Stansfield S, Patel C, North F, Head J, White I, Brunner E & Feeney A. Health inequalities among British civil servants: the Whitehall II study. Lancet, 1991; 337:1387–93.
20. Mackenbach JP. Income inequality and population health. British Medical Journal, 2002; 324:1–2.
21. Berzon R, Hays RD, Shumaker SA. International use, application and performance of health-related quality of life instruments. Quality of Life Research 1993; 2(6):367–368.
22. Schmidt S, Bullinger M. Current issues in cross-cultural quality of life instrument development. Archives of Physical Medicine and Rehabilitation, 2003; 84:29–33.
23. Campbell I (Ed.) The quality of American life. Russel-Sage, New York, 1981.
24. Glatzer W, Zapf W. Anthropological Perspectives. The importance of culture in the assessment of quality of life. In: Spilker B (Ed.): Quality of Life and Pharmaeconomics in Clinical Trials, second edition 1996, Lippincott-Raven Publishers Philadelphia, 1984; pp. 523–529.
25. Hutchinson JE. Quality of Life in ethnic groups. In: Spilker B (Ed.): Quality of Life and Pharmaeconomics in Clinical Trials, second edition 1996, Lippincott-Raven Publishers Philadelphia, 1996; pp. 587–595.
26. Johnson TM (1996) Cultural Considerations. In: Spilker B (Ed.): Quality of Life and Pharmaeconomics in Clinical Trials, second edition 1996, Lippincott-Raven Publishers Philadelphia, 1996; pp. 511–517.
27. Guarnaccia PJ. Anthropologial perspectives: The Importance of Culutre in the Assessment of Quality of Life. In: Spilker B (Ed.): Quality of Life and Pharmaeconomics in Clinical Trials, second edition, Lippincott-Raven Publishers Philadelphia, 1996; pp. 523–528.
28. Brislin RW, Lonner WJ, Thorndike RM. Cross-cultural research methods. New York: John Wiley, 1973.
29. Hui C, Triandis HC. Measurement in cross-cultural psychology: a review and comparison of strategies. Cross-Cultural Psychology 1985, 16:131–152.
30. Cantrill H. The pattern of human concerns. New York: University Press, 1966.
31. Aaronson NK, Cull A, Kaasa S, Sprangers M, Acquadro C, Jambon B, Ellis D, Marquis P (1996) Language and Translation Issues. In: Spilker B (Ed.): Quality of Life and Pharmaeconomics in Clinical Trials, second edition, Lippincott-Raven Publishers Philadelphia, 1996; pp. 575–587.
32. The European Organisation for Research and Treatment of Cancer (EORTC): Modular approach to quality of life assessment in oncology: an update. In: Spilker B (Ed.): Quality of Life and Pharmaeconomics in Clinical Trials, second edition, Lippincott-Raven Publishers Philadelphia, 1996; pp. 179–191.
33. The European Group for Quality of Life and Health Measurement. European Guide to the Nottingham Health Profile. ESCUBASE, Montpellier, 1992.
34. Aaronson NK, Acquadro C, Alonso J, Apolone G et al. International Quality of Life Assessment (IQOLA) Project. Quality of Life Res 1, 349–351.
35. Kind P. The EuroQOL instrument: an index of health-related quality of life. In: Spilker B (Ed.): Quality of Life and Pharmaeconomics in Clinical Trials, second edition 1996, Lippincott-Raven Publishers Philadelphia, 1996; pp. 191–203.
36. The WHOQOL Group. The Development of the World Health Organization Quality of Life Assessment Instrument (the WHOQOL). In: Orley J, Kuyken W (Eds.) Quality of Life Assessment: International Perspectives. Springer Berlin Heidelberg 1994, pp. 41–61.
37. Bergner M, Bobbit RA, Carter WB, Gilson BS. The Sickness Impact Profile: Development and Final Revision of a Health Status Measure. Med Care, 1981, 19, 780–805.

38. Anderson RT, Aaronson NK, Wilkin D. Critical review of the international assessment of health-related quality of life. Quality of Life Research, 1993; 2(6):369–395.
39. Cella & Bonomi. The Functional Assessment of Cancer Therapy (FACT) and Functional Assessment of HIV Infection (FAHI) quality of life measurement system. In Quality of Life and Pharmacoeconomics in Clinical Trials 2nd edition Spilker B (ed) Philadelphia: Lippincott-Raven; 1996.
40. Cella DF, Tulsky DS, Gray G. The Functional Assessment of Cancer Therapy (FACT). Scale: Development and Validation of the General Measure. J Clin Oncol. 1993; 11:572–597.
41. Fumimoto, H; Kobayashi, K; Chang,-C-H; Eremenco,-S; Fujiki,-Y; Uemura,-S; Ohashi,-Y; Kudoh,-S. Cross-cultural validation of an international questionnaire, the General Measure of the Functional Assessment of Cancer Therapy scale (FACT-G), for Japanese. Qual-Life-Res. 2001; 10(8): 701–9.
42. Schipper H, Olweny CLM, Clinch JJ. A Mini-Handbook for Conducting Small-Scale Clinical Trials in Developing Countries. In: Spilker B (Ed.): Quality of Life and Pharmaeconomics in Clinical Trials, second edition 1996, Lippincott-Raven Publishers Philadelphia, 1996; pp. 669–681.
43. Bullinger M, Power MJ, Aaronson NK, Cella DF, Anderson RT. Creating and Evaluating Cross-Cultural Instruments. In: Spilker B (Ed.): Quality of Life and Pharmaeconomics in Clinical Trials, second edition 1996, Lippincott-Raven Publishers Philadelphia, 1996; pp. 659–669.
44. Ware Jr. JE, Gandek BL, Keller SD, and the IQOLA Project Group. Evaluating nstruments Used Cross-Nationally: Methods from the IQOLA Project. In: Spilker B (Ed.): Quality of Life and Pharmaeconomics in Clinical Trials, second edition 1996, Lippincott-Raven Publishers Philadelphia, 1996; pp. 681–693.
45. Hunt SM, McEwen J, McKenna SP, Williams J, Papp E. The Nottingham Health Profile: Subjective Health Status and Medical Consultations. Sc Sci Med, 1981; 19A, 221–229.
46. Chwalow AJ. Cross-cultural validation of existing quality of life scales. Patient Education and Counseling, 1995; 26(1–3):313–318.
47. Sartorius N, Kuyken W (1994) Translation of Health Status Instruments. In: Orley J, Kuyken W (Eds.) Quality of Life Assessment: International Perspectives. Springer Berlin Heidelberg, 1994; pp. 3–19.
48. Guillemin F, Bombardier C, Beaton D. Cross-cultural adaption of health-related quality of life measures: literature review and proposed guidelines. Journal of Clinical Epidemiology, 1993; 46(12):1417–1432.
49. Acquadro C, Jambon B, Ellis D, Marquis P. Language and Translation Issues. In: Spilker B (Ed.): Quality of Life and Pharmaeconomics in Clinical Trials, second edition, Lippincott-Raven Publishers Philadelphia, 1996; pp. 575–587.
50. Guyatt GH The philosophy of health-related quality of life translation. Quality of Life Research, 1993; 2(6):461–465.
51. Kohlmann T, Bullinger M, & Kirchberger-Blumstein I. Die deutsche Version des Nottingham Health Profile (NHP). Übersetzungsmethodik und psychometrische Validierung. Soz Praventivmed, 1997; 42(3), 175–185.
52. Sartorius N. A WHO method for the assessment of health-related quality of life (WHOQoL). In Quality of life assessment: Key issues in the 1990s, ed. S. R. Walker & R. M. Rosser. Boston: Kluwer, 1993; pp. 201–207.
53. Skevington, S., Sartorius, N. Armin, M. Developing methods for assessing quality of life in different cultural settings. The history of the WHOQOL instruments: Social Psychiatry and Psychiatric Epidemiol., 2004; 39(1):1–8.
54. Orley J & Kyuken, W. (Eds). Quality of Life Assessment: International Perspectives. Heidelberg, Springer Verlag, 1994.
55. Ware JE, Gandek B, Kosinski MA, Aaronson NK, Apolone G, Brazier J, Bullinger M, Kaasa S, Leplege A, Prieto L, Sullivan M & Thunedborg K. The equivalence of SF-36 summary health scores estimated using standard and country-specific agorithms in 10 countries: Results from the IQoLA Project. Social Science and Medicine, 1998; 51:1167–1170.
56. Ware JE, Kosinski MA, Gandek BL, Aaronson NK, Apolone G, Bech P, Brazier J, Bullinger M, Kaasa S, Leplege A, Prieto L & Sullivan M. The factor structure of the SF-36 health survey in

10 countries: Results from the IQoLA Project. International quality of life assessment. Journal of Clinical Epidemiology, 1998; 51:1159–1165.

57. Fukuhara S, Ware JE, Kosinski MA, Wada S & Gandek B. Psychometric and clinical tests of validity of the Japanese SF-36 Health Survey. J Clin Epidemiol, 1998; 51:1045–1054.
58. Hall GC & Maramba GG. In search of cultural diversity: recent literature in cross-culutral and ethnic minority psychology. Cultur Divers Ethn Minor Psychol, 2001; 7:12–26.
59. The WHOQOL-Group. The World Health Organization Quality of Life Assessment (WHOQOL): Development and General Psychometric Properties. Social Science and Medicine, 1998; 46: 1569–85.
60. The WHOQOL-Group. Development of The World Health Organization WHOQOL-BREF Quality of Life Assessment. Psychological Medicine, 1998; 28:551–8.
61. Saxena S, Carlson D, Billington R. & WHOQoLGroup. The WHO quality of life assessment instrument (WHOQoL-Bref): the importance of its items for cross-cultural research. Quality of Life Research, 2001; 10:711–721.
62. Bullinger M, Schmidt S, Petersen C, The-DISABKIDS-Group. Assessing quality of life of children with chronic health conditions and disabilities: A European approach. International Journal of Rehabilitation Research, 2002; 25:197–206.
63. Ravens-Sieberer U. Der Kindl-R Fragebogen zur Erfassung der gesundheitsbezogenen Lebensqualität bei Kindern und Jugendlichen – Revidierte Form. In Diagnostische Verfahren zu Lebensqualität und Wohlbefinden, ed. J. Schumacher, A. Klaiberg & E. Brähler. Göttingen: Hogrefe, 2003; pp. 184–188.
64. Hu L, Bentler PM. Cutoff criteria for fit indexes in covariance structure analysis: Conventional criteria versus new alternatives. Structural Equation Modeling, 1999; 6:1–55.
65. Schermelleh-Engel K, Moosbrugger H, Müller H. Evaluating the fit of structural equation models: Test of significance and descriptive goodness-of-fit measures. Methods of Psychological Research – Online, 2003; 8:23–74.
66. Schmidt S, Power M, Bullinger M. Cross-cultural analysis of relationships of health indicators across Europe: First results based on the Eurohis project. Rep. EUR/02/5041391., World Health Organisation – Regional Office for Europe, 2002.
67. WHO. The European Health Report 2002. Copenhagen: WHO Regional Office. 2002.
68. Murray CJ & Lopez AD. Global mortality, disability, and the contribution of risk factors: Global Burden of Disease Study. Lancet, 1997; 349:1436–42.
69. Murray C & Lopez AD. Regional patterns of disability-free life expectancy and disability-adjusted life expectancy: global Burden of Disease Study. Lancet, 1997; 349:1347–52.
70. Davey Smith G, Charsley K, Lambert H, Paul S, Fenton S & Ahmad W. Ethnicity, health and the meaning of socio-economic position. In Understanding health inequalities, ed. H. Graham. Buckingham/Philadelphia: Open University Press, 2000.
71. Davey Smith, G., Gunnell, D. & Ben-Shlomo, Y. (2001). Life-course approaches to socio-economic differentials in cause-specific adult mortality. In Poverty, inequality and health, ed. D. A. Leon & G. Walt. New York: Oxford University Press, 2001; pp. 88–124.
72. Marmot MG. & Shipley MJ. Do socioeconomic differences in mortality persist after retirement? 25 Year follow up of civil servants from the first Whitehall study. British Medical Journal, 1996; 313:1177–80.
73. Cavelaars AE, Kunst AE, Geurts JJ, Helmert U, Lundberg O, Mielck A, Matheson J, Mizrahi A, Mizrahi A, Rasmussen N, Spuhler T & Mackenbach JP. Morbidity differences by occupational class in men in 7 European countries. International Journal of Epidemiology, 1998; 27:222–230.
74. Mackenbach JP, Kunst AE, Cavelaars AE, Groenhof F & Geurts JJ. Socioeconomic inequalities in morbidity and mortality in western Europe. The EU working group on socioeconomic inequalities in health. Lancet, 1997; 349:1655–1659.
75. Cooper, RS. Health and social status of Blacks in the US. Annual Epidemiology, 1993; 3:137–44.
76. Salant T. & Lauderdale DS. Measuring culture: a critical review of acculturation and health in Asian immigrant populations. Social Science and Medicine, 2003; 57:71–90.

77. Sartorius N, Jablensky A & Shapiro R. Cross-cultural differences in the short-term prognosis of schizophrenic psychoses. Schizophrenia Bulletin, 1978; 4:102–13.
78. Draguns JG & Tanaka-Matsumi J. Assessment of psychopathology across and within cultures: issues and findings. Behaviour Research and Therapy, 2003; 41:755–776.
79. Cheng ATA. Expressed emotion: a cross-culturally valid concept? The British Journal of Psychiatrists, 2002; 181:466–467.
80. Hupkens CL, van den Berg J & van der Zee, J. National health interview surveys in Europe: An overview. Health Policy, 1999; 47:145–168.
81. Meltzer H. Development of a common instrument for mental health. In Developing common instruments for health surveys, ed. A. Nosikov & C. Gudex. Amsterdam: IOS Press, 2003; pp. 35–60.
82. Gudex C & Lafortune G. An inventory of health and disability-related surveys in OECD countries. Paris: Organisation for Economic Co-operation and Development (OECD Labour Market and Social Policy), 2000.
83. Diener E, Diener M & Diener C. Factors predicting the subjective well-being of nations. Journal of Personality and Social Psychology, 1995; 69:851–864.
84. Diener, E., Suh, E. M., Smith, H. & Shao, L. (1995b). National differences in reported subjective well-being: Why do they occur? Social Indicators Research, 1995; 34:7–32.
85. EORG. (2003). The mental health status of the European population. Brussel: The European Opinion Research Group; written for The SANCO Directorate-General, 2003.
86. Mesquita B & Walker R. Culutral differences in emotions: a context for interpreting emotional experiences. Behaviour Research and Therapy, 2003; 41:777–793.
87. Bullinger M, & Naber D. Assessing the Quality of Life in Mental Illness. In F Henn, N Satorius, H Helmchen & M Lauter (Eds.), Contemporary Psychiatry Vol. 1, Part 2: General Psychiatry, 2000; pp. 135–150.
88. Perkins, M R; Devlin, N J; Hansen, P. The validity and reliability of EQ-5D health state valuations in a survey of Maori. Qual-Life-Res. 2004; 13(1):271–4.
89. Bowden A, Fox-Rushby J. A systematic and critial review of teh process of translation and adaptation of generic health-related quality of life measures in Africa, Asia, Eastern Europe, the Middel East, South Africa. Social Science & Medicine, 2003; 57:1289–306.
90. Herdman M, Fox-Rushby J, Badia X. A model of equivalence in the cultural adaption of HRQoL instruments: the universalist approach. Qual Life Res, 1998; 7:323–35.
91. Naber D, Moritz S, Lambert M et al. Improvement of schizophrenic patients' subjective well-being under atypical antipsychotic drugs. Schizophr Res, 2001; 50:79–88.
92. Naber D, Riedel M, Klimke A et al. Randomized double blind comparison of olanzapine and clozapine on subjective well-being and clinical outcome in patients with schizophrenia. Acta Psychatr Scand, 2005; 111:106–115.
93. De Haan L, Weisfelt M, Dingemans PM et al. Psychometric properties of the subjective well-being under neuroleptics scale and the subjective deficit syndrome scale. Psychopharmacology, 2002; 162:24–28.
94. Spilker B & Revicki DA. Taxonomy of quality of life. In Quality of life and pharmaeconomics in clinical trials, ed. B Spilker. Philadelphia: Lippincott-Raven, 1996; pp. 25–31.
95. Anderson RT, Aaronson NK, Leplège AP, Wilkin D. International Use and Application of Generic Health-Related Quality of Life Instruments. In: Spilker B (Ed.): Quality of Life and Pharmaeconomics in Clinical Trials, second edition 1996, Lippincott-Raven Publishers Philadelphia, 1996; pp. 613–633.
96. Bernhard J, Hürny CDT, Coates A, Gelber R. Applying Quality of Life Principles in International Cancer Clinical Trials. In: Spilker B (Ed.): Quality of Life and Pharmaeconomics in Clinical Trials, second edition 1996, Lippincott-Raven Publishers Philadelphia, 1996; pp. 693–707.
97. Mathias SD, Fifer SK, Patrick DL (1994) Rapid translation of quality of life measures for international clinical trials: avoiding errors in the minimalist approach. Quality of Life Research, 1994; 3(6):403–412.
98. Cella D, Wiklund I, Shumaker SA, Aaronson NK. Integrating health-related quality of life into cross-national clinical trials. Quality of Life Research, 1993; 2(6):433–440.

CHAPTER 6

MEASURING THE VALUE OF HEALTH-RELATED QUALITY OF LIFE

GRAEME HAWTHORNE

Department of Psychiatry, The University of Melbourne, Australia

Abstract: Health-related quality of life is concerned with the relationship between the effect of treatment on the patient's life with society's value for this effect. It is only through balancing these two concerns that transparent decisions concerning the best choice of treatment at the intervention level and, at the policy level, the allocation of scarce health resources can incorporate both the patient's and society's views. Where these decisions are important, the appropriate form of evaluation is cost-utility analysis (CUA). With rising health care costs, the ageing of populations and the determination of governments to cap health care expenditure, the mental health field will increasingly be asked for evidence of its cost-effectiveness. In the absence of such evidence, patients access to mental health services in the future may be hindered and the choice of treatment restricted as decision-makers may be disinclined to increase or even maintain funding.

This paper reviews the axioms of cost-utility analysis and the role of multi-attribute (MAU) utility instruments. Seven leading instruments are reviewed, and examples of their use in cost-utility analysis in the mental health field are presented. It is concluded that none of the existing instruments fully meet the axioms of either utility or measurement theory, and that the instruments provide HRQoL estimates that are so different that study outcomes are likely to be as much a function of the instrument chosen for a particular study as the effectiveness of the intervention itself.

It is recommended that mental health professionals undertaking CUAs review available instruments carefully, use two MAU-instruments in any particular study, and report both sets of results. The shortcomings of existing MAU-instruments should not be taken as a reason to avoid economic evaluation; at the moment they are the only practical way of capturing the balance described above. Few CUA studies have been carried out in the mental health field, there is thus an opportunity to undertake studies providing the evidence needed by clinicians and decision-makers for transparent decisions regarding treatment options and the future funding of mental health care

Keywords: AQoL, Burden of disease, Cost-effectiveness, Cost-utility analysis, CUA, Economic evaluation, EQ5D, Health-related quality of life, HUI3, 15D, Multi-attribute utility instrument, Quality-adjusted life years, QALY, QWB, Rosser Index, SF6D

M.S. Ritsner and A.G. Awad (eds.), Quality of Life Impairment in Schizophrenia, Mood and Anxiety Disorders, 99–132.

INTRODUCTION

Although some of the key ideas underpinning quality of life (QoL) can be traced back to Aristotle[1], in the medical field the term 'quality of life' was first used by Long[2] in 1960 when his paper entitled "On the quantity and quality of life" was published in the Medical Times. This paper explored the ethical issues behind the allocation of health care resources to patients in extremis. Six years later in an editorial for Annals of Internal Medicine titled "Medicine and the quality of life" Elkinton posed the central questions that QoL researchers have been trying to answer ever since:

> How does the physician protect the proper quality of life of the individual patient? How can the quality of life be improved...? Into which programs of preventive and therapeutic medicine should the resources of society be put to achieve the maximum in health and quality of life for all members of that society?[3]

The roots of QoL research in health care were concerned with life quality, length and the allocation of resources. During the course of the past forty years there has been growing interest in answering these questions, and the measurement of QoL as an outcome in health care interventions is now well accepted. This interest can be attributed to five interrelated health and health care changes[4]. Health care technologies have reduced early mortality and prolonged the lives of those who would otherwise have died (usually from an infectious disease)[5]; there has been a shift in economically developed societies from exogenous to endogenous chronic diseases[6], such as mental health conditions; there has been increasing recognition that interventions should respect patients' concerns and incorporate their experiences into medical decision-making[7,8]; many health services are now designed to prevent deterioration in QoL[9]; and there is increasing conflict between potentially useful interventions and the (limited) resources available to fund them[10].

The issue of resource constraint is critically important: across the world health care costs are increasing as a proportion of Gross Domestic Product (GDP). As in most countries, in my own country, Australia, this is driven by the demand for health care, the use of more expensive technologies and the changing demographic profile of society[11,12]. Table 1 illustrates this for the OECD countries: between 1980 and 2000 there was an increase across the OECD of 2.8% in the cost of health care as a proportion of GDP. Very few countries experienced a decline in the cost of health care. The implications are that there is and will be increased competition within the health care sector for these limited resources. Although mental health conditions are one of the largest contributors to the burden of disease (for example, mental health was the largest contributor to non-fatal burden of disease in Australia in 1996[13]), as resourcing issues become more important mental health professionals will increasingly be asked to justify their costs and the benefits of treatments relative to other health areas. In this context, Elkinton's questions go right to the heart of the matter: how can this be best done?

This question poses three key issues for medical clinicians and researchers: How do we define and measure quality of life? Whose values should be taken into account when distributing society's health care resources? And how should these be brought

Table 1. Health care costs, OECD countries, 1980–2000, percentage of Gross Domestic Product

	1980	1985	1990	1995	2000
Australia	6.8	7.2	7.5	8.0	8.8
Austria	7.5	6.5	7.0	9.7	9.4
Belgium	6.3	7.0	7.2	8.2	8.6
Canada	7.1	8.2	9.0	9.2	8.9
Czech Republic			4.7	7.0	6.7
Denmark	8.9	8.5	8.3	8.1	8.3
Finland	6.3	7.1	7.8	7.4	6.7
France	7.0	7.9	8.4	9.4	9.2
Germany	8.7	9.0	8.5	10.3	10.4
Greece	6.6		7.4	9.6	9.9
Hungary				7.4	7.1
Iceland	6.2	7.2	7.9	8.4	9.2
Ireland	8.3	7.5	6.1	6.7	6.3
Italy			7.7	7.1	7.9
Japan	6.5	6.7	5.9	6.8	7.6
Korea		4.1	4.4	4.2	4.8
Luxembourg	5.2	5.2	5.4	5.6	5.8
Mexico			4.8	5.6	5.6
Netherlands	7.2	7.1	7.7	8.1	7.9
New Zealand	5.9	5.1	6.9	7.2	7.7
Norway	7.0	6.6	7.7	7.9	8.5
Poland			4.9	5.6	5.7
Portugal	5.6	6.0	6.2	8.2	9.4
Slovak Republic					5.5
Spain	5.3	5.4	6.5	7.4	7.2
Sweden	9.0	8.6	8.3	8.1	8.4
Switzerland	7.4	7.8	8.3	9.7	10.4
Turkey	3.3	2.2	3.6	3.4	6.6
United Kingdom	5.6	5.9	6.0	7.0	7.3
United States	8.8	10.1	11.9	13.3	13.3

Source: OECD (2006)

together? The first of these questions lies within the field of measurement, and the second in ethics and economics. The inclusion of health-related QoL (HRQoL)* assessment in evaluative research[14] is an attempt to bring these fields together to answer Elkinton's questions – questions which are particularly important in the treatment of mental health conditions where, historically, treatment has often been as much about social control as health[15].

It is widely accepted that it is the patient's or participant's self-report of HRQoL that should be elicited. This position is consistent with that of the World Health Organization, which defined QoL as:

*Health-related QoL (HRQoL) refers to those aspects of life that are directly affected by health. QoL is a much broader term that embraces all of life.

> ... an individual's perceptions of their position in life in the context of the culture and value system in which they live and in relation to their goals, expectations, standards and concerns. It is a broad ranging concept affected in a complex way by the person's physical health, psychological state, level of independence, social relationships, and their relationship to salient features of their environment [16].

The problem with this broad-ranging definition is that it is extraordinarily difficult to actually measure it with any great precision because it is assumed that QoL is a stable construct, but one which is subject to within-the-person standards that may vary over time or in response to changed circumstances. This is particularly the case in mental health because this definition assumes individuals have the necessary insight into their own lives to provide accurate self-reports. Those living with mental health conditions may lack this insight due to affective bias, poor self-awareness, and cognitive or reality distortions [17]. In spite of these concerns, the evidence suggests that those with mental health conditions are capable of providing appropriate self-report assessments [18–20].

It is also accepted that judgements concerning health resource allocation under conditions of resource constraint should be transparent: before the development of cost-utility analysis (CUA) the perspective of the patient was simply ignored. In 1978 Sackett & Torrance [21] argued that transparency could be achieved through eliciting the values of the general public for different health states. Once these values were elicited they could be used in the planning and financing of health interventions. Utility theory, which had been developed in the 1950s, was brought to medicine when Torrance argued that:

> The applicability of cost-effectiveness analysis as a technique for the evaluation and comparison of alternative health care programs is limited by the difficulty of measuring program effects (outcomes, consequences) in commensurable units across programs of different types. One approach to overcoming this limitation is to determine the ultimate impact of each program on the health states of each individual affected by the program and to use a social preference function defined over the relevant health states as the common unit of measure [22].

The problem was how to validly and reliably measure social preferences for subjective within-the-person self-report health states. Early work in the 1970s used the holistic utility method in which vignettes describing the health state of interest were valued [23]. The limitations were that each different health state needed to be separately valued and that the vignettes didn't reflect patient self-report. Regarding the question of whose values, Rosser & Kind [24] asked patients, nurses, clinicians and volunteers. A key finding was that there were significant differences in valuations by respondent group with psychiatric patients and psychiatric nurses providing the highest valuations (i.e. the worst impacts on life of selected health conditions) – suggesting that valuations of health states are directly affected by the experience of illness and neuroticism. They speculated that this difference may be due to perceptual distortion by those with mental illnesses. Subsequent work has confirmed that patients, clinicians and the general public do indeed hold different values regarding health states, and the consensus that has emerged is that it is the general public's values which should be used (although there is continuing debate over this [25,26]).

Torrance et al.[22], drawing on multi-attribute utility theory (MAU), explicitly decomposed health states for both valuing and for patient self-report. Figure 1 shows the Physical Function dimension of their original descriptive system for assessing the health outcomes for neonatal intensive care children. Their descriptive system consisted of three dimensions: Physical Function, Role Function and Social-Emotional Function.

Each child could be classified at any one of the levels, and on each of the dimensions. These assignations to health state levels provided an estimate of the child's health state. The next step was to obtain social utilities (preferences) for the different health states. A random sample of parents was drawn and asked to interpret selected health states using the time trade-off technique (fully explained below). This involved consideration of the health state and valuing it on a scale where 1.00 represented the best possible health state for 70 years and 0.00 death shortly after birth. To combine valuations of the different dimensions into an index score (i.e. to re-compose the dimensions into a comprehensive health state), a weighted multiplicative model was used.

Although this is an incomplete description of their study, all subsequent MAU-instruments can be traced back to this seminal work. It offered a model for identifying the HRQoL of the patient in such a way that the patient's perspective, weighted by societal values, could be used alongside clinical (mortality/morbidity) evidence in resource allocation. In other words, it provided a method with which Elkinton's questions could be answered.

***X_1 PHYSICAL FUNCTION*: MOBILITY AND PHYSICAL ACTIVITY**		
Level	**Code**	**Description**
1	P1	Being able to get around the house, yard, neighborhood or community WITHOUT HELP from another person; AND having NO limitation in physical ability to lift, walk, run, jump or bend.
2	P2	Being able to get around the house, yard, neighborhood or community WITHOUT HELP from another person; AND having SOME limitations in in physical ability to lift, walk, run, jump or bend.
3	P3	Being able to get around the house, yard, neighborhood or community WITHOUT HELP from another person; AND NEEDING mechanical aids to walk or get around
4	P4	NEEDING HELP from another person in order to get around the house, yard, neighborhood or community; AND having SOME limitations in physical ability to life, walk, run, jump or bend.
5	P5	NEEDING HELP from another person in order to get around the house, yard, neighborhood or community; AND NEEDING mechanical aids to walk or get around.
6	P6	NEEDING HELP from another person in order to get around the house, yard, neighborhood or community; AND NOT being able to use or control the arms and legs.

Adapted from: Torrance et al (1982; 1045)[22]

Figure 1. Health State Classification System: Physical Function

Today there are 7 well-known generic MAU-instruments, and several other specialised measures. This paper reviews the generic instruments with particular attention to their use in mental health care.

UNDERSTANDING UTILITIES AND CUA

Medical evaluative research has traditionally been concerned with summative evaluation where the criteria have been mortality or morbidity[14,27,28]. For the reasons outlined above, there is increasing interest in the allocation of limited health care resources to interventions that provide the best health for the community. This necessarily involves making comparisons across different health conditions and interventions. Where this comparative information is sought for resource allocation decisions, economic evaluation is used[10]. For an introduction and overview of economic evaluation in psychiatry see Singh et al.[29].

Economic evaluation offers three important mechanisms: it can describe the cost of the burden of mental health conditions, it can predict the level of resources that will be needed in the future for treating them, and it can provide information about the best use of resources available for mental health care interventions. Although cost-benefit analysis is the most comprehensive form of economic evaluation[10], increasingly it is CUA that is being used in health care evaluative research. MAU instruments are designed explicitly to be used in these cost-utility analyses.

The Axioms of Utility Measurement

The basic axiom of CUA is simple: life years are weighted by the value of a given health state so that the values—referred to as 'utilities'—act as an exchange rate between the quantity and quality of life. In this context, 'utilities' are assumed to be preferences for a given health state. Consider the following. Most people would prefer to be healthy over a given time rather than suffer constant depression. Utility measurement refers to valuing these preferences on a life-death scale with endpoints of 0.00 and 1.00, where 0.00 is death equivalent and 1.00 is perfect (very good) HRQoL. For example, the measured utility for depression may be 0.60. If treatment improves this to 0.70, then the value of the treatment is $0.70–0.60 = 0.10$. If this utility gain is maintained over time, say for 10 years, then the gain is $0.10 \times 10 = 1.00$ Quality Adjusted Life Year (QALY). This is shown graphically in Figure 2. The lines show the utility values for usual care of chronic depression (Treatment A, lower line) and treated depression (Treatment B, upper line). Although both patient groups have the same life-length, the treated patients have a higher QoL across their lives. If the years of life are weighted by the utility values, the area between the two lines represents the QALYs gained from the experimental treatment[29].

Because utilities fall on the life-death scale, they are (in theory) common across all health states and therefore can be used to compare the effect of interventions in different health fields, or different interventions within the same field. For example, the QALYs gained from Treatment A for depression could be compared with those

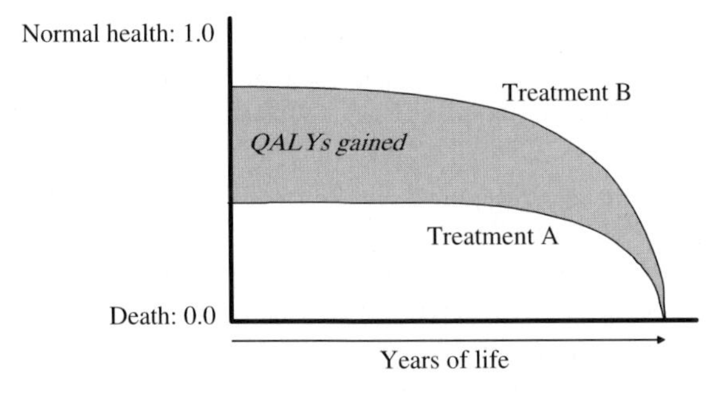

Figure 2. Understanding QALYS

gained from Treatment B for incontinence. Where treatment costs (including costs to the patient) are known, the treatment providing the lowest cost-per-QALY gained is preferred as this ensures society gains the greatest benefit from the health care dollar. For example, consider two treatments for schizophrenia. If for Treatment A the cost per QALY gained is US$45,000 whereas it is US$55,000/QALY gained for Treatment B, then Treatment A is preferred because the difference (US$10,000) could be used to either treat more patients or be allocated to another health area.

To allow for this comparison, utility measures must be generic and must allow for respondents to report they have excellent HRQoL (full health equivalent state: 1.00). Additionally they must allow those who have appalling HRQoL to report this (death equivalent state: 0.00). If an instrument does not permit this full range of responses, it cannot accurately measure the HRQoL of people who fall outside its range. For example, if an instrument only allows measurement between 0.50 to 1.00, then it is incapable of reporting the effect of treatment for people who are in a desperate health state (say, close to death). Under these circumstances, any claim to generalizability for the instrument is foregone.

The instrument must be applicable to HRQoL states deemed worse than death (i.e. the respondent indicates they would rather die now than continue living in their current HRQoL state). These negative health states are needed to allow for people who commit suicide or euthanasia; they have clearly made the decision that death is preferable to living in their current health state and any possible future health states. When determining negative utility boundaries, the developers of the EQ5D and HUI3 adopted Torrance's symmetry argument. This states that since a person can 'lose' HRQoL value from 1.00 (full health) to 0.00 (death equivalent), he/she must be able to 'gain' an equivalent amount from –1.00 to 0.00[30]. However, since negative utility values do not possess the same interval properties as positive utility scores[31,32], there are difficulties. For example, improving the HRQoL of a person from –0.35 to –0.25 (i.e. bringing them closer to a HRQoL death-equivalent state) does not have the same meaning as improving their HRQoL state from 0.25 to 0.35; yet both these would have a utility gain of +0.10. This is implausible. It seems likely, then, that negative values should have lower boundaries close to 0.00 (death equivalent)[32].

Implicit in axioms and mathematical modelling of utilities is that utility measurement must be at the interval level, where interval level refers to measurement scales that have equal-intervals between the measurement points. There are two forms of interval measurement that MAU-instruments must have if they are to do their job correctly. One is known as the "weak" interval property and the other the "strong" interval property[33]. The weak interval property is where a gain of 0.10 means the same thing across the range of instrument scores. For a person who has severe depression, his/her utility score might be 0.15; as a result of treatment this rises to 0.25; i.e. the value of the treatment is $0.25–0.15 = 0.10$. Similarly, the value of the treatment is also 0.10 for a person with schizophrenia with an initial utility of 0.60, and who gains a utility of 0.70 after treatment ($0.70–0.60 = 0.10$). The strong interval property is where there is a direct relationship between gains in utility and gains in life-length. Since QALY calculation represents the time spent in a given state multiplied by the quality of that state, this implies that a 0.20 utility gain multiplied by 5 years in the improved health state equals 1.00 QALY (from 0.20×5). But a gain of 1 QALY could also be the product of a 0.40 utility gain over 2.5 years (or any other combination).

Measuring Utilities Using MAU-instruments

There are two steps to measuring utilities using MAU-instruments. First, the health state of interest is described. Second, the value or utility of the health state is assigned.

When a person completes a MAU-instrument, his/her numerical responses provide a description of his/her health. For example, consider two people completing an imaginary instrument with four dimensions each of which has four levels. This instrument's 'descriptive system' would be: physical, mental, social and cognitive health dimensions, and the response levels are: 1 = normal, 2 = some impairment, 3 = major impairment, 4 = gross impairment. Person A, who is in the best of health, selects the best response to each item (i.e. '1': normal,). Her health state would be described as '1,1,1,1'. Person B reported major depression (level 3: major impairment on the mental health dimension), normal physical heath (level 1), some social impairment (level 2), and normal cognitive function (level 1). Her health state would be '1,3,2,1'.

Valuing these health states is called 'scaling'. Five procedures have been used: time trade-off (TTO), standard gamble (SG), visual analog rating scale (VAS), magnitude estimation (ME) and person trade-off (PTO).

- *Time trade-off (TTO).* A person with severe depression can have a treatment which will restore her to full health; but a side effect is she will live a shorter life. She is asked to choose how many years of her life she would be willing to 'give up' in order to be in full health. If, in her untreated condition, her life expectancy was 10 years and after the treatment this was 5 years she may reject the treatment.

If after the treatment it was 9 years, she may accept it; if her life expectancy was 6 years, she may not. Her choices would continue back-and-forth like this until she indicated that she was indifferent to whether she had the treatment or not. If the point of indifference was that 8 years of full health was the equivalent of 10 years with severe depression, then the QoL value for her current health state is 8/10 or 0.80.

- *Standard gamble (SG).* A person with depression is presented with a treatment option that has two possible outcomes: either full health for the remainder of his life, or death. He is free to choose either the treatment or to remain with lifelong depression. If the probability of full health is 1.00 (i.e. his depression will be cured and there is no chance of death), then obviously he will choose to have the treatment. If the probability of full health is 0.90 and death 0.10, he may still choose the treatment. However there would be a point, for example at 0.80 for full health and 0.20 for death, where he is not clear as to whether he would want the treatment or would choose to remain in his current health state. This point of indifference is the 'value' of his health state.
- *Visual analog scale (VAS).* The respondent is asked to consider a health state (depression) and then to rate this on a scale, where the endpoints are typically 0.00 (death equivalent) and 1.00 (full health equivalent). Unlike the TTO or SG, with the VAS there is no uncertainty: the respondent is not asked to 'trade' anything. Consequently many consider that VAS scores do not represent utilities because they provide a simple ranking of health states. Where VAS scores are used, a transformation is generally required, based on TTO or SG[34–36].
- *Magnitude estimation (ME).* The respondent is asked to consider the distance of the health state of interest (e.g. depression) from 1.00 (full health). Once several of these rating exercises have been carried out, the respondent is then asked to rank these in order[37]. Because there is no uncertainty in the procedure, it is uncertain if ME represents utility.
- *Person trade off (PTO).* The respondent is asked to estimate the number of people that would have to be treated to make an intervention worthwhile. For example, a respondent might be asked to choose between extending the life of 10,000 people who were in full health by 1 year against a treatment which extended the life of N people with depression, also for 1 year. The number of people with depression would be varied until the respondent indicated they were indifferent between the two choices[37].

When these techniques are used to obtain the utility weights used in an MAU-instrument, in theory each health state described by the descriptive system can be scaled (as was done with the original Rosser Index[38]), but this is impractical because MAU-instruments typically generate thousands of different health states. Instead, a limited number of health states are scaled and the values for other health states are then inferred using econometric or decision analytic techniques. During scoring, the health state descriptors (1,2,3, etc.) are replaced with the appropriate values. For example, if the value of suffering mild pain is '0.70' and the response levels on an item measuring pain were '1' (no pain), '2' (mild pain), and '3' (severe pain), then a person who selected ‘2’ would have this level replaced with the value ‘0.70’ during scoring of the instrument.

Once item-level values have been assigned, these are combined into an index on a life-death scale. Three procedures have been used.

- *Additive models*. The substituted importance values are summed and the resulting score represents the utility index. The limitation is that each instrument item or dimension must contribute a fixed amount. Under this model, a respondent can obtain a very poor utility score only if they report poor scores on all items or dimensions. Consider an instrument measuring two dimensions: physical and mental health. In an additive model, each may contribute 0.50 towards the utility score. In this model, appalling mental health (leading to suicide) could never, by itself, lead to a utility value lower than 0.50 because 0.50 (a person in good physical health) $+0.00$ (mental health) $= 0.50$. Thus additive models cannot explain people who commit suicide if their physical health is good, or euthanasia if their mental health is good.
- *Econometric models*. The items are treated as explanatory variables to derive a regression equation predicting utilities. This method, however, suffers the same limitation as the additive model.
- *Multiplicative models*. These involve multiplying items or dimension scores together. This overcomes the limitation of the additive model as it allows any dimension to carry a person to a death equivalent value. Consider the case above. Here the person’s value for mental health would be 0.50 (physical health) $\times 0.00$ (mental health) $= 0.00$.

Given these assumptions, preference independence is required to avoid double-counting, which is where the same underlying health condition contributes more than once to the MAU-instrument utility index. For example, if a person suffers major depression this should be counted in their utility score once, although the effect of this health state may be measured in several different aspects of their life; i.e. on several different scales. Where these effects are measured using unidimensional scales that are orthogonal to each other there is no difficulty. Where the scales, however, are correlated the effect of depression will be counted several times over thereby biasing the utility measurement. It is for this reason that MAU-instruments are required to possess structural independence (i.e. where the scales are unidimensional and orthogonal)[39]. For example, if depression is counted on dimensions measuring social, physical and psychological dimensions as well as its effects being directly measured, then there is loss of preference independence as the scores on the social dimension may be a function of mental health scores.

DESCRIPTION OF MAU-INSTRUMENTS

MAU-instruments (in alphabetical order) are the Assessment of QoL (AQoL)[40–42], EQ5D (formerly the EuroQol)[43,44], Health Utility Index 3 (HUI3)[45–47], 15D[48–50], Quality of Well-Being (QWB)[51,52], Rosser Index[38], and the SF6D[53,54].

Assessment of QoL (AQoL)

The Australian AQoL used the WHO's definition of health, and items describe 'handicap' as distinct from impairment and disability[55]. The descriptive system has 15 items and 12 are used in computing the index[56]. Each item has 4 levels. There are five dimensions: Illness (not used in utility computation), Independent Living, Social Relationships, Physical Senses and Psychological Well-being[40]. Designed for self-completion, Nord[56] reported the AQoL took 5-10 minutes. A stratified sample ($n = 350$ respondents; response rate 72%) representative of the Australian adult population completed TTOs based on a 10 year timeframe to provide the utility weights[42]. A multiplicative model is used to compute the utility index[41]. The upper boundary is 1.00, and the lower boundary is –0.04: it permits health state values worse than death. Permission to use the AQoL must be obtained, but there is no cost for its use. Further information can be obtained at: *http://www.psychiatry.unimelb.edu.au/qol/.*

EQ5D (formerly the EuroQol)

The EQ5D (formerly the EuroQoL), developed by a team from 7 European countries[43,57], was based on the QWB[58], the Sickness Impact Profile[59], the Nottingham Health Profile[60], the Rosser Index[38], and group members' opinions. Designed for use in cross-cultural comparisons it has 5 items, each with 3 response levels, measuring Mobility, Self-care, Usual Activities, Pain/Discomfort and Anxiety/Depression. It takes 1-2 minutes to self-complete[56]. The original utility weights were from a British population random sample ($n = 3395$ respondents, response rate 56%) based on the TTO for 42 marker health states using a 10 year timeframe[61]. Recently, US weights have been published. The mean difference in health state values between the British and US weights was 10% of a life-death scale. Consequently it was recommended that when used in the US, US-derived weights should be used[63–65].

The intermediate health state values (i.e. those for which direct TTO weights were not obtained) were regression modelled[61,66,67]. The index is computed using an econometric regression model. The upper boundary is 1.00, and the lower boundary is –0.59: it permits health state values worse than death. Although the EQ5D is in the public domain for public health research, the EQ5D management group ask that researchers register their use of it. There are no costs for its use, unless it is used by commercial organisations. The EQ5D has been translated in many languages. Further information on the EQ5D can be obtained from: *http://www.eur.nl/bmg/imta/eq-net/EQ5d.htm.*

Health Utilities Index, Mark 3 (HUI3)

The Canadian Health Utilities Index (HUI3), for general population use, is based on the HUI2 which was designed for survivors of childhood cancer. To render it generic and overcome reported difficulties, it was revised into the HUI3[45]. The HUI1 has been superseded. The HUI3 measures 'within the skin' functional capacity[46], a perspective adopted to enhance its use in clinical studies[68]. Social aspects of HRQoL are not measured. Items have 4–6 response levels. Twelve of the 15 items form 8 attributes (Vision, Hearing, Speech, Ambulation, Dexterity, Emotion, Cognition and Pain). Designed for self-completion, Nord[56] reported it took 2 minutes to complete, although 5–10 minutes is more likely given it has 15 items. The utility weights were elicited using the VAS, and scores then transformed based on four 'corner' health states valued with the SG where a 60 year timeframe was used. These results were based on stratified sampling ($n = 256$; response rate 22%) of the Hamilton, Ontario, population[69]. A multiplicative function combines the attributes into the utility score[47,69]. The upper boundary is 1.00, and the lower boundary is –0.36, permitting health states worse than death. Users must be registered and the instrument is available at a cost of CAN$4,000 per trial (at the time of writing). Copies of the HUI3 and application forms can be found at: *http://www.healthutilities.com/hui3.htm/.*

15D

The Finnish 15D was defined by Finnish health concerns, the WHO definition of health and medical and patient feedback[48,70]. It is concerned with impairment and disability of 'within the skin' functions. There are 15 items, each with 5 levels, measuring Mobility, Vision, Hearing, Breathing, Sleeping, Eating, Speech, Elimination, Usual Activities, Mental Function, Discomfort & Symptoms, Depression, Distress, Vitality and Sexual Function[48]. Nord[56] reported it took 5–10 minutes for self-completion. The weights came from five random samples of the Finnish population ($n = 1290$ respondents; response rate 51%) using VAS questions; responses were combined using a simple additive model[49,50]. The upper boundary is 1.00, and the lower boundary is +0.11: death-equivalent and worse than death health states are not allowed. Permission must be obtained to use the instrument, however there are no costs for its use. The 15D has been translated into a number of European languages. Although there is no website devoted to the 15D, details can be obtained from *http://195.101.204.50:443/public/15D.html.*

Quality of Well-Being (QWB or IWB)

The American QWB was designed to bridge the gap between clinical measurement, functional status and health planning policy[71] and was an adaptation of US health surveys[72]. It has three dimensions (Mobility, Physical Activity, and Social Activity) with 3–5 levels each. There are an additional 27 illness symptoms. Combined, these provide an index of 'Well-life expectancy' of which there are 43 functioning levels[51,71,73]. This would seem to support Anderson et al.'s[74] description of it as measuring dysfunction. Mental and social health are not measured. The QWB was designed for interview administration (15–35 minutes), although a shorter version has been developed which takes about 15 minutes[75]. Interviewer training is required[76]. The preference weights were elicited using VAS scores which were obtained from a sample of the San Diego population. A linear transformation was then used to place these on a 0.00–1.00 scale[73,75]. An additive model is used to compute the index. Extensive efforts to validate that VAS provides interval properties led to the release of a revised version[51,73,77]. The upper boundary is 1.00, and the lower boundary is 0.00 (death equivalent) and health states worse than death are not permitted. Permission must be obtained to use the QWB and there are no costs for its use. Further information on the QWB can be obtained at: *http://medicine.ucsd.edu/fpm/hoap/instruments.html.*

Rosser Index

The British Rosser Index was designed for use in hospital settings. The original version had two dimensions measuring disability and distress, and measured 29 health states. Values were elicited using magnitude estimation from a convenience sample of 70 respondents[38]. A revised version was released in the early 1990s based on SG procedures and included an additional dimension of discomfort[38]. Administration requires a trained interviewer. The upper boundary is 1.00, and the lower boundary –1.49; i.e. health states worse than death are permitted. The Rosser Index has given rise to two variants: the Health Measurement Questionnaire (HMG)[78] and the Utility-based Quality of Life-Heart Questionnaire (UBQ-H)[79]. Permission must be obtained for using the instrument, however there are no costs for its use. No website was identified for the Rosser Index.

SF6D

Two different algorithms have been published by Brazier for deriving preference-based values from the SF-36[53,54]. The more recent algorithm supersedes the earlier version, so only the more recent algorithm is described here. The advantage of the SF6D is that wherever SF-36 raw scores are available, the SF6D preference measure can be used.

The SF6D[53] uses 10 items from the SF-36: three from the physical functioning scale, one from physical role limitation, one from emotional role limitation, one from social functioning, two bodily pain items, two mental health items and one vitality item. These form 6 dimensions: Physical Functioning (PF: 6 levels), Role Limitation (RL: 4 levels), Social Functioning (SF: 5 levels), Pain (PA: 6 levels), Mental Health (MH: 5 levels) and Vitality (VI: 5 levels). Utility weights were computed from VAS scores, which were modelled using SG values for two link health states. Values were obtained from a random sample of the British population ($n = 611$; response rate = 45%). An additive econometric model is used to compute the utility index. The endpoints for the SF6D are 1.00, and 0.30 for the worst possible health state. No website for the SF6D was identified.

COMPARISON OF INSTRUMENTS

Hawthorne and Richardson[80] outlined the axioms of utility measurement which MAU-instruments should conform to in order to possess basic validity. These axioms can be used as a checklist in instrument selection. They are:

- The use of a preference measurement to weight instrument items.
- Instruments must measure the dimensions of HRQoL deemed to be important. These are usually defined as physical, mental, social and somatic sensations (e.g. pain).
- There must be coverage of the full spectrum of HRQoL values, from full health states to values representing states worse than death.
- The combination rule for the utility index must prevent double-counting.
- There must be evidence of both the weak and strong interval measurement.
- Instruments must be sensitive to the health states of interest. This requirement is covered in the next section. For general sensitivity comparisons between the instruments the reader is referred to validation papers by Barton et al.[81,82], Conner-Spady et al.[83], Hawthorne et al.[56,80,84], Kopec and Willison[85], Marra et al.[86], or Pickard et al.[87].

An additional requirement is that:

- There must be evidence of valid and reliable measurement.

Use of a Preference Measurement Technique to Weight Instrument Items

Instruments using the SG or TTO may be regarded as possessing preference weights since both involve decisions under uncertainty. In the SG, the life outcome is uncertain (the probability of full health versus death). In the TTO, life-length is uncertain (how many life-years a person is willing to sacrifice).

There are doubts over whether ME delivers preferences because the procedure requires the respondent to estimate the divergence of a given health state from the 'full' health state (which is assigned a value of 1.00). Once several given health states have been so assigned, the respondent is then asked to rank these in order[37].

As reported above, there is doubt whether the VAS delivers preference measurement. Consequently it has been argued that the VAS has no place in economic theory[34] and that untransformed VAS scores should not be used[88,89]. It is recommended that VAS data should always be transformed, based on TTO or SG[34–36]; the transformation function that has been used was developed by Torrance et al.[22]. The preference measurement of instruments weighted with VAS scores therefore rests upon the validity of this transformation. For the EQ5D, Dolan et al.[90] reported that the explanatory power of the transformations used was $R^2 = 0.46$, which was considered to be very good. However Sintonen[49] reported that when applied to the 15D VAS data it assigned 12–25% of the adult population to values worse than death, a result he stated was 'implausible'. Bleichrodt & Johannesson[91] noted that individual transformations were unstable; Robinson et al.[88] reported difficulties with the transformations; as did Torrance et al.[89].

Instruments weighted with a preference measure are the EQ5D (which used the SG) and the AQoL (the TTO). The Rosser Index relies upon ME. The HUI3 and the SF6D both rely upon transformed VAS scores; the extent to which these can claim preference weighting is dependent upon the validity of the transformations. Nord[92] has questioned the validity of the linear transformations for the QWB, arguing that one of the primary reasons its use in Oregon was so heavily criticised was that it lacked cardinal values. Given that the 15D uses untransformed VAS ratings there are doubts that it meets this requirement, although Martin[79] averred that this gave the opportunity to quickly establish new weights for different populations—a procedure which Sintonen argued should be followed for each population from which study participants were drawn[48].

Instruments must Measure the Dimensions of HRQoL Deemed to be Important

Important areas of HRQoL are usually defined as physical, mental, social and somatic sensations (e.g. pain). Unless instruments measure all these they cannot claim to be 'generic'. It should be remembered that the measurement of utilities was explicitly developed to enable cross-condition, health state and health care comparisons. By definition MAU-instruments are supposed to be generic.

Generally there are no published formal tests of content validity[80]. Where this is mentioned, instrument developers have reported 'face' validation, i.e. that instrument content as judged by the instrument developers 'looks about right'. For example, it has been argued the very restricted Rosser Index descriptive system makes it insensitive and provides a narrow band of responses[93–96]. In a study of the EQ5D descriptive system it was reported that it only covers 39% of the concepts regarded by the public as salient to health[97]. Feeny et al.[45] reported that the HUI3 was valid because all levels of scores had been assigned at least once in population surveys. These various assertions do not engender confidence that the universe of HRQoL is actually measured by any of the instruments, a point which has been noted in the literature.

In three recent review articles Hawthorne et al.[41,80] mapped the content of MAU-instruments against the dimensions of 14 HRQoL instruments published between 1971 and 1993. Table 2 summarizes their work. This shows that even in the better instruments coverage of the universe is limited. Some instruments offer very narrow measurement (for example, the Rosser Index and EQ5D), others have in-depth or duplicated measurement in particular areas (for example, the QWB, 15D and HUI3), and some offer very broad but sketchy coverage (for example, the AQoL and SF6D). Duplicated measurement may bias the obtained utility values. Two examples illustrate the problems. Despite its broad coverage, the QWB primarily measures pain and physical disability[51] yet does not include either social or mental health[74], and analysis of the HUI3 showed it was a measure of physical impairment which did not adequately measure physical, social or mental dimensions[98].

Table 2. Content of descriptive systems of MAU-instruments (a)

HRQoL dimensions (b)	*AQoL*	*EQ5D*	*HUI3*	*15D*	*QWB (c)*	*Rosser Kind (d)*	*SF6D*
Relative to the body							
Anxiety/depression/distress	*	*		**	**	*	**
Bodily care	*	*					*
Cognitive ability			*	*	*		
General health							
Memory			*				
Mobility	*	*	*	*	*		**
Pain	*	*	*	*	********		*
Physical ability/vitality/disability			*	*	*****	*	*
Rest and fatigue	*			*	**		**
Sensory functions	**		****	*****			
Social expression							
Activities of daily living	*	*		*			*
Communication	*		**	*	*		
Emotional fullfilment			*				
Environment					*		
Family role	*						
Intimacy/Isolation	*						
Medical aids use					**		
Medical treatment							
Sexual relationships				*	*		
Social function	*				*		*
Work function							*

Note: a = Table shows only those items used in calculation of utility scores. Each asterisk represents an item. Based on item content examination.
b = Dimensions of HRQoL defined by a review of 14 HRQoL instruments, 1971-1993.
c = Excludes intoxication.
d = Areas subsumed within the two items: mobility, employment, housework.
Source: Adapted from Hawthorne et al.[80]

There must be Coverage of the Full Spectrum of HRQoL Values

This refers to instruments providing utility values from full health states to values representing states worse than death. There are two issues. First, instruments must have combination rules permitting very poor HRQoL, irrespective of how this is caused. Second, the range of utility scores must cover the full spectrum.

Regarding combination rules, multiplicative models are to be preferred for the reasons outlined above. Instruments with multiplicative models are the HUI3 and AQoL. The EQ5D and SF6D rely upon regression models which are essentially additive in nature, and the 15D is an additive instrument.

The Rosser Index, EQ5D and HUI3 allow large negative values. Given the difficulty with the symmetry argument, these values are problematic. Hawthorne & Richardson[80] calculated that the effect of restricting the lower boundary for the

HUI3 and EQ5D to 0.00, in population studies, would raise mean utility values by 9% and 14% respectively. This suggests the net effect of the symmetry argument is to overstate the value of interventions where people are in very poor health states. This problem does not apply to the QWB and AQoL which have lower boundaries at or near to 0.00.

The lower endpoints for the 15D (+0.11) and SF6D (+0.30) raise other questions. Hawthorne & Richardson[80] reported these boundaries resulted in very different QALY estimates: a 1 QALY gain from the AQoL, EQ5D or HUI3, where a person was returned from the lowest quartile to full health for 1 year, implied a 0.50 and 0.37 QALY gain on the 15D and SF6D respectively. These contradictory results suggest that at least one of these of instrument groups is wrong.

For allowing the full range of scores, the QWB or AQOL instruments would be preferred, as would the 15D.

The Utility Combination Rule must Prevent Double-counting

During construction of the Rosser Index, care was taken to ensure orthogonality between the dimensions[38]. Brazier et al.[34] reported that for the QWB there is multicollinearity between the scales and symptoms. In papers describing the EQ5D there is no mention of this issue[43,44]. Based on clinicians' opinions, structural independence was claimed for the HUI3[68]; the factor analysis of the HUI3 published by Richardson & Zumbo[98], which revealed a lack of independence between the attributes, challenges this claim. Sintonen claimed independence for the 15D, although no evidence was provided[49].

For the SF6D Brazier et al.[53] reported that since an econometric model was used preference independence, structural independence and double-counting were unimportant. The form of the SF6D for the prediction of SG scores is

$$y_{ij} = g\left(\beta' x_{ij} + \theta' r_{ij} + \delta' z_j\right) + \varepsilon_{ij}$$

which is essentially an additive model. Brazier's argument seems extraordinary given that orthogonality to prevent double-counting caused by multicollinearity has been axiomatic of both psychometric and decision-making theory for over 50 years [39,99].

For the AQoL, during construction exploratory factor analysis was used to ensure orthogonality[40]; the structure has since been confirmed by structural equation modelling[84].

There must be Evidence of Both Weak and Strong Interval Measurement

For meeting this criteria, all MAU-instruments rely on the presumed interval properties of the TTO, SG, or VAS. No instrument construction or validation paper has reported any formal testing of these properties and it has not been convincingly demonstrated that these properties are embedded with the TTO, SG, magnitude estimation or VAS[34,38].

The weak interval property

VAS responses may be functions of adaptation, context, endpoints or anchorpoints, end-aversion and rating effects. These imply VAS may produce ordinal rather than interval data[33,88,89,100]. For the TTO and SG even less is known as these issues do not appear to have ever been properly investigated. Although Cook et al.[100] challenged the claim of interval data for all three techniques, there were methodological difficulties with the paper[101]. Subject to these caveats, Hawthorne & Richardson[80] asserted it was likely the SG and TTO possessed interval properties given they allowed incremental probabilities (SG) or time fractions (TTO).

The strong interval property

This means that any given incremental value in HRQoL utility is directly equivalent to the same incremental value in life-length or life-probability. This is a fundamental requirement for the correct calculation of QALYs. There is no evidence available for any of the MAU-instruments that they meet this requirement.

Finally, although the EQ5D may theoretically meet the weak interval properties, there is evidence that it fails this requirement at the empirical level due to the 'N3 term'[102]. In the EQ5D scoring process, any person who endorses a level-3 response (the worst possible level) automatically incurs a coefficient loss of −0.269 utilities. The effect of the 'N3 term' on EQ5D scores is shown in Figure 3, which presents data from the 2004 South Australian Health Omnibus Survey. As shown

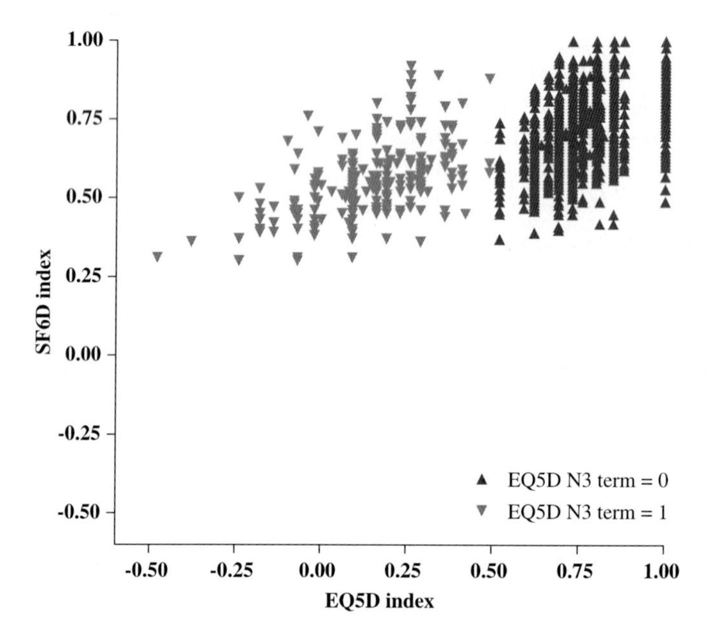

Figure 3. Data distribution issues for the EQ5D
Source: 2004 South Australian Health Omnibus Survey.

there is a large gap in scores in the region of 0.4–0.5, implying that there are areas of the utility range where scores are virtually impossible to attain.

Valid and Reliable Measurement

The validity and reliability of various MAU-instruments has been assessed through either tests of concurrent validity where monotonic relationships are sought, or test-retest. Additionally, there are issues around the stability of the utility values used in the different instruments due to sample bias.

Monotonicity refers to a relationship in which the instrument of interest group or mean scores progressively increase in line with a criterion measure. For example, if a sample of people suffer symptoms of depression from "feeling a bit blue" to "major depression", then an instrument measuring this underlying health condition should report manifest scores that systematically increase with the level of actual depression. This does not imply, of course, that there will always be a 1:1 relationship between the two measures, for there will be individual variation.

Hawthorne et al.[41] examined monotonicity for the EQ5D, 15D, HUI3, AQoL and SF6D (Version 1) against health status as defined by their sample strata of community random sample, outpatients and inpatients; they also examined the same instruments by combined utility quartile[84] and by instrument predictive power[80]. In general their findings support monotonicity for all the instruments, although they did observe that the instruments formed two groups: those which correctly classified >50% of cases (AQoL, 15D and SF6D) and those which predicted <50% (EQ5D and HUI3).

Data on the Rosser Index are mixed. Although Rosser Index scores have been shown to match empirical and population general health data quite well when predicting the healthy/unhealthy dichotomies[78,94], in a replication study it was shown that there are several health states where monotonicity is violated leading to difficulties with assigning logical QALY values[37].

For the QWB there is mixed evidence regarding monotonicity. Kaplan et al.[103] reported very high correlations with a number of chronic conditions, where the average was $r = 0.96$. Based on the revised version, similar correlations with chronic conditions have been reported[77,104]. For example, Kaplan et al.[105] reported a monotonic relationship between QWB scores and HIV-status; similarly monotonicity has been reported for functional status of children suffering cancer[106]. Against this the QWB has been criticised for producing QALY values that are non-monotonic. Thus a person wearing glasses is worse off than someone confined to a wheelchair, or curing five people with pimples would equate with saving one life[92,107]. In a study of heart disease, non-monotonicity was reported for half the QWB scales[108].

The Hawthorne et al. results for the EQ5D (see above) were particularly interesting as they indicated that the EQ5D assigned too many cases to a utility value of 1.00, a finding consistent with earlier work by Brazier et al.[109]. Both research groups suggested this may have been due to the insensitivity of the EQ5D at the healthy end of the range and the consequent limited capacity to discriminate between those with full health and some health problems. At the other end of the range (very poor health states) Nord et al.[94], in a study comparing Norwegian and Australian populations, reported that the EQ5D assigned excessively low values for some health states; a finding consistent with that of Hawthorne et al.[41] who found that the EQ5D assigned 4% of a population sample to health states worse than death. In a comparison with the SF-36, Brazier[109] pointed out that the EQ5D correlated poorly with physiological symptoms, and Andersen et al.[110] reported that the EQ5D assigned non-monotonic values for people with fractures: a person with a fractured arm was assigned worse utility than someone with a fractured vertebrae.

Sintonen[49] tested the 15D for monotonicity in five population-based samples, reporting that up to 2.5% of respondents valued health states inconsistently, rising to 20% who valued 'death' higher than being 'unconscious'.

For the AQoL, several papers have suggested monotonicity, including Hawthorne et al.'s work[41,80], and in Alzheimer's disease[111], cochlear implants[112], depression[113], psychosis[20], suicidal ideation[114] and stroke[115].

Test-retest reliability estimates have been reported for the QWB, 15D, EQ5D, HUI3 and AQoL. For the QWB, Kaplan et al.[103] reported test-retest reliability at $r = 0.93$–0.98. In a study of chronic obstructive pulmonary disease, at 14-day separation, Stavem[116] reported that the EQ5D and 15D test-retest reliability using Spearman correlations were $r = 0.73$ and $r = 0.90$ respectively. This result for the 15D is more encouraging than that reported by Sintonen[50], who did not give a statistical estimate but stated that the agreement was not very good. In a study of stroke patients, Dorman et al.[117] reported test-retest reliability estimates for the EQ5D of ICC = 0.83; and in a Dutch population study of the EQ5D where test-retest was carried out at 10-month interval the correlation was $r = 0.90$[118]. Studies of the HUI3[46,119], based on a community random sample with telephone follow-up, reported test-retest reliability where $r = 0.77$. For the AQoL, Hawthorne[120], using random population sampling and mail/telephone comparisons reported the test-retest ICC = 0.83. An earlier study reported test-retest reliability for the AQoL descriptive system where $\alpha = 0.80$[121].

Finally, and importantly, there are issues concerning the stability of the utility weights used in the various instruments. This concern stems from the fact that utility weights for most of the instruments—with the notable exception of the EQ5D where the British sample size was 3395—were obtained from either small (e.g. 70 cases for the Rosser Index) or convenience samples (e.g. the 1290 respondents for the 15D). In most cases, this was because of the cost of data collection: face-to-face interviews where SG or TTO questions are asked are costly. Because the SG or TTO is extremely tedious, all the instrument designers eroded their sample sizes further by breaking their health states up into sub-interview routines and then administering each sub-interview to a strata within the sample. This is commonly referred to as a 'sort' procedure. The extreme case where this occurred was with the SF6D[53]. The weights for the SF6D were obtained from a representative sample of 836 Englishpersons of which 611 interviews were actually used. Based on a sort procedure, each respondent was asked to value 6 health states out of a possible 249 health states. Altogether 3,518 valuations were made: an average of 15 responses for each health state (the range was from 8 for health state 5,3,5,6,4,6 to 19 for health state 1,3,1,5,4,2) . Similar procedures were followed for the HUI3[69], AQoL[122], and 15D[49], although in each case the numbers were greater than for the SF6D. For example, for the HUI3 the numbers for each health state varied from 19 to 246; for the AQoL the range was 70 through 225). These difficulties for each instrument were compounded by the relatively low response rates (typically about 50% although the AQoL's was higher).

These wafer-thin estimates raise fundamental questions concerning the transparency of utility scores, their stability and the generalizability of the instruments. Other than for the EQ5D, none of the instrument developers have reported validation of the obtained utility results or published an analysis of these data. Given this, it is highly likely the utility values for all instruments, other than the EQ5D, are biased and lack transparency. Because of the restricted response rates and small sample sizes utility weights may be less than stable; a problem compounded by the fact that all instrument weights have been derived using means rather than medians. Clearly, under these circumstances, claims for generalizability to many health conditions, including mental health, should be interpreted cautiously.

USING MAU-INSTRUMENTS IN CUA OF MENTAL HEALTH INTERVENTIONS

CUA is a form of cost-effectiveness analysis (CEA). In a CEA study, two or more interventions are compared where the comparison is made on the basis of the cost-per-health-outcome-unit gained. For example, Araya et al.[123] reported on a stepped care improvement program versus usual care for women with depression who were being treated in primary care. The outcome measure was symptom-free days over a 6-month period. Costs collected included primary care consultation, psychiatric consultation, group therapy sessions, psycho-education training and patient support. The incremental cost-effectiveness ratio (additional costs above usual care per extra depression-free day) was 216 (95%CI: 130-343) Chilean pesos. To compare this study with Adams et al.'s[124] study of the effectiveness of the diagnostic utility of endocrine and neuroimaging for the treatment of adolescent psychosis, would be extremely difficult because the outcome units would be different (depression-free days versus abnormal screening tests).

CUA overcomes this difficulty because the outcomes are expressed in a common metric, the QALY (discussed above). Where the QALY is the outcome, then the cost-per-QALY gained as a result of intervention A is directly comparable to that for intervention B. For example, Pinkerton et al.[125] conducted a CUA of a small-group HIV risk reduction intervention for adults with a psychiatric diagnosis. Seven small group sessions were conducted over a month, followed by two 'booster' sessions later. Sexual behaviours were assessed every 3 months for a year after completion of the small group sessions. Costs collected included all HIV-associated costs based on a societal perspective. The QALYs saved were modelled from earlier work. The CUA showed that the cost-per-QALY saved ratio was US$136,295. The researchers compared this finding with the cost-per-QALY gained from annual mammography (US$62,000 - $190,000 per QALY saved), sickle disease treatment for new-born babies (US$110,000), and pneumonia vaccination for children (US$160,000 – $170,000). It was concluded that the intervention was cost-effectiveness.

This example gives rise to several important issues to be considered when conducting a CUA.

The Perspective of the Study

Costs can be viewed from any of three perspectives: costs to the service provider, costs to the patient, and to the society. Generally, the narrowest perspective is that of the service provider since this only includes the costs of providing treatment; on the other hand this perspective may be preferred by service providers or medical insurers who have to plan for facilities, staff and re-imbursements. The broadest is the societal perspective because this captures all costs: the costs of treatment, the consumption of personal care products or medical aids, the loss of wages, legal costs, family breakdown etc. In the mental health field, this broader perspective is preferred because mental health conditions are often chronic and have multiple effects on intimate relationships, friendships, the capacity to work and so on. The societal perspective is also that recommended by the Panel on Cost-Effectiveness in Health and Medicine[126]. In a study comparing levodopa and pramipexole for Parkinson's disease, Noyes et al.[127] asked participants to keep a health service use diary of all health care visits, procedures used, diagnostic tests, hospitalisations, emergency department visits, outpatient surgeries and days off work. Additionally, medication use, care-receiving, home help and rehabilitation services were all collected and costed. Although the researchers acknowledged that there were some limitations to these data, they argued that as far as possible they were using a societal perspective. The EQ5D was used as the outcome measure over a 4-year period, and changes in EQ5D utility scores over time were used to compute QALYs. The results showed the cost-per-QALY gained for levodopa was US$42,989 compared with US$46,218 for pramipexole.

Issues in Cost Variability

Costs within a country may vary by insurance arrangements. For example, many managed care/insurance plans limit the services covered with the consequence that non-covered services may be avoided because they involve out-of-pocket costs to the patient. Costs may also vary between countries, depending upon the regulatory framework. For example, in the USA there are generally limits on long-term care whereas in Japan long-term health care benefits are available. To overcome cost variability, CUA usually uses standard or average costs. For example, in a study of escitalopram versus venlafaxine for major depression across 8 European countries health care costs were based on country-specific costs for six countries, which were then averaged across all participating countries [128]. The outcomes were measured with the EQ5D: the mean scores at baseline were reported to be 0.52 for the escitalopram cohort and 0.54 for venlafaxine. At 8-week follow-up there was no statistically significant difference in EQ5D scores (0.78 for escitalopram and 0.77 for venlafaxine): both treatment arms of the study had gained equally. The healthcare costs, though, were 110€ and 161€ respectively, and sick leave costs were 655€ and 712€ respectively. The authors concluded that treatment with escitalopram led to per patient cost savings of 51€ when compared with venlafaxine.

Sensitivity Analysis

Because of the uncertainty in costs CUA findings are usually subject to sensitivity analysis. This involves varying costs (or the outcomes) by between 5-10% and re-running the analysis. In Noyes et al.'s study of Parkinson's disease described above sensitivity analyses were conducted by varying imputed missing data for the EQ5D, by varying the cost of the pharmaceuticals and by applying a 3% discount rate over the 4-years of the study. The resulting extreme 'best' case scenario for pramipexole was US$29,759/QALY when compared with the base estimate of US$46,218/QALY.

The Presence of Co-morbidities, Confounders or Changes in Treatment Regimen

Co-morbidities are usually ignored in CUA, and the changes in the utilities of whichever measure is being used are accepted uncritically. Yet there is considerable evidence that co-morbidities, confounders and treatment variation may play a major role in study outcomes. In an open multicentre randomized trial CEA of atypical (SGA; second-generation antipsychotics) versus typical (FGA; first-generation antipsychotics) medication for schizophrenia, Jones et al. [129] reported that there was no statistically significant difference after 12-months between the two treatment arms of the study in scores on the Quality of Life Scale, although patients assigned to typical medications had lower costs (£18, 800 versus £20, 100) indicating a clear advantage for typical medication. However, the results of the study may have been confounded since more participants in the typical medication study arm were referred owing to adverse effects, so this factor was entered as a co-variate during data analysis. The study findings may also have been affected by the numbers dropping out of the study (17% of participants), that 41% of participants did not remain in their allocated treatment arm throughout the study, and differences in the prescribing practices of participating psychiatrists (most of whom prescribed sulpiride to patients in the typical medication treatment arm).

MAU-instrument Sensitivity to Mental Health Conditions

The computation of QALYs assumes that a 10% increase in utility is equivalent to a 10% increase in life length and that from this it is possible to compute the number of patients needed to be treated to achieve a 1-QALY gain. As outlined above, one of the axioms of CUA is that by using QALYs as the outcomes it is possible to compare across different health conditions and interventions. This assumes that a QALY generated on one MAU-instrument has the same meaning as a QALY generated on another instrument. This assumption is, unfortunately, not supported by studies that have directly compared different MAU-instruments[56,85,86,130]. For example, in a study of incontinence drawing upon a population-weighted sample of 3015 participants from South Australia, Hawthorne[131] compared the AQoL, EQ5D, HUI3, 15D and SF6D. He reported that study findings were dependent upon which measure was used as much as upon treatment success.

Re-analysis of the same dataset examining depression classification based on completion of the PRIME-MD[132,133] shows the same finding, as shown in Table 3. The table shows the utility scores for those with no symptoms, minor (other) depressions and major depression for each of the MAU-instruments. There are few differences between the AQoL, EQ5D and HUI3. According to these instruments, a person with major depression would, on average, be experiencing a HRQoL that was about 30% lower than that of a person with no symptoms. The scores on the 15D and SF6D are different. For example, those classified as suffering minor depression have a utility score on the 15D that is higher than that for those with no symptoms of depression on any of the other instruments. The most sensitive (i.e. responsive) instrument to depression is the 15D, as shown by it obtaining the highest relative efficiency[134]. The least sensitive instrument was the EQ5D. However, for CUA the important factor is not so much statistical significance as the number of patients needed to treat to obtain a 1-QALY gain. When assessed by this criterion, the 15D performed particularly poorly: the number of patients needed to treat for a 1-QALY gain is double that of the AQoL, EQ5D or HUI3. These findings suggest that there are substantial differences between MAU-instruments and that these differences may have a profound impact on the calculation of QALYs arising from an intervention and any subsequent cost-utility analyses.

The Use of Proxies in Mental Health Studies

The use of proxies has been advocated for reasons of current affective state, adaptation, lack of insight, neuroticism and emotional adjustment, cognitive impairment and possible effects of neuroleptic therapy[135–140]. Yet the research into patient self-report versus proxy report is not supportive of this position. In general, it suggests that patient self-report is to be preferred[18,141–145], although there are some equivocal studies[146]. Voruganti et al.[19], for example, reported that based on correlation analysis the relationships between patients with schizophrenia and clinicians' assessments concurred on structured measures whereas there was a lower

Table 3. Depression and utility as assessed by 5 MAU-instruments

		N	*MAU-instrument*									
			AQoL		*EQ5D*		*HUI3*		*15D*		*SF6D*	
			Mean	*sd*	*Mean*	*sd*	*Mean*	*sd*	*Mean*	*sd*	*Mean*	*sd*
Depression classification	No symptoms	2521	0.85	0.17	0.86	0.17	0.86	0.17	0.95	0.07	0.84	0.12
	Minor depression (a)	253	0.68	0.22	0.72	0.24	0.71	0.23	0.88	0.09	0.70	0.14
	Major depression	241	0.53	0.25	0.53	0.32	0.53	0.30	0.79	0.12	0.60	0.12
Statistics	$ANOVA_{ADJ}$(b)		F = 306.69		F = 246.77		F = 283.47		F = 374.22		F = 303.56	
Relative efficiency (c)			1.24		1.00		1.15		1.52		1.23	
Number needed to treat (d)	Minor depression		5.88		7.14		6.67		14.29		7.14	
	Major depression		3.13		3.03		3.03		6.25		4.17	

Notes a = Dysthymia and other depressions
b = Scores log adjusted prior to analysis
c = $RE = F_{INTEREST}/F_{SMALLEST}$
d = Number needed to treat for a 1-QALY gain

level of agreement on more global measures. Similarly, Sainfort et al.[147] reported that there was general agreement on the clinical aspects and less on social assessments. This is well illustrated by Herrman et al.[20] who examined the relationship between patient self-report for those with long-term psychosis and the reports of their case managers on the AQoL. The findings showed that case managers systematically underestimated utility, particularly in the area of social relationships where case managers assessments were 14% lower (i.e. worse) than those of the patients. Overall, the difference in utility was 0.05 – which was within the range considered to be a minimally important difference[148]. Herrman et al. concluded that patient reports were to be preferred.

CONCLUSION

MAU-instruments were developed to value HRQoL through combining the respondent's health state with societal preferences for that health state, thus enabling the quantification of treatment effects as the QALY. Although they were developed for use in CUA, almost all of their application to date has been in epidemiological or evaluation research[20,149–154] and burden of disease studies[114,155–158]. A recent review of the use of QALYs in cost-effectiveness studies (of which CUA is a sub-category) identified 60 published studies of rigorous methodology of which just 4 concerned psychiatry[159]. This may reflect difficulties with study design, a general unawareness of MAU-instruments, a reluctance to rely upon self-report, a lack of familiarity with economic evaluation or even a belief that economic evaluation is not needed in psychiatry.

Increasing demand for mental health services combined with rising health care costs will inexorably lead to the situation where mental health professionals will be required to justify their health care expenditures against other possible health care uses of those resources. This was introduced into Australia in 1993 for the registration of new pharmaceuticals[160,161], and other countries are following Australia's lead. It is against this background that MAU-instruments were developed and are being used to provide evidence of the value of health interventions to patients; information which can be placed alongside clinical outcomes to assist clinicians in making judgements about which treatments to use and decision-makers when considering the allocation of scarce resources.

This chapter has reviewed available MAU-instruments and issues in their use in economic evaluation. As shown, none of the available instruments fully meet the axioms of utility theory, they are different to each other, and QALYs generated from one instrument will not be equivalent to those generated from another instrument. The best advice is for researchers to use two different MAU-instruments and to report both sets of results.

Overall, the paucity of CUA in mental health research needs to be redressed. If it is not, in the future it may hinder patients access to mental health services and restrict the range of treatments available as decision-makers may be disinclined to increase or even maintain funding for interventions that do not have evidence of cost-effectiveness.

As shown in this chapter, economic evaluation using cost-utility analysis and MAU-instruments offers a way for mental health research to collect this information and use it to answer the challenges set by Elkinton 40 years ago. It should do so in the interests of the QoL of its patients.

ACKNOWLEDGMENTS

I would like to acknowledge the valuable assistance of Professor Helen Herrman with the initial conceptualization of this paper, and of Professor Robert Goldney for kindly agreeing to let me use his data (Table 3).

REFERENCES

1. Ostenfeld E. Aristotle on the good life and quality of life. In: Nordenfelt L, editor. Concepts and Measurement of Quality of Life in Health Care. Dordrecht: Kluwer Academic Publishers; 1994. p. 19–34.
2. Long PH. On the quantity and quality of life. Med. Times 1960;88:613–9.
3. Elkinton JR. Medicine and the quality of life. Ann. Intern. Med. 1966;64(3):711–4.
4. Imhof A. The implications of increased life expectancy for family and social life. In: Wear A, editor. Medicine in Society: Historical Essays. Cambridge: Cambridge University Press; 1992.
5. Nordenfelt L, editor. Concepts and Measurement of Quality of Life in Health Care. Dordrecht: Kluwer Academic; 1994.
6. Walker S, Rosser R. Quality of life assessment: key issues in the 1990s. Dordrecht: Kluwer Academic Publishers; 1993.
7. Williams B. Patient satisfaction: a valid concept? Soc. Sci. Med. 1994;38(4):509–516.
8. Sitzia J, Wood N. Patient satisfaction: a review of issues and concepts. Soc. Sci. Med. 1997;45(12):1829–1843.
9. Bowling A. Measuring Health: A Review of Quality of Life Measurement Scales. Milton Keynes: Open University; 1991.
10. Drummond M, O'Brien B, Stoddart G, et al. Methods for the Economic Evaluation of Health Care Programmes. 2nd ed. Oxford: Oxford University Press; 1998.
11. Productivity Commission. Economic Implications of an Ageing Australia. Canberra: Commonwealth of Australia; 2005.
12. Productivity Commission. Impacts of Medical Technology in Australia. Melbourne: Productivity Commission; 2005.
13. Mathers C, Vos T, Stevenson C. The Burden of Disease and Injury in Australia. Canberra: Australian Institute of Health and Welfare; 1999.
14. Suchman EA. Evaluative Research: Principles and Practice in Public Service & Social Action Programs. New York: Russell Sage Foundation; 1967.
15. Foucault M. Madness and civilization; a history of insanity in the Age of Reason. New York: Vintage Books; 1965.
16. WHOQoL Group. Study protocol for the World Health Organization project to develop a quality of life assessment instrument (WHOQOL). Qual. Life Res. 1993;2:153–159.
17. Atkinson M, Zibin S, Chuang H. Characterizing quality of life among patients with chronic mental illness: a critical examination of the self-report methodology. Am. J. Psychiatry 1997;154:99–105.
18. Awad G, Voruganti P. Cost-utility analysis in schizophrenia. J. Clin. Psychiatry 1999;60(Suppl 3): 22–28.
19. Voruganti L, Heslegrave R, Awad A, et al. Quality of life measurement in schizophrenia: reconciling the quest for subjectivity with the question of reliability. Psychol. Med. 1998;28: 165–172.

20. Herrman H, Hawthorne G, Thomas R. Quality of life assessment in people living with psychosis. Soc. Psychiatry Psychiatr. Epidemiol. 2002;37(11):510–518.
21. Sackett D, Torrance G. The utility of different health states as perceived by the general public. J. Chronic Dis. 1978;31:697–704.
22. Torrance G, Boyle M, Horwood S. Application of multi-attribute theory to measure social preferences for health states. Oper. Res. 1982;30:1043–1069.
23. Patrick DL, Bush JW, Chen MM. Toward an operational definition of health. J. Health Soc. Behav. 1973;14(1):6–23.
24. Rosser R, Kind P. A scale of valuations of states of illness: is there a social consensus. Int. J. Epidemiol. 1978;7(4):347–58.
25. Ubel PA, Richardson J, Menzel P. Societal value, the person trade-off, and the dilemma of whose values to measure for cost-effectiveness analysis. Health Econ. 2000;9(2):127–36.
26. Ubel PA, Loewenstein G, Jepson C. Whose quality of life? A commentary exploring discrepancies between health state evaluations of patients and the general public. Qual. Life Res. 2003;12:599–607.
27. Shortell R, Richardson W. Health Program Evaluation. St Louis: Mosby; 1978.
28. Ovretveit J. Evaluating Health Interventions. Buckingham: Open University Press; 1998.
29. Singh B, Hawthorne G, Vos T. The role of economic evaluation in mental health care. Aust. N. Z. J. Psychiatry 2001;35:104–117.
30. Torrance G. Measurement of health state utilities for economic appraisal: a review. J. Health Econ. 1986;5:1–30.
31. Hawthorne G, Richardson J, Day N, et al. Life and Death: Theoretical and Practical Issues in Utility Measurement. Melbourne: Centre for Health Program Evaluation; 2000. Working Paper 102.
32. Richardson J, Hawthorne G. Negative Utilities and the Evaluation of Complex Health States: Issues Arising from the Scaling of a Multiattribute Utility Instrument. Melbourne: Centre for Health Program Evaluation; 2000. Working Paper 113.
33. Richardson J. Cost utility analysis: what should be measured? Soc. Sci. Med. 1994;39(1):7–21.
34. Brazier J, Deverill M, Green C, et al. A review of the use of health status measures in economic evaluation. Health Technol. Assess. 1999;3(9):1–165.
35. Bennett K, Torrance G. Measuring health state preferences and utilities: rating scale, time trade-off, and standard gamble techniques. In: Spilker B, editor. Quality of Life and Pharmacoeconomics in Clinical Trials. Philadelphia: Lippincott-Raven Publishers; 1996. p. 253–265.
36. Robinson A, Dolan P, Williams A. Valuing health states using VAS and TTO: what lies behind the numbers? Soc. Sci. Med. 1997;45(8):1289–1297.
37. Gudex C, Kind P, van Dalen H, et al. Comparing Scaling Methods for Health State Valuations - Rosser Revisited. Discussion Paper. York: Centre for Health Economics, University of York; 1993. Report No.: 107.
38. Rosser R. A health index and output measure. In: Walker S, Rosser R, editors. Quality of Life Assessment: Key Issues in the 1990s. Dordrecht: Kluwer Academic Publishers; 1993.
39. von Winterfeldt D, Edwards W. Decision Analysis and Behavioural Research. Cambridge: Cambridge University Press; 1986.
40. Hawthorne G, Richardson J, Osborne R. The Assessment of Quality of Life (AQoL) instrument: a psychometric measure of health related quality of life. Qual. Life Res. 1999;8:209–224.
41. Hawthorne G, Richardson J, Day N. A comparison of the Assessment of Quality of Life (AQoL) with four other generic utility instruments. Ann. Med. 2001;33(5):358–370.
42. Hawthorne G, Richardson J, Day N, et al. Construction and Utility Scaling of the Assessment of Quality of Life (AQoL) Instrument. Melbourne: Centre for Health Program Evaluation; 2000. Working Paper 101.
43. EuroQol Group. EuroQol: a new facility for measurement of health-related quality of life. Health Policy 1990;16:199–208.
44. Kind P. The EuroQoL instrument: an index of health-related quality of life. In: Spilker B, editor. Quality of Life and Pharmacoeconomics in Clinical Trials. 2nd ed. Philadelpha: Lippincott-Raven Publishers; 1996. p. 191–201.

45. Feeny D, Furlong W, Torrance G. Health Utilities Index Mark 2 and Mark 3 (HUI2/3) 15-item questionnaire for self-administered, self-assessed usual health status. Hamilton: Centre for Health Economics and Policy Analysis, McMaster University; 1996.
46. Feeny D, Torrance G, Furlong W. Health utilities index. In: Spilker B, editor. Quality of Life and Pharmacoeconomics in Clinical Trials. 2nd ed. Philadelpha: Lippincott-Raven Publishers; 1996.
47. Torrance GW, Furlong W, Feeny D, et al. Multi-attribute preference functions. Health Utilities Index. Pharmacoeconomics 1995;7(6):503–20.
48. Sintonen H, Pekurinen M. A fifteen-dimensional measure of health-related quality of life (15D) and its applications. In: Walker S, Rosser R, editors. Quality of Life Assessment. Dordrecht: Kluwer Academic Publishers; 1993.
49. Sintonen H. The 15D measure of health-related quality of life: feasibility, reliability and validity of its valuation system. Melbourne: National Centre for Health Program Evaluation; 1995. Working Paper 42.
50. Sintonen H. The 15D measure of health-related quality of life: reliability, validity and sensitivity of its health state descriptive system. Melbourne: National Centre for Health Program Evaluation; 1994. Working Paper 41.
51. Kaplan R, Anderson J, Ganiats T. The Quality of Well-Being Scale: rationale for a single quality of life index. In: Walker S, Rosser R, editors. Quality of Life Assessment: Key Issues in the 1990s. Dordrecht: Kluwer Academic Publishers; 1993.
52. Kaplan RM, Alcaraz JE, Anderson JP, et al. Quality-adjusted life years lost to arthritis: effects of gender, race,and social class. Arthritis Care Res. 1996;9(6):473–82.
53. Brazier J, Roberts J, Deverill M. The estimation of a preference-based measure of health from the SF-36. J. Health Econ. 2002;21:271–292.
54. Brazier J, Usherwood T, Harper R, et al. Deriving a preference-based single index from the UK SF-36 Health Survey. J. Clin. Epidemiol. 1998;51(11):1115–1128.
55. Hawthorne G, Richardson J. An Australian MAU/QALY Instrument: Rationale and Preliminary Results. In: Sansoni J, editor. Proceedings Health Outcomes and Quality of Life Measurement Conference; 1995; Canberra: The Australian Health Outcomes Collaboration. The University of Wollongong; 1995.
56. Nord E. A Review of Synthetic Health Indicators. Oslo: National Institute of Public Health for the OECD Directorate for Education, Employment, Labour and Social Affairs; 1997.
57. Rabin R, de Charro F. EQ-5D: a measure of health status from the EuroQol Group. Ann. Med. 2001;33:337–343.
58. Kaplan RM, Anderson JP. A general health policy model: update and applications. Health Serv. Res. 1988;23(2):203–35.
59. Bergner M, Bobbitt RA, Carter WB, et al. The Sickness Impact Profile: development and final revision of a health status measure. Med. Care 1981;19(8):787–805.
60. Hunt S, McKenna S, McEwen J, et al. The Nottingham Health Profile: subjective health status and medical consultations. Soc. Sci. Med. 1981;15A:221–229.
61. Dolan P. Modeling valuations for EuroQol health states. Med. Care 1997;35(11):1095–1108.
62. Shaw JW, Johnson JA, Coons SJ. US valuation of the EQ-5D health states: development and testing of the D1 valuation model. Med. Care 2005;43(3):203–20.
63. Havranek EP, Steiner JF. Valuation of health states in the US versus the UK: two measures divided by a common language? Med. Care 2005;43(3):201–2.
64. Johnson JA, Luo N, Shaw JW, et al. Valuations of EQ-5D health states: are the United States and United Kingdom different? Med. Care 2005;43(3):221–8.
65. Fryback DG. A US valuation of the EQ-5D. Med. Care 2005;43(3):199–200.
66. MVHGroup. The Measurement and Valuation of Health: Final Modelling of Valuation Tariffs. York: Centre for Health Economics; 1995.
67. Dolan P, Gudex C, Kind P, et al. Valuing health states: a comparison of methods. J. Health Econ. 1996;15:209–231.
68. Furlong WJ, Feeny DH, Torrance GW, et al. The Health Utilities Index (HUI) system for assessing health-related quality of life in clinical studies. Ann. Med. 2001;33:375–384.

69. Furlong W, Feeny D, Torrance G, et al. Multiplicative Multi-attribute Utility Function for the Health Utilities Index Mark 3 (HUI3) System: A Technical Report. Working Paper. Hamilton: McMaster University, Centre for Health Economics and Policy Analysis; 1998. Report No.: 98–11.
70. Sintonen H. The 15D instrument of health-related quality of life: properties and applications. Ann. Med. 2001;33:328–336.
71. McDowell I, Newell C, editors. Measuring health: a guide to rating scales and questionnaires. New York: Oxford University Press; 1987.
72. Cadet B. History of the construction of a health indicator integrating social preference: the Quality of Well-Being Scale. In: 7th Meeting, International Network on Health Expectancy (REVES); 1994 23–25 February; Canberra; 1994.
73. Kaplan R, Bush J, Berry C. Health status: types of validity and the Index of Well-being. Health Serv. Res. 1976;11(4):478–507.
74. Anderson J, Kaplan R, Berry C, et al. Interday reliability of function assessment for a health status measure: The Quality of Well-Being scale. Med. Care 1989;27(11):1076–1084.
75. Kaplan R, Ganiats T, Sieber W, et al. The Quality of Well-being Scale. Medical Outcomes Trust Bulletin 1996:2–3.
76. Bombardier C, Raboud J. A comparison of health-related quality-of-life measures for rheumatoid arthritis research. The Auranofin Cooperating Group. Control. Clin. Trials 1991;12(4 Suppl): 243S-256S.
77. Coons S, Kaplan R. Quality of life assessment: understanding its use as an outcome measure. Hosp. Formul. 1993;28:486–498.
78. Kind P, Gudex C. Measuring health status in the community: a comparison of methods. J. Epidemiol. Community Health 1994;48:86–91.
79. Martin A, Glasziou P, Simes R. A Utility-Based Quality of Life Questionnaire for Cardiovascular Patients: Reliability and Validity of the UBQ-H(eart) Items. Sydney: NHMRC Clinical Trials Centre, University of Sydney; 1996.
80. Hawthorne G, Richardson J. Measuring the value of program outcomes: a review of utility measures. Exp. Rev. Pharm. Outcomes Res. 2001;1(2):215–228.
81. Barton GR, Bankart J, Davis AC, et al. Comparing utility scores before and after hearing-aid provision: results according to the EQ-5D, HUI3 and SF-6D. Appl. Health Econ. Health Policy 2004;3(2):103–5.
82. Barton GR, Bankart J, Davis AC. A comparison of the quality of life of hearing-impaired people as estimated by three different utility measures. Int. J. Audiol. 2005;44(3):157–63.
83. Conner-Spady B, Suarez-Almazor ME. Variation in the estimation of quality-adjusted life-years by different preference-based instruments. Med. Care 2003;41(7):791–801.
84. Hawthorne G, Richardson J, Day N. A comparison of five multi-attribute utility instruments.Aust. Stud. Health Serv Admin 2001;89: 151–179.
85. Kopec JA, Willison KD. A comparative review of four preference-weighted measures of health-related quality of life. J. Clin. Epidemiol. 2003;56(4):317–25.
86. Marra CA, Woolcott JC, Kopec JA, et al. A comparison of generic, indirect utility measures (the HUI2, HUI3, SF-6D, and the EQ-5D) and disease-specific instruments (the RAQoL and the HAQ) in rheumatoid arthritis. Soc. Sci. Med. 2005;60(7):1571–82.
87. Pickard AS, Johnson JA, Feeny DH. Responsiveness of generic health-related quality of life measures in stroke. Qual. Life Res. 2005;14(1):207–19.
88. Robinson A, Loomes G, Jones-Lee M. Visual analog scales, standard gambles and relative risk aversion. Med. Decis. Making 2001;21(1):17–27.
89. Torrance GW, Feeny D, Furlong W. Visual analog scales: do they have a role in the measurement of preferences for health states? Med. Decis. Making 2001;21(4):329–34.
90. Dolan P, Gudex C, Kind P, et al. Social tariff for EUROQoL: results from a UK general population survey. York: Centre for Health Economics, University of York; 1995. Discussion Paper 138.
91. Bleichrodt H, Johannesson M. An experimental test of a theoretical foundation for rating-scale valuations. Med. Decis. Making 1997;17(2):208–16.

92. Nord E. Unjustified use of the Quality of Well-Being Scale in priority setting in Oregon. Health Policy 1993;24(1):45–53.
93. Hollingworth W, Mackenzie R, Todd CJ, et al. Measuring changes in quality of life following magnetic resonance imaging of the knee: SF-36, EuroQol or Rosser index? Qual. Life Res. 1995;4(4):325–34.
94. Nord E, Richardson J, Macarounas-Kirchmann K. Social evaluation of health care versus personal evaluation of health states. Evidence on the validity of four health-state scaling instruments using Norwegian and Australian surveys. Int. J. Technol. Assess. Health Care 1993;9(4):463–78.
95. Mulkay M, Ashmore M, Pinch T. Measuring the quality of life: a sociological intervention concerning the application of economics to health care. Sociology 1987;21:541–564.
96. Elvik R. The validity of using health state indexes in measuring the consequences of traffic injury for public health. Soc. Sci. Med. 1995;40(10):1385–98.
97. DeptHealth. Research Group on the measurement and valuation of health. In: Methodology Workshop, EUROQoL Conference; 1995; London: Economics & Operational Research Division, Department of Health; 1995.
98. Richardson C, Zumbo B. A statistical examination of the Health Utility Index-Mark III as a summary measure of health status for a general population survey. Soc. Ind. Res. 2000;51(2):171–191.
99. Cattell R. Factor Analysis: an Introduction and Manual for the Psychologist and Social Scientist. New York: Harper & Row; 1952.
100. Cook K, Ashton C, Byrne M, et al. A psychometric analysis of the measurement level of the rating scale, time trade-off, and standard gamble. Soc. Sci. Med. 2001;53:1275–1285.
101. Hawthorne G, Osborne RH, Elliott P. Commentary on: a psychometric analysis of the measurement level of the rating scale, standard gamble and time-trade off, by Cook et al. Soc. Sci. Med. 2003;56:895–897.
102. Brazier J, Roberts J, Tsuchiya A, et al. A comparison of the EQ-5D and SF-6D across seven patient groups. Health Econ. 2004;13(9):873–84.
103. Kaplan R, Bush J, Berry C. The reliability, stability and generalizability of a health status index. In: Proceedings of the Social Statistics Section; 1978: American Statistical Association; 1978. p. 704–709.
104. Kaplan RM. Quality of life assessment for cost/utility studies in cancer. Cancer Treat. Rev. 1993;19 Suppl A:85–96.
105. Kaplan RM, Anderson JP, Patterson TL, et al. Validity of the Quality of Well-Being Scale for persons with human immunodeficiency virus infection. HNRC Group. HIV Neurobehavioral Research Center. Psychosom. Med. 1995;57(2):138–47.
106. Bradlyn AS, Harris CV, Warner JE, et al. An investigation of the validity of the quality of Well-Being Scale with pediatric oncology patients. Health Psychol. 1993;12(3):246–50.
107. O'Connor R. Issues in the Measurement of Health-Related Quality of Life. Melbourne: National Centre for Health Program Evaluation; 1993. Working Paper 30.
108. Visser M, Fletcher A, Parr G, et al. A comparison of three quality of life instruments in subjects with angina pectoris: the Sickness Impact Profile, the Nottingham Health Profile, and the Quality of Well Being Scale. J. Clin. Epidemiol. 1994;47(2):157–163.
109. Brazier J, Jones N, Kind P. Testing the validity of the EuroQoL and comparing it with the SF-36 health survey questionnaire. Qual. Life Res. 1993;2:169–180.
110. Andersen L, Kristiansen I, Falch J, et al. Cost-effectiveness of Alendronate for the prevention of osteoporotic fractures in Norwegian women. Working Paper. Oslo: Folkehelsa; Statens Institutt for Folkehelsa; 1995. Report No.: 11/1995.
111. Wlodarczyk JH, Brodaty H, Hawthorne G. The relationship between quality of life, Mini-Mental State Examination, and the Instrumental Activities of Daily Living in patients with Alzheimer's Disease. Arch. Gerontol. Geriatr. 2004;39(1):25–33.
112. Hogan A, Hawthorne G, Kethel L, et al. Health-related quality-of-life outcomes from adult cochlear implantation: a cross-sectional study. Cochlear Imp. Int. 2001;2(2):115–128.

113. Hawthorne G, Cheok F, Goldney R, et al. The excess cost of depression in South Australia: a comparative study of two methods of calculating burden. Aust. N. Z. J. Psychiatry 2003;37(3):362–373.
114. Goldney RD, Fisher LJ, Wilson DH, et al. Suicidal ideation and health-related quality of life in the community. Med. J. Aust. 2001;175(10):546–549.
115. Sturm JW, Osborne RH, Dewey HM, et al. Brief comprehensive quality of life assessment after stroke: the Assessment of Quality of Life instrument in the North East Melbourne Stroke Incidence study (NEMESIS). Stroke 2002;33(12):2888–94.
116. Stavem K. Reliability, validity and responsiveness of two multiattribute utility measures in patients with chronic obstructive pulmonary disease. Qual. Life Res. 1999;8(1–2):45–54.
117. Dorman P, Slattery J, Farrell B, et al. Qualitative comparison of the reliability of health status assessments with the EuroQol and SF-36 questionnaires after stroke. United Kingdom Collaborators in the International Stroke Trial. Stroke 1998;29(1):63–68.
118. van Agt HM, Essink-Bot ML, Krabbe PF, et al. Test-retest reliability of health state valuations collected with the EuroQol questionnaire. Soc. Sci. Med. 1994;39(11):1537–1544.
119. Boyle MH, Furlong W, Feeny D, et al. Reliability of the Health Utilities Index – Mark III used in the 1991 cycle 6 Canadian General Social Survey Health Questionnaire. Qual. Life Res. 1995;4:249–257.
120. Hawthorne G. The effect of different methods of collecting data: mail, telephone and filter data collection issues in utility measurement. Qual. Life Res. 2003;12:1081–1088.
121. Hawthorne G, Osborne R, McNeil H, et al. The Australian Multi-attribute Utility (AMAU): Construction and Initial Validation. Melbourne: Centre for Health Program Evaluation; 1996. Working Paper 56.
122. Hawthorne G, Richardson J, Osborne R, et al. The Assessment of Quality of Life (AQoL) Instrument: Construction, Initial Validation and Utility Scaling. Melbourne: Centre for Health Program Evaluation; 1997. Working Paper 76.
123. Araya R, Flynn T, Rojas G, et al. Cost-effectiveness of a primary care treatment program for depression in low-income women in Santiago, Chile. Am. J. Psychiatry 2006;163(8):1379–87.
124. Adams M, Kutcher S, Antoniw E, et al. Diagnostic utility of endocrine and neuroimaging screening tests in first-onset adolescent psychosis. J. Am. Acad. Child Adolesc. Psychiatry 1996;35(1):67–73.
125. Pinkerton SD, Johnson-Masotti AP, Otto-Salaj LL, et al. Cost-effectiveness of an HIV prevention intervention for mentally ill adults. Ment Health Serv Res 2001;3(1):45–55.
126. Manning W. Panel on cost-effectiveness in health and medicine recommendations: identifying costs. J. Clin. Psychiatry 1999;60(Suppl 3):54–58.
127. Noyes K, Dick AW, Holloway RG. Pramipexole v. levodopa as initial treatment for Parkinson's disease: a randomized clinical-economic trial. Med. Decis. Making 2004;24(5):472–85.
128. Fernandez JL, Montgomery S, Francois C. Evaluation of the cost effectiveness of escitalopram versus venlafaxine XR in major depressive disorder. Pharmacoeconomics 2005;23(2):155–67.
129. Jones PB, Barnes TR, Davies L, et al. Randomized controlled trial of the effect on Quality of Life of second- vs first-generation antipsychotic drugs in schizophrenia: Cost Utility of the Latest Antipsychotic Drugs in Schizophrenia Study (CUtLASS 1). Arch. Gen. Psychiatry 2006;63(10):1079–87.
130. Holland R, Smith R, Harvey I, et al. Assessing quality of life in the elderly: a direct comparison of the EQ-5D and AQoL. Health Econ. 2004;13(8):793–805.
131. Hawthorne G. Measuring incontinence in Australia. Melbourne: Department of Psychiatry, The University of Melbourne for the Department of Health and Ageing, Australian Government, Canberra. May 2006.
132. Spitzer RL, Williams JB, Kroenke K, et al. Utility of a new procedure for diagnosing mental disorders in primary care. The PRIME-MD 1000 study. JAMA 1994;272(22):1749–56.
133. Spitzer RL, Kroenke K, Linzer M, et al. Health-related quality of life in primary care patients with mental disorders. Results from the PRIME-MD 1000 Study. JAMA 1995;274(19): 1511–7.

134. Fayers P, Machin D. Quality of Life: Assessment, Analysis and Interpretation. Chichester: Wiley; 2000.
135. Wood J, Taylor S, Lichtman R. Social comparison in adjustment to breast cancer. J. Pers. Soc. Psychol. 1985;49:1169–1183.
136. Diener E, Suh E, Lucas R, et al. Subjective well-being: three decades of progress. Psychol. Bull. 1999;125:276–302.
137. Kring A, Kerr S, Smith D, et al. Flat affect in schizophrenia does not reflect diminished subjective experience of emotion. J. Abnorm. Psychol. 1993;102:507–517.
138. Jenkins C. Assessment of outcomes of health intervention. Soc. Sci. Med. 1992;35:367–375.
139. Awad G, Hogan T, Voruganti L, et al. Patients' subjective experiences on antipsychotic medications: implications for outcome and quality of life. Int. Clin. Psychopharmacol. 1995;10S:123–132.
140. Coucill W, Bryan S, Bentham P, et al. EQ-5D in patients with dementia: an investigation of interrater agreement. Med. Care 2001;39(8):760–71.
141. Becchi A, Rucci P, Placentino A, et al. Quality of life in patients with schizophrenia—comparison of self-report and proxy assessments. Soc. Psychiatry Psychiatr. Epidemiol. 2004;39(5):397–401.
142. Ankri J, Beaufils B, Novella JL, et al. Use of the EQ-5D among patients suffering from dementia. J. Clin. Epidemiol. 2003;56(11):1055–63.
143. Bullinger M, Azouvi P, Brooks N, et al. Quality of life in patients with traumatic brain injury-basic issues, assessment and recommendations. Restor Neurol Neurosci 2002;20(3–4):111–24.
144. Tamim H, McCusker J, Dendukuri N. Proxy reporting of quality of life using the EQ-5D. Med. Care 2002;40(12):1186–95.
145. Naglie G, Tomlinson G, Tansey C, et al. Utility-based Quality of Life measures in Alzheimer's disease. Qual. Life Res. 2006;15(4):631–43.
146. Scocco P, Fantoni G, Caon F. Role of depressive and cognitive status in self-reported evaluation of quality of life in older people: comparing proxy and physician perspectives. Age Ageing 2005;35(2):166–71.
147. Sainfort F, Becker M, Diamond R. Judgements of quality of life of individuals with severe mental disorders: patient self-report versus provider perspectives. Am. J. Psychiatry 1996;153:497–502.
148. Hawthorne G, Osborne R. Population norms and meaningful differences for the Assessment of Quality of Life (AQoL) measure. Aust. N. Z. J. Public Health 2005;29(2):136–142.
149. Patterson TL, Shaw W, Semple SJ, et al. Health-related quality of life in older patients with schizophrenia and other psychoses: relationships among psychosocial and psychiatric factors. Int. J. Geriatr. Psychiatry 1997;12(4):452–61.
150. Depp CA, Davis CE, Mittal D, et al. Health-related quality of life and functioning of middle-aged and elderly adults with bipolar disorder. J. Clin. Psychiatry 2006;67(2):215–21.
151. Lonnqvist J, Sintonen H, Syvalahti E, et al. Antidepressant efficacy and quality of life in depression: a double-blind study with moclobemide and fluoxetine. Acta Psychiatr. Scand. 1994;89(6):363–369.
152. Sapin C, Fantino B, Nowicki ML, et al. Usefulness of EQ-5D in assessing health status in primary care patients with major depressive disorder. Health Qual Life Outcomes 2004;2:20.
153. Pyne JM, Patterson TL, Kaplan RM, et al. Assessment of the quality of life of patients with major depression. Psychiatr. Serv. 1997;48(2):224–30.
154. Kasckow JW, Twamley E, Mulchahey JJ, et al. Health-related quality of well-being in chronically hospitalized patients with schizophrenia: comparison with matched outpatients. Psychiatry Res. 2001;103(1):69–78.
155. Saarni SI, Harkanen T, Sintonen H, et al. The Impact of 29 Chronic Conditions on Healthrelated Quality of Life: A General Population Survey in Finland Using 15D and EQ-5D. Qual. Life Res. 2006; 15(9):1403–1414. 2006.
156. Goldney RD, Fisher LJ. Double depression in an Australian population. Soc. Psychiatry Psychiatr. Epidemiol. 2004;39(11):921–6.
157. Manuel DG, Schultz SE, Kopec JA. Measuring the health burden of chronic disease and injury using health adjusted life expectancy and the Health Utilities Index. J. Epidemiol. Community Health 2002;56(11):843–50.

158. Goldney R, Fisher LJ, Dal Grande E, et al. Have education and publicity about depression made a difference? A comparison of prevalence, service use and excess costs in South Australia: 1998 and 2004. Aust. N. Z. J. Psychiatry 2007:41(1);38–53.
159. Rasanen P, Roine E, Sintonen H, et al. Use of quality-adjusted life years for the estimation of effectiveness of health care: A systematic literature review. Int. J. Technol. Assess. Health Care 2006;22(2):235–41.
160. PBAC. Background Document on the use of economic analysis as a basis for inclusion of pharmaceutical products on the Pharmaceutical Benefits Scheme. Canberra: Australian Government Publishing Service; 1993.
161. PBAC. Guidelines for the Pharmaceutical Industry on Preparation of Submissions to the Pharmaceutical Benefits Advisory Committee. 2nd draft. Canberra: Pharmaceutical Benefits Advisory Committee, Commonwealth of Australia; 1995.
162. Thomas S, Nay R, Moore K, et al. Continence Outcomes Measurement Suite Project (Final Report). Canberra: Australian Government, Department of Health and Ageing; 2006 April.

CHAPTER 7

COMPARISON OF INSTRUMENTS FOR MEASURING THE QUALITY OF LIFE IMPAIRMENT SYNDROME IN SEVERE MENTAL DISORDERS

Q-LES-Q versus QLS and LQOLP

MICHAEL S. RITSNER

Department of Psychiatry, The Rappaport Faculty of Medicine, Technion - Israel Institute of Technology and Acute Department, Sha'ar Menashe Mental Health Center, Israel

Abstract: In the present chapter, we compared the psychometric properties of the Quality of Life Enjoyment and Satisfaction Questionnaire (Q-LES-Q)[1] with the Quality of Life Scale (QLS)[2] and the Lancashire Quality of Life Profile (LQOLP)[3] in the same patients with schizophrenia, schizoaffective, and mood disorders. These instruments were chosen since they are mental illness-related, and Q-LES-Q is a self-report evaluation scale, whereas QLS is observer-rated, and LQOLP has domains similar to Q-LES-Q, which enables a comparison of the instruments. The compared instruments proved to be mental health-related, but none were mental-disorder-specific. Despite the acceptable psychometric properties and the correlation of the general indices, similar domains proved to be instrument-specific and were not sufficiently compatible. These discrepancies should be considered when comparing evaluations from similar domains in these scales

Keywords: Psychiatric patients, Quality of life scales, Reliability, Validity, Comparison

INTRODUCTION

Health related quality of life (HRQL) impairment has become an important syndrome of severe mental disorders. It has been intensively investigated in clinical trials, in compliance with psychopharmacological treatment, and regarding adaptation to social environments. Today there is no universal instrument that can be recommended for all studies. Specific features and psychometric properties of self-reported, observer-rated, and combined (observer and self-reported) instruments have been reviewed[3–7]. Differences between these instruments in terms of the underlying HRQL concepts and data collection procedures are substantial. Using self-reported or observer-rated instruments makes comparisons rather difficult[8,9]. In addition they are impacted upon by different determinants[10,11].

M.S. Ritsner and A.G. Awad (eds.), Quality of Life Impairment in Schizophrenia, Mood and Anxiety Disorders, 133–142.

Several studies have compared several measurements in the same patient populations. Lehman et al.[12], in comparing the convergent validity of the Quality of Life Interview and the Quality of Life Scale[2] in patients with severe mental illness, found low but significant convergence between both. Oliver et al.[3,13] compared the psychometric properties of the Lancashire Quality of Life Profile (LQOLP) and the Life Experiences Checklist (LEC) between psychiatric patients and healthy subjects, and found that the scales reflect low internal consistency in both groups. The observed discriminatory power of both measures was attributed to differences in the scales' underlying concepts, with LEC focusing on objective statements of current life experiences, and LQOLP focusing on satisfaction with life domains. Since research concerning direct comparisons of different HRQL measures within the same patient populations is rare, further comparison studies and statistical testing with widely used HRQL measures are necessary to facilitate the choice of appropriate research instruments and to enhance the comparability of HRQL data.

In the present chapter, we compare the Quality of Life Enjoyment and Satisfaction Questionnaire[1] with the observer-rated QLS[2] and the LQOLP[3] in the same patients with schizophrenia, schizoaffective, and mood disorders. We chose these instruments since they are mental illness-related, and Q-LES-Q is a self-report evaluation scale, whereas QLS is observer-rated, and LQOLP has domains similar to Q-LES-Q, which enables a comparison of the instruments.

THE QUALITY OF LIFE ENJOYMENT AND SATISFACTION QUESTIONNAIRE

Q-LES-Q is a self-report measure consisting of 93 items grouped into ten summary scales. Responses are scored on a 5-point scale ("not at all or never"' to "frequently or all the time"), with higher scores indicating more enjoyment and satisfaction with specific life domains. In the present study a global Q-LES-Q_{index} and the following domains were evaluated: physical health, subjective feelings, leisure activities, social relationships, general activities, and satisfaction with medication and life. Reported internal consistency for the Q-LES-Q domains (Cronbach's α) ranged from 0.84 to 0.96, and the test-retest reliability ranged from 0.47 to 0.89[6,14–16].

Since 1993 Q-LES-Q has been widely used in HRQL studies for assessing patients with bipolar disorder[17], depression[18,19], seasonal affective disorder[20], subthreshold affective symptoms[21], premenstrual dysphoric disorder[22], major psychoses[14,15,23], and posttraumatic stress disorder[24]. Recently a parsimonious subset of items from Q-LES-Q that can accurately predict the basic Q-LES-Q domain mean scores was sought and evaluated in 339 inpatients meeting DSM-IV criteria for schizophrenia, schizoaffective, and mood disorders[25]. It was found that 18-items predicted basic Q-LES-Q domains (physical health, subjective feelings, leisure time activities, social relationships) and general index scores with high accuracy. Q-LES-Q-18 indicated that the test-retest ratings had high reliability, validity, and stability. Thus, Q-LES-Q-18, a brief, self-administered questionnaire, may aid in monitoring the quality of life outcomes of schizophrenia, schizoaffective, and mood disorder patients.

THE QUALITY OF LIFE SCALE

QLS[2], a 21-item observer-rated scale based on a semistructured interview designed to assess deficit symptoms, is rated by a clinician on a 7-point scale (0-1 severe impairment to 5-6 normal or unimpaired functioning). QLS assesses four domains: (a) intrapsychic foundations, (b) interpersonal relations, (c) instrumental role functioning, and (d) common objects and activities that are often incorporated into the construct of intrapsychic foundations. However, the scale items comprising QLS were derived from analysis reflecting manifestations of the deficit syndrome in schizophrenia. Therefore the scale has been criticized for reflecting the presence of negative and deficit symptoms rather than presenting a clear HRQL evaluation[26]. Indeed, in a discussion of the conceptual and methodological requirements needed for measuring HRQL in psychiatry, it was concluded that there should be a clear distinction between psychopathological symptoms and HRQL domains[26–28].

Recently, a predictive model approach was developed that reduces the number of items collected for scales that yield a total summary score. More specifically, it pioneered the use of a condensed seven-item scale (QLS_7) that includes acquaintances, social initiatives, extent, motivation, anhedonia, objects, and empathy[29]. QLS_7 predicted the QLS_{21} total summary score with high accuracy (r=0.98), but with some limitations. The patient sample used to develop QLS_7 included 31.3% drug naïve and 32.3% first-episode patients, a sample deemed to be not representative of the broad heterogeneity of schizophrenia populations. In addition, the relationship of QLS_7 to the severity of symptoms, side effects, emotional distress, and perceived quality of life was not addressed.

In attempting to address these limitations, we tested and validated a condensed QLS_5, based on QLS_{21}, which is briefer and thus easier to administer than the complete rating scale[30]. The analyses suggest that QLS_5 has been shown to be a valid predictor of the QLS_{21} total scores. Psychometric properties (inter-rater, the test-retest reliability, and sensitivity to change) for QLS_5 were also high and comparable to QLS_{21}. In addition, QLS_5 does not reflect the presence of psychiatric symptoms as do the QLS_{21} and QLS_7-abbreviated versions[29]. The most reliable items in QLS_5 are social initiatives, adequacy, acquaintances, time utilization, and motivation. Although it may be argued that lack of motivation is symptomatic of schizophrenia, and therefore indicative of the deficit syndrome associated with schizophrenia, it should be noted that the one item rated in all versions of the scale is the degree of motivation, and not the lack thereof. Whereas lack of motivation may be disease-related, motivation, or enthusiasm per se is indeed a component of the HRQL construct. Thus, the five-item condensed Quality of Life Scale for schizophrenia maintains the validity of the full QLS, and has the advantage of a shorter administration time. Utilization of the revised QLS_5 in routine care and clinical trials may potentially facilitate evaluation of treatment outcomes in schizophrenia.

QLS is widely used in clinical trials, course of illness, and rehabilitation studies[31,32]. This scale has acceptable psychometric qualities: test-retest reliability is good for nearly all items of the scale, categories, and overall score. Internal consistency, alpha-coefficients were 0.8-0.9 for the global score and convergent validity is good[33]. Cramer et al.[34] reported that QLS appeared to be substantially more sensitive to subtle changes and treatment effects clinical trials compared to Lehman self reported Quality of Life Interview.

COMPARISON OF Q-LES-Q WITH QLS

Q-LES-Q and the complete QLS were simultaneously administered to 133 schizophrenia outpatients (time from discharge, mean=8.5 months, SD=6.4). These patients were assessed during a naturalistic comparative investigation of second-generation (SGAs; risperidone or olanzapine) versus first-generation antipsychotic agents (FGAs; see details)[35,36]. Pearson's correlation coefficients between domains of both instruments ranged from 0.35 to 0.64 (all $p<0.001$). Correlations were performed for measures of Q-LES-Q and QLS with distressing and protective factors (Table 1). Interestingly, our results indicate that QLS is more strongly associated with illness and symptom severity, side effects, and general functioning than Q-LES-Q. Correlations with emotional distress, somatization, and two protective factors (self-efficacy and social support) were found to be similar.

In another study, longitudinal data about 124 schizophrenia patients treated with antipsychotics for 12 month were analyzed[37]. Table 2 presents the treatment-group effect on changes in the quality of life from baseline to end-of study. Improvement was indicated in the general Q-LES-Q_{index} ($p=0.029$), subjective feelings ($p=0.050$),

Table 1. Correlation of the two instruments with related factors among 133 schizophrenia outpatients

HRQL measures	Symptom severity (PANSS)	Illness severity (CGI)	General functioning (GAF)	Side effects (ESRS)	Emotional distress (TBDI)	Somatization (BSI)	Self-efficacy (GSES)	Social support (MSPSS)
General Q-LES-Q index	−0.20	−0.05	0.14	−0.15	−0.47***	−0.19*	0.63***	0.50***
Physical health	−0.12	0.00	0.10	−0.16	−0.39***	−0.29***	0.57***	0.31***
Subjective feelings	−0.15	0.05	0.10	−0.11	−0.43***	−0.15	0.62***	0.40***
Leisure time activities	−0.10	−0.07	0.04	−0.08	−0.26**	−0.02	0.54***	0.36***
Social relationships	−0.15	0.01	0.05	−0.09	−0.26**	0.02	0.49***	0.57***
General activities	−0.30***	−0.20*	0.29***	−0.24**	−0.57***	−0.34***	0.44***	0.51***
Satisfaction with medication	−0.37***	−0.30***	0.38***	−0.21*	−0.45***	−0.18*	0.26**	0.37***
Life satisfaction	−0.29***	−0.22*	0.26**	−0.15	−0.62***	−0.31***	0.33***	0.45***
QLS, total score	−0.47***	−0.31***	0.45***	−0.21*	−0.39***	−0.19*	0.39***	0.37***
Interpersonal relations	−0.38***	−0.26**	0.39***	−0.19*	−0.32***	−0.16	0.32***	0.40***
Instrumental role	−0.38***	−0.28***	0.41***	−0.12	−0.38***	−0.15	0.31***	0.31***
Intrapsychic foundations	−0.51***	−0.33***	0.44***	−0.22*	−0.39***	−0.21*	0.41***	0.31***
Common objects and activities	−0.42***	−0.29***	0.43***	−0.24**	−0.34***	−0.18*	0.42***	0.23**

Table 2. Changes (%) in quality of life scores during 12 months among schizophrenia patients treated with antipsychotic agents

Quality of life measures	Change from baseline (n=124, %)				The treatment-group effect[a]	
	Mean	SD	95%	CI	F	P
Q-LES-Q general index	6.4	21.9	2.5	10.3	**3.1**	**0.029**
Physical health	11.1	31.2	5.6	16.6	2.1	0.15
Subjective feelings	10.7	32.3	5.0	16.5	**2.7**	**0.050**
Leisure time activities	9.8	40.2	2.7	17.0	**3.2**	**0.026**
Social relationships	9.0	33.1	3.1	14.9	0.6	0.64
General activity	3.4	34.5	−2.7	9.6	0.3	0.84
Life satisfaction	16.7	58.3	6.4	27.1	0.2	0.88
QLS, total score	8.8	30.3	3.4	14.2	0.4	0.75
Interpersonal relations	6.8	34.6	0.7	12.9	0.2	0.88
Instrumental role	11.1	35.4	4.8	17.4	0.3	0.85
Intrapsychic foundations	13.0	35.4	6.8	19.3	0.4	0.72
Common objects and activities	11.8	44.9	3.9	19.8	0.6	0.63

[a] ANCOVA tests of 4 treatment-group effect on change in HRQL controlled for sex and baseline values (df=3, 124)
Treatment-groups: RP – risperidone, OL – olanzapine, FGAs – first-generation antipsychotic agents, and CA – combined agents.

and leisure time activities (p=0.026), but no changes were found in the QLS domains (all p>0.05).

Thus, it seems that QLS strongly reflects the severity of the psychopathology. It shows lower scores in deficit than in non-deficit patients due to its construct and also because deficit symptomatology is partially redundant. Observer-rated QLS did not detect any changes in HRQL levels that were associated with antipsychotic therapy during the follow-up period, a finding supported by others[38]. Consistent with Gourevitch et al.[39], we suggest that QLS might be an inappropriate or at least unspecific measure of HRQL in schizophrenia patients.

THE LANCASHIRE QUALITY OF LIFE PROFILE

LQOLP[3,13] combines observer and self-report measures. More specifically, LQOLP is a structured self-report interview conducted by a trained non-clinical interviewer, and elicits patients' ratings of their HRQL. LQOLP is a structured interview based on the Lehman's Quality of Life Interview[40]. Responses are scored on a 7-point Life Satisfaction Scale ("How satisfied are you with…") ranging from "can't be worse" to "can't be better". The reported internal consistency for the social relations, leisure and participation, and work/education domains was high (0.80), but fell to 0.50 in some samples. Test-retest reliability ranged from 0.34 to 0.89[3,13,41].

COMPARISON OF Q-LES-Q WITH LQOLP

We sought to compare the psychometric properties concerned with the validity and reliability of both the Q-LES-Q and self-report items of LQOLP within a population of patients with severe mental disorders, and between psychiatric patients and non-patient control subjects. Therefore, we simultaneously administered Q-LES-Q and LQOLP to 148 schizophrenia patients and 175 non-patients.

The compositions of Q-LES-Q and LQOLP differ in format. There are 15 domains covered by each instrument, five of which are concordant, five instrument-specific domains of LQOLP (religion, finances, living situation, family relations, and legal and safety issues), and four were specific domains in Q-LES-Q (subjective feelings, household duties, general activities, and satisfaction with medication). Cronbach's α coefficients for the concordant life domains were higher for Q-LES-Q than for LQOLP. Internal consistency for the Work and Legal and Safety domains was low ($r<0.60$; LQOLP). Despite the domain discrepancies, the general quality of life index of each instrument indicated high internal consistency[41].

Pearson's correlation coefficients for the concordant domains of both instruments in 148 schizophrenia patients were moderate for the general index ($r=0.61$, $p<0.001$), physical health ($r=0.58$, $p<0.001$), work ($r=0.41$, $p<0.001$), leisure activities ($r=0.46$, $p<0.001$), social relationships ($r=0.45$, $p<0.001$), and satisfaction with life ($r=0.50$, $p<0.001$).

Both instruments yielded significantly higher HRQL mean scores among the control subjects than among patients (all $p<0.001$). Correlations were performed for general indices of the LQOLP and Q-LES-Q, and related factors. As Table 3 shows, with both instruments there was a negative correlation with the severity of emotional distress, depressive symptoms, most of the PANSS factors, and side effects. On the other hand a positive correlation with self-constructs and social support was observed.

Thus, the identified five concordant domains, five instrument-specific domains for LQOLP and four instrument-specific domains for Q-LES-Q. Q-LES-Q provides better psychometric properties than LQOLP in both samples. Both instruments show a good capacity to evaluate HRQL and discriminate between the patient and non-patient controls. Within the patient sample, both measures exhibited similar negative correlations with the main distressing and protective factors. Despite the acceptable psychometric properties and the correlation of general indices, similar HRQL domains proved to be instrument-specific and were not sufficiently compatible. These discrepancies should be considered when comparing evaluations of similar domains assessed by both scales.

CONCLUDING REMARKS

We have attempted to compare three different instruments for measuring HRQL impairment syndrome in the same patients with severe mental disorders. Whether mentally ill persons can judge or report their own HRQL continues to be controversial. Those who believe that the mentally ill are unable to reliably observe the

Table 3. Correlations of general quality of life impairment scores with related factors

Related factors	Quality of life enjoyment and satisfaction questionnaire	Lancashire quality of life profile
PANSS, total	−0.25**	−0.22**
Negative factor	−0.16*	−0.13
Positive factor	−0.17*	−0.20*
Activation factor	−0.11	−0.14
Dysphoric mood	−0.43***	−0.28***
Autistic preoccupations	−0.22**	−0.22**
Depression severity (MADRS)	−0.49***	−0.38***
Side effects (DSAS)	−0.49***	−0.26**
Emotional distress (TBDI)	−0.79***	−0.53***
Self-efficacy (GSES)	0.68***	0.53***
Self-esteem (RSES)	0.73***	0.46***
Social support (MSPSS)	0.54***	0.57***

Two-tailed t-test: *p < 0.05; **p < 0.01; ***p < 0.001
PANSS - Positive and Negative Syndrome Scale; MADRS - Montgomery and Äsberg Depression Rating Scale; DSAS - Distress Scale for Adverse Symptoms; TBDI - Talbieh Brief Distress Inventory; BSI - Brief Symptom Inventory; CSES - General Self-Efficacy Scale; RSES - Rosenberg Self-Esteem Scale; MSPSS - Multidimensional Scale of Perceived Social Support scale.

world they live ignore subjective measures and rely on proxy objective evaluations. The issue of self-report instruments within the psychiatric population has been extensively discussed in the literature[27,28,42,43]. Voruganti et al.[44] examined schizophrenia patients' ability to appraise their quality of life by examining the reliability and validity of self-rated quality of life estimates and found that the patients' self-reports were highly consistent and that their HRQL ratings correlated significantly with clinician's estimates.

Probably the most fundamental feature of the HRQL construct is its subjective nature. On this basis, patients have been given a voice in evaluating interventions. Giving the patients a voice implies a shift of reference from the health care professional to the patient. Integration of patients' experiences remains controversial as a result of the difficulties in expressing emotions and perceptions in patients with deficit syndrome or severe thought disorders[45,46]. The patients' appraisal of their HRQL, in whatever domains, is considered valid by definition, given a valid measure[47]. This view is strongly supported by the findings of the present study.

Comparing Q-LES-Q with QLS and LQOLP proved to be mental health-related, but none were mental-disorder-specific. Despite the acceptable psychometric properties and the correlation of general indices, similar domains proved to be instrument-specific and not sufficiently compatible. These discrepancies should be considered when comparing evaluations of similar domains assessed by these scales.

REFERENCES

1. Endicott J, Nee J, Harrison W, Blumenthal R. Quality of Life Enjoyment and Satisfaction Questionnaire: a new measure. Psychopharmacol Bull 1993; 29: 321–326.
2. Heinrichs DW, Hanlon TE, Carpenter WT Jr. The Quality of Life Scale: an instrument for rating the schizophrenic deficit syndrome. Schizophr Bull 1984; 10:388–398.
3. Oliver J, Huxley P, Bridges K, Mohammed H. Quality of Life and Mental Health Service. London & New York: Routledge, 1996.
4. Van Nieuwenhuizen C, Schene AH, Boevink WA, Wolf JRLM. Measuring the quality of life of clients with severe mental illness: a review of instruments. Psychiatr Rehabil J 1997; 20:33–41.
5. Awad AG, Voruganti LN, Heslegrave RJ. Measuring quality of life in patients with schizophrenia. Pharmacoeconomics. 1997; 11:32–47.
6. Bishop SL, Walling DP, Dott SG, Folkes CC, Bucy J. Refining quality of life: validating a multidimensional factor measure in the severe mentally ill. Qual Life Res 1999; 8:151–160.
7. Hawthorne G, Richardson J. Measuring the value of program outcomes: a review of multiattribute utility measures. Expert Rev. Pharmacoeconomics Outcomes Res. 2001; 1: 215–228.
8. Trauer T, Duckmanton RA, Chiu E. A study of the quality of life of the severely mentally ill. Int J Soc Psychiatry 1998; 44:79–91.
9. Lobana A, Mattoo SK, Basu D, et al. Quality of life in schizophrenia in India: comparison of three approaches. Acta Psychiatr Scand 2001; 104:51–55.
10. Fitzgerald PB, Williams CL, Corteling N, et al. Subject and observer-rated quality of life in schizophrenia. Acta Psychiatr Scand 2001; 103:387–392.
11. Ruggeri M, Bisoffi G, Fontecedro L, et al. Subjective and objective dimensions of quality of life in psychiatric patients: a factor analytical approach: The South Verona Outcome Project 4. Br J Psychiatry 2001; 178:268–275.
12. Lehman AF, Postrado LT, Rachuba LT. Convergent validation of quality of life assessments for persons with severe mental illnesses. Qual. Life Res. 1993: 2: 327–333.
13. Oliver JP, Huxley PJ, Priebe S, Kaiser W. Measuring the quality of life of severely mentally ill people using the Lancashire Quality of Life Profile. Soc Psychiatry Psychiatr Epidemiol 1997; 32:76–83.
14. Ritsner M, Modai I, Endicott J, et al. Differences in quality of life domains and psychopathologic and psychosocial factors in psychiatric patients. J Clin Psychiatry 2000; 61:880–889.
15. Ritsner M, Ponizovsky A, Endicott J, et al. The impact of side-effects of antipsychotic agents on life satisfaction of schizophrenia patients: a naturalistic study. Eur Neuropsychopharmacol 2002; 12:31–38.
16. Rossi A, Rucci P, Mauri M, Maina G, Pieraccini F, Pallanti S, Endicott J; (For the EQUIP group). Validity and reliability of the Italian version of the Quality of Life, Enjoyment and Satisfaction Questionnaire. Qual Life Res. 2005; 14(10):2323–8.
17. Ozer S, Ulusahin A, Batur S, et al. Outcome measures of interepisode bipolar patients in a Turkish sample.Soc Psychiatry Psychiatr Epidemiol. 2002; 37:31–37.
18. Seidman SN, Rabkin JG. Testosterone replacement therapy for hypogonadal men with SSRI-refractory depression. J Affect Disord. 1998; 48: 157–161.
19. Russell JM, Koran LM, Rush J, et al. Effect of concurrent anxiety on response to sertraline and imipramine in patients with chronic depression. Depress Anxiety. 2001;13:18–27.
20. Michalak EE, Tam EM, Manjunath CV, et al. Quality of life in patients with seasonal affective disorder: summer vs winter scores. Can J Psychiatry. 2005; 50:292–295.
21. Goracci A, Martinucci M, Scalcione U, et al. Quality of life and subthreshold affective symptoms. Qual Life Res. 2005; 14:905–909.
22. Pearlstein TB, Halbreich U, Batzar ED, et al. Psychosocial functioning in women with premenstrual dysphoric disorder before and after treatment with sertraline or placebo. J Clin Psychiatry 2000; 61:101–109.
23. Ritsner M, Kurs R, Gibel A, et al. Predictors of quality of life in major psychoses: a naturalistic follow-up study. J Clin Psychiatry 2003; 64:308–315.

24. Rapaport MH, Endicott J, Clary CM. Posttraumatic stress disorder and quality of life: results across 64 weeks of sertraline treatment. J Clin Psychiatry. 2002; 63: 59–65.
25. Ritsner M, Kurs R, Gibel A, et al. Validity of an abbreviated quality of life enjoyment and satisfaction questionnaire (Q-LES-Q-18) for schizophrenia, schizoaffective, and mood disorder patients. Qual Life Res. 2005; 14:1693–1703.
26. Katschnig, H. How useful is the concept of quality of life in psychiatry? Curr Opin Psychiatry 1997; 10: 337–345.
27. Ritsner M, Kurs R. Impact of antipsychotic agents and their side effects on the quality of life in schizophrenia. Expert Review and Pharmacoeconomic Outcomes Research 2002; 2: 89–98.
28. Ritsner M, Kurs R. Quality of life outcomes in mental illness: schizophrenia, mood and anxiety disorders. Expert Review and Pharmacoeconomic Outcomes Research 2003; 3:189–199.
29. Bilker, W.B., Brensinger, C., Kurtz, M.M., et al. Development of an abbreviated schizophrenia quality of life scale using a new method. Neuropsychopharmacology 2003; 28: 773–777.
30. Ritsner M, Kurs R, Ratner Y, Gibel A. Condensed version of the Quality of Life Scale for schizophrenia for use in outcome studies. Psychiatry Res. 2005; 135:65–75.
31. Hamilton SH, Revicki DA, Edgell ET, et al. Clinical and economic outcomes of olanzapine compared with haloperidol for schizophrenia. Results from a randomised clinical trial. Pharmacoeconomics. 1999;15: 469–480.
32. Kaneda Y, Ohmori T. Effects of quetiapine on gonadal axis hormones in male patients with schizophrenia: a preliminary, open study. Prog Neuropsychopharmacol Biol Psychiatry. 2003; 27:875–878.
33. Kaneda Y, Imakura A, Fujii A, Ohmori T. Schizophrenia Quality of Life Scale: validation of the Japanese version. Psychiatry Res. 2002;113:107–113.
34. Cramer JA, Rosenheck R, Xu W, et al. Quality of Life in Schizophrenia: A Comparison of Instruments. Schizophr Bull 2000; 26: 659–666.
35. Ritsner, M., Perelroyzen, G., Kurs, R., et al. Quality of life outcomes in schizophrenia patients treated with atypical and typical antipsychotic agents: A naturalistic comparative study. International Clinical Psychopharmacology 2004; 24:582–591.
36. Ritsner M, Perelroyzen G, Ilan H, Gibel A. Subjective response to antipsychotics of schizophrenia patients treated in routine clinical practice: a naturalistic comparative study. J Clin Psychopharmacol. 2004; 24:245–54.
37. Ritsner MS, Gibel, A. The effectiveness and predictors of response to antipsychotic agents to treat impaired quality of life in schizophrenia: a 12-month naturalistic follow-up study with implications for confounding factors, antidepressants, anxiolytics, and mood stabilizers. Progress in Neuro-Psychopharmacology & Biological Psychiatry 2006; 30:1442–1452.
38. Voruganti L, Cortese L, Oyewumi L, et al. Comparative evaluation of conventional and novel antipsychotic drugs with reference to their subjective tolerability, side-effect profile and impact on quality of life. Schizophr. Res 2000; 43:135–145.
39. Gourevitch R, Abbadi S, Guelfi JD. Quality of life in schizophrenics with and without the deficit syndrome. Eur Psychiatry. 2004; 19:172–174.
40. Lehman AF. (1995) Measuring quality of life in a reformed health system. Health Affairs 1995; 14: 90–101.
41. Ritsner M, Modai I, Kurs R, et al. Subjective Quality of Life Measurements in Severe Mental Health Patients: Measuring Quality of Life of Psychiatric Patients: Comparison Two Questionnaires. Quality Life Research 2002; 11: 553–561.
42. Awad AG, Voruganti LN. Impact of atypical antipsychotics on quality of life in patients with schizophrenia. CNS Drugs. 2004; 18:877–893.
43. Lambert M, Naber D. Current issues in schizophrenia: overview of patient acceptability, functioning capacity and quality of life. CNS Drugs. 2004 (Suppl 2): 5–17; discussion 41–43.
44. Voruganti L, Heslegrave R, Awad AG, Seeman MV. Quality of life measurement in schizophrenia: reconciling the quest for subjectivity with the question of reliability. Psychol Med 1998; 28:165–172.

45. McKenna, S.P. Measuring quality of life in schizophrenia. Eur. Psychiatry 1997 (Suppl. 3), 267s–274s.
46. Wilkinson, G., Hesdon, B., Wilde, D., et al. Self-report quality of life measure for people with schizophrenia: the SQLS. Br. J. Psychiatry 2000; 177: 42–46.
47. Bernhard J, Lowy A, Mathys N, et al. Health related quality of life: a changing construct? Qual Life Res. 2004; 13:1187–1197.

CHAPTER 8

INTEGRATIVE BOTTOM-UP APPROACH TO HRQOL MEASUREMENT

RALF PUKROP AND ANDREAS BECHDOLF

Department of Psychiatry and Psychotherapy, University of Cologne, Kerpenerstr. 6250937 Köln, Germany

Abstract: HRQOL is characterized by definitional uncertainty and a vast amount of heterogeneous assessment tools, and there is no really satisfying integrative theoretical model. The present chapter describes two studies using a bottom-up approach to establish an integrative model of HRQOL. The first study compared seven (inter)nationally validated HRQOL questionnaires in mentally healthy subjects (n = 479), patients with major depression (n = 171), and patients with schizophrenia (n = 139) to explore convergent and divergent aspects of 7 instruments, 45 subscales, and a great number of single items. Multivariate analyses have been primarily conducted by a nonparametric multidimensional scaling technique (faceted similarity structure analysis). A set of seven reliable QOL domains could be extracted by simultaneous analysis of all assessment tools. This set consists of one 'G-factor' of general well-being vs. depressed mood and six specific QOL dimensions. This basic structure represents a core module which holds for nonclinical and clinical samples, and which can be completed by specific modules for particular subpopulations (e.g. persons with family, partnership, profession). Reliability, validity, and sensitivity to change (from admission to discharge and 4-month-follow-up) of this Modular System for QOL were investigated in a second study in healthy controls (N = 346), patients with depression (N = 114), and patients with schizophrenia (N = 91) using the SF-36 as a comparison standard. Results and major conclusions will be discussed with an emphasis on the clear impact of depressive symptoms and current mood on HRQOL, and the implications given by this relationship

Keywords: HRQOL, Depression, Schizophrenia, Facet analysis

INTRODUCTION

The concept of HRQOL is characterized by definitional uncertainty and lacks clear causal specificity and theoretical extensive (see chapter 1 for a more detailed discussion). This is reflected on the operational level by an immense number of assessment tools (see chapters 5 and 6 for an overview). For example, a literature review restricted to quality of life in schizophrenia and to the interval from January 2005 to May 2006 revealed 113 original studies using 38 different HRQOL

M.S. Ritsner and A.G. Awad (eds.), Quality of Life Impairment in Schizophrenia, Mood and Anxiety Disorders, 143–155.

instruments. One way to overcome this heterogeneity may be a top-down approach developing more elaborated models to integrate the diverse empirical findings and theoretical concepts (chapter 1); another way is to start from bottom-up and search for integrative potentials in already existing HRQOL models and item pools that all claim to be content valid with regard to QOL.

We followed the second approach and conducted two studies with independent samples: the first study compared seven HRQOL instruments to discover convergent and divergent aspects[1]. The second one tested the set of reliable and non-redundant QOL domains extracted in the first study for its psychometric and clinical usefulness[2]. Finally, the most relevant issues for an integrative approach to QOL in psychiatric populations will be discussed.

STUDY 1: SELECTION OF HRQOL INSTRUMENTS AND DESCRIPTION OF SUBJECTS

Seven HRQOL measures have been selected meeting some necessary criteria: each single instrument should have sufficient psychometric qualities and should be (inter)nationally established, subjective evaluations and more objective behavioral or functional aspects of life quality should be covered by the whole set of instruments, and all instruments had to be multidimensional (measuring different QOL domains). All instruments have been applied as self-ratings.

1. Lancashire Quality of Life Profile (LQLP[3]) based on the Quality of Life Interview[4]: The LQLP has been developed for chronically mentally ill patients in community care settings and assesses objective conditions and subjective satisfaction (12 dimensions/81 items; 10 subjectively evaluated QOL domains have been included in the present study leaving out 'religious belief' and the integrated 'Affect Balance Scale').
2. Short-Form 36 (SF-36[5]): The SF-36 measures behavior-related functioning and subjective well-being (8 dimensions/36 items).
3. Sickness Impact Profile (SIP[6]): The SIP is a behaviorally oriented instrument measuring health-related dysfunctions in acute and chronic conditions (12 dimensions/136 items; for the present study a physical and a psychosocial sum scale score have been used).
4. Nottingham Health Profile (NHP[7]): The NHP assesses the subjective experience of health problems (6 dimensions/38 items).
5. Satisfaction Questionnaire (SQ): The SQ is a selection of items that emerged as the most relevant in the popular surveys by Andrews and Withey[8], and by Campbell et al.[9] (10 items).
6. Psychological General Well-Being Schedule (PGWBS[10]): The PGWBS is an instrument for the assessment of subjective well-being and discomfort (6 dimensions/22 items). The PGWBS has not been applied in the clinical samples.
7. Questionnaire for Everyday-living (QEL[11]): The QEL has been developed for German-speaking patients with acute and chronic conditions and assesses subjective experience of health-related everyday-living problems (4 dimensions/39 items).

These questionnaires have been completed by n = 479 healthy controls, n = 171 patients with major depression (n = 171), and patients with schizophrenia (n = 139). Clinical subjects were recruited in open wards and day-hospitals making their QOL assessments two weeks before discharge to minimize the impact of acute psychopathology. Further sample characteristics have been published elsewhere[1].

STUDY 1: COMPARISON ON THE LEVEL OF INSTRUMENTS (TOTAL SCORES)

To obtain single scores for every questionnaire either total scores or subscale scores for general QOL assessments were used. These scores correlated between 0.32 and 0.76 (median 0.54) in the nonclinical control sample, between 0.48 and 0.72 (median 0.62) in patients with MD, and between 0.33 and 0.74 (median 0.54) in the schizophrenia sample. Thus, the different instruments share about 30% of variance (including error variance) independent of the population investigated. The assumption of a common core despite clearly different operationalizations is also supported by principal components analyses (PCA). Calculating a PCA for the seven QOL total scores yielded one component with an eigenvalue > 1 for all samples. This general component explained 63%, 68%, and 65% of QOL variance in the nonclinical, depression, and schizophrenia sample, respectively. Thus, QOL instruments with quite different conceptualizations demonstrate sufficient convergence to postulate an identical construct (whatever this construct may be). However, it should also be noted, that there is substantial divergence between instruments. The degree of divergence even implies sufficient potential to produce different results and to lead to completely different conclusions depending on the instrument chosen for a single study. Thus, the selection of an instrument should always be explicitly justified.

STUDY 1: COMPARISON ON SUBSCALE LEVEL

It is widely agreed that QOL is a multidimensional construct. However, our seven selected instruments cover a range of 4 to 12 dimensions or QOL domains, and there is even more variation of QOL subdimensions in the literature. Thus, going beyond the level of total scores, it is important to identify convergent structures of QOL across different instruments if this research area gets more integrated. Usually, latent structures and common sources of variance are investigated by factor analysis techniques. There are a lot of problems, however, associated with parametric statistics: QOL measures are heavily skewed, large number of variables imply many factors that are hard to interpret, and results are dependent on metric conditions such as the absolute size of correlation coefficients or the linear relationship between variables. Therefore, we have applied a faceted similarity structure analysis (FSSA) which is not affected by these limitations.

Faceted Similarity Structure Analysis (FSSA)

Because FSSA is not well known, this nonmetric multivariate analysis technique will be described in some detail. Forty-five subscales from seven QOL measures have been included in the analysis. All scores have been rescaled from 0 to 100 so that a higher score indicates higher QOL. First, monotonicity coefficients[12] have been used to calculate the pairwise relationships between these 45 subscales. This coefficient is not affected by skewed raw score distributions, and its size is not dependent on metric qualities of variables. The resulting similarity matrix can be inspected to search for scales that do not fit with other QOL subdimensions because of consistent negative relationships.

In a second step of the analysis, the similarity matrix is graphically represented by transferring the empirical similarities into spatial distances using nonmetric multidimensional scaling techniques[13]. High emprical similarities (monotonicity coefficients) correspond to small distances in space, and low coefficients correspond to large spatial distances. The result of this second step is a (two-dimensional) spatial configuration of all 45 variables. The appropriateness of a certain dimensionality can be quantified by an alienation coefficient that should not clearly exceed the conventional limit of 0.15. Thirdly, predictions about the partitioning of this spatial configuration are tested as regional hypotheses. The most important criterion is the replicability of a postulated spatial configuration. Because of space limitations, the FSSA method cannot be explained in more detail here, but the interested reader is referred to the special literature[12–16]. The most important results of our FSSA on QOL subdimensions in clinical and nonclinical samples are outlined as follows:

First Law of Attitude

First, the similarity matrix for 45 QOL subscales has been inspected for the nonclinical sample, and for 39 subscales (without PGWBS) for the MD and schizophrenia samples. A first strong hypothesis that can be tested by this matrix is sometimes labelled the first law of attitude[15,16]. QOL items usually meet the definition of attitude items given by Guttman[16]: 'An item belongs to the universe of attitude items if and only if its domain asks about behavior in a cognitive/affective/instrumental modality towards an object, and its range is ordered from very positive to very negative toward that object'. The common object in QOL studies toward that cognitive, affective, or instrumental behavior is directed and that is evaluated from very positive to very negative is the subjects' life. If this definition is met, the relationships of all QOL variables (scales) should be monotonous, and coefficients should have a positive or at least no negative sign. This is a well known fact about intelligence test items that are positively related and can be aggregated to measure a G-factor of intelligence.

In all three samples there were no more than 6 negative correlations < -0.10 out of 990 coefficients (741 in the clinical samples). Thus, less than 1% of the investigated relationships between QOL subdimensions is negative. This confirmation of the first law of attitude supports the assumption of a common underlying construct, i.e. a G-factor of QOL, for nonclinical subjects, and patients with depression or schizophrenia. If one scale consistently correlates negatively with other scales, this would imply that the scale contents do not necessarily contribute to subjective QOL. There would be no constant attitude object. Another possibility would be that certain dimensions exclusively contribute to QOL in specific populations. However, this is not the case, and we can confirm the first law of attitude and the existence of a common attitude object for nonclinical and clinical subjects.

Regional Hypotheses (radex)

After the similarity matrix had been transferred into a two-dimensional spatial configuration, the alienation coefficients for all three samples indicated the appropriateness of this dimensionality. All coefficients were < 0.20, and a third dimension did not reduce them substantially. The two-dimensional representation of 45 variables (QOL subdimensions) and their 990 interrelationships for the nonclinical sample is given in Figure 1. For example, one can easily identify a cluster of physical subscales on the left side of the spatial representation. However, the regional hypothesis for this representation, i.e. the partitioning of the space, can be formulated much more systematically. The separation lines in Figure 1 define a regional partitioning of the space that is called a radex[14–16].

A radex is defined by two principles (which are called facets in FSSA): A nominal principle plays a polarizing role and induces regions of different directions that are

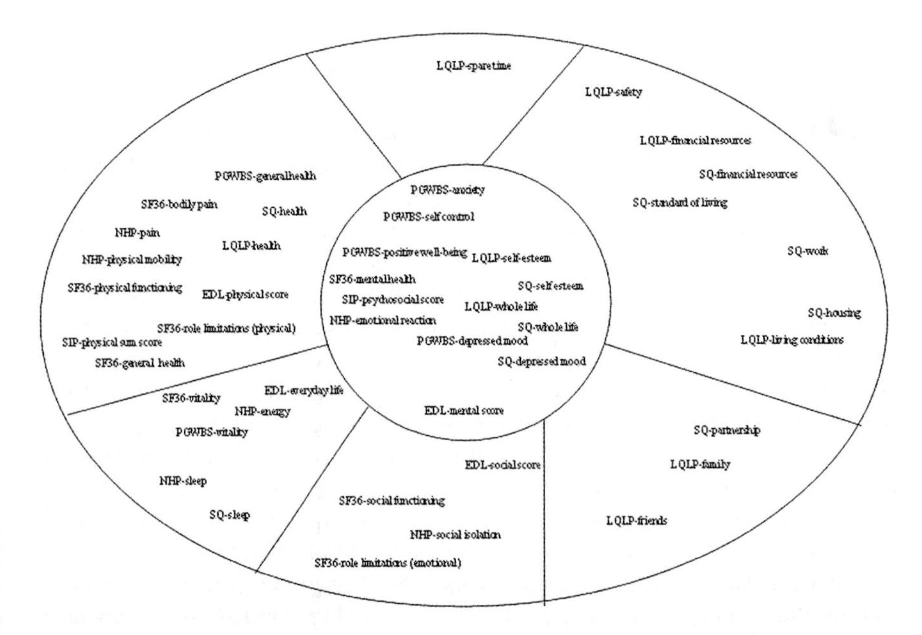

Figure 1. Spatial representation of interrelationships between 45 subscales from 7 QOL instruments

arranged in a circular manner around the center of the configuration (the center is not necessarily a mathematical or geometric center). The elements of this facet are different life areas (e.g. physical, social or mental QOL domains). These areas are not ordered in any way, and therefore they define a nominal facet. A second principle or facet plays a modularizing role and induces different distance classes with regard to the center. To keep it simple, the radex in Figure 1 has only two distance classes (central and peripheral) that are two elements of an ordinal facet. Thus, the radex hypothesis is a combination of a nominal facet (QOL domains) and an ordinal facet (distance classes) and can be summarized as follows: QOL subdimensions in the inner circle are highly correlated independent of the section or domain they belong to. The further a QOL subdimension is located in the periphery the more dependent relationships to other QOL variables become on the affiliation to the same section.

A closer look at Figure 1 will help to understand these abstract considerations: The central region is characterized by QOL subdimensions from different instruments measuring evaluations of one's life as a whole and self-related psychological (or psychopathological) conditions such as self-esteem or depression. In positive terms, this QOL domain is characterized by general psychological well-being or mental health; in negative terms, it is best described by the degree of depression with that somebody is looking at his life. Because these QOL aspects are located in the center of the configuration, they must be considered as making the most significant contribution to overall QOL.

Moreover, six different regions can be separated that are arranged around this QOL center. The exact number is not the most important aspect here, but the replicability and reliability of the partitioning is a necessary criterion. First, there is a clearly defined region of physical performance or impairments including pain assessments. Functional measures (SIP, SF-36, NHP) are overrepresented in that region, but satisfaction judgments of general health (LQLP, SQ) are also located here. Second, close to that physical health region, a section of vitality/energy/sleep has been established. Third, adjacent to the physical health region in clockwise direction, a spare time region can be identified. Because this QOL subscale is not closely related to any other domain, it makes sense to separate the subjective evaluation of spare time activities from other QOL areas. The spare time region is located between (closely related to) physical abilities on the one side, and material resources on the other side. This fourth region of material resources covers financial aspects, safety and living conditions. A fifth region is defined by intimate social relationships (partner, family, friends). Finally, the sixth region is characterized by subscales measuring the quality of social relationships in a more general sense and may be called psychosocial QOL.

When this analysis was repeated for the two clinical samples, the basic structure with a center of general well-being/depression and six peripheral regions with identical contents could be replicated. In the depression sample 5 subscale variables, and in the schizophrenia sample 4 variables slightly shifted to neighbouring regions.

Thus, based on the empirical relationships between QOL domains measured by quite different QOL instruments it is possible to infer a stable integrative structure of the whole construct not dependent on the assumptions underlying single models or questionnaires. The following characteristics of this bottom-up validation of QOL should be kept in mind: 1. The QOL model is not based on metric properties usually not met by QOL measures. 2. Since each model (or instrument) included in the analysis already claims content validity, an integrative model establishing a set union of all QOL domains considered by single instruments, should have even increased this aspect of validity. 3. Because all subdimensions are positively related and refer to a constant attitude object (the subject's life), it is legitimate to summarize neighbouring regions such as physical health and vitality, or intimate relationships and psychosocial QOL. Based on this rationale, it may be even useful to aggregate the information in one total QOL score. Moreover, if additional aspects or domains of QOL are suggested that are not represented in the present model, it will be easy to test if these domains contribute to overall QOL (and fit the first law of attitude), and if they define an additional region that can be reliably separated. 4. Most relevant for a clinical context, is that the basic integrative structure of QOL is replicable across different populations of nonclinical subjects, and subjects with depression and schizophrenia. One important implication of this replicability is the following: the investigation of *differences* in QOL levels between populations only makes sense when the structure of QOL is *similar* in these populations.

STUDY 1: COMPARISON ON ITEM LEVEL

To measure the integrative model for QOL it is either possible to use an existing instrument that covers all suggested QOL domains or to create a new instrument by extracting a suitable number of items for each region in Figure 1. We completed our first study by designing such a new assessment tool. For this reason, psychometric properties of all questionnaires, scales and single items have been thoroughly investigated in clinical and nonclinical samples. Based on item difficulty, face validity with regard to psychiatric patients, item convergent validity (item-scale-correlation corrected for overlap), item discriminant validity (correlation with other scales in comparison to corrected correlation with own scale), retest reliability (after 14 days within a nonclinical subsample), internal consistency, factorial validity and FSSA results, 49 items could be extracted that sufficiently met the listed criteria and that were representative of all QOL domains identified in Figure 1. Finally, these 49 items from different questionnaires have been analyzed simultaneously by FSSA as described above to test whether the structure found on the subscale level was replicable on item level[1]. Results showed that the central region and all peripheral regions could be basically confirmed for the nonclinical, depression and schizophrenia sample. However, three modifications should be noted: 1. especially PGWBS items formed a clear region of its own in the nonclinical sample that was called affective QOL (the PGWBS had not been assessed in the clinical samples); 2. items measuring self-esteem and self-confidence were located in the psychosocial region indicating that the quality of social relationships is substantially linked to self-related concepts; 3. because not every participant could answer questions on partnership, family, and occupation, these items were removed from the QOL core module to specific non-obligatory QOL modules.

Thus, the final integrative questionnaire, called the *Modular System for Quality of Life (MSQOL)*, has been composed as follows:

- 'general QOL' from depressed to happy or satisfied (8 items)
- 'physical health' from limited to excellent health (7 items)
- 'vitality' from exhausted and tired to full of energy (8 items)
- 'material resources' from materially satisfied to dissatisfied (4 items)
- 'affective QOL' from stressed and worried to relaxed and easy-going (9 items)
- 'psychosocial QOL' from socially integrated and well-accepted to lack of self-confidence and feelings of rejection (9 items)
- 'spare time QOL' from satisfying leisure to unsatisfied withdrawal from leisure activities (3 items)

These seven domains (48 items) together form a core module of QOL, because they can be applied in every population. Four specific modules have been added that only make sense for specific subpopulations and therefore are not obligatory and can be analyzed separately:

- 'partnership QOL' (4 items)
- 'family QOL' (3 items)
- 'children relationships' (2 items)
- 'occupational QOL' (2 items)

All items have been reformulated and have been given an identical response format. The core module and the specific modules are completed by a demographic module and a question for the current mood of the subject. Because QOL questionnaires usually ask for the last 4 weeks and tap something in between states and traits, it is important to estimate the impact of current states on more stable QOL estimations.

STUDY 2: VALIDATION OF THE INTEGRATIVE BOTTOM-UP APPROACH TO HRQOL

The integrative MSQOL has been tested and validated in a second study with new samples[2]: healthy controls (n = 346), patients with major depression[(MD) (n = 114)], and patients with schizophrenia (n = 91). The MSQOL and the SF-36 as a control questionnaire have been administered to inpatients at admission, discharge (n = 95 MD; n = 79 schizophrenia), and 4-month-follow-up (n = 77 MD; n = 47 schizophrenia) in the clinical samples, and at two occasions with an interval of 4 months in the nonclinical sample (n = 330). Patients with severe depression or who were not able to complete self-report forms because of other reasons even when supported by the research staff have been excluded from the study. Further sample characteristics and quantitative details of the analyses are described elsewhere[2]. The most important results can be summarized as follows:

Reliability

Median internal consistencies for MSQOL core dimensions were: 0.88 (nonclinical), 0.87 (MD), 0.83 (schizophrenia). Median internal consistencies for the SF-36 dimensions were: 0.78 (nonclinical), 0.79 (MD), 0.83 (schizophrenia); retest reliabilities after 4 months in the nonclinical sample were 0.55 (MSQOL) and 0.42 (SF-36). It should be noted that both instruments ask to evaluate the last four weeks. Therefore, these retest coefficients are adequate and offer sufficient range for measuring change.

Group Validity

At admission patients with schizophrenia showed significantly higher QOL levels than MD patients in all MSQOL and SF-36 scales (except for 'material resources'), and both clinical samples scored significantly lower than healthy controls. At discharge the difference to healthy controls remained significant for nearly all QOL dimensions in both instruments, but there were no systematic differences between the two patient groups anymore. These results were not affected by age and gender differences. The observation that MD patients demonstrate lower (at admission) or at least similar (at discharge) QOL levels than schizophrenia samples despite objectively worse conditions going along with schizophrenia (such as financial resources, social relationships, living conditions, age of onset, or number of hospitalizations) is consistent with other reports[17,18] and needs to be explained (see conclusions).

Selectivity of QOL Deficits

When each QOL dimension is compared with the mean of all other dimensions, selectively low QOL levels can be identified. Both patient groups had selectively low QOL levels on 'General QOL' and 'psychosocial QOL'. The MD group showed additional selective deficits on 'Physical Health'.

Responsiveness

Patients with depression revealed significant improvements from admission to discharge on nearly all QOL dimensions. No further change could be observed from discharge to follow-up. However, when controlling for depressive symptoms (Hamilton Depression Rating Scale) all QOL improvements showed substantial interactions with psychopathological improvements. Moreover, QOL improvements in six out of eight SF-36 dimensions and in three out of seven MSQOL core dimensions even became insignificant when controlling for depressive symptom change.

Patients with schizophrenia showed the same pattern of change with a general improvement on all QOL scales from admission to discharge and no further change at 4-month-follow-up. However, the relationship of these improvements to change in psychopathology was quite different to the MD sample. When improvement in positive and negative symptoms (Positive and Negative Syndrome Scale) was controlled, nearly all QOL changes remained significant. Accordingly, there were just a few interactions between change in schizophrenia psychopathology and in QOL measures.

Discriminant Validity

After the alpha level had been adjusted, positive and negative symptom scores were unrelated to QOL as measured by MSQOL and SF-36 at admission and discharge in the schizophrenia sample. This picture was quite different for MD patients: nearly all QOL dimensions correlated significantly with depressive symptomatology at admission and discharge. Correlations between depression and QOL were higher at discharge than at admission. Other clinical and demographic variables (age at onset, number of hospitalizations, age, gender) did not show any systematic relationships with subjective QOL in both clinical samples. However, there was one important exception for all three samples including nonclinical controls: current mood state during QOL self-report had a substantial and general impact on QOL assessment independent of the population, psychopathology, QOL instrument, life domain and time of measurement, since nearly all QOL dimensions (MSQOL and SF-36) were significantly correlated at any occasion with the MSQOL current mood measure. When demograhic and clinical characteristics, psychopathology and current mood were entered as predictors for QOL subdimensions in multiple regression analyses, current mood was the most powerful and consistent predictor for the majority of QOL scales.

MAJOR CONCLUSIONS

1. Convergence and divergence of QOL assessments: There is strong evidence that subjective QOL assessments refer to a constant construct. Total scores of instruments developed within diverse approaches show considerable empirical overlap and all load on the same principal component. Moreover, when QOL is conceptualized as an attitude toward one's own life then QOL items from completely different questionnaires confirm the first law of attitude (i.e. do not show substantial negative relationships). An important implication of these findings is to have a rationale for a QOL total score or 'G-factor'. These considerations seem to be valid for nonclinical as well as patients with MD or schizophrenia.

   On the other hand, correlation coefficients as low as 0.32 between total scores from different instruments have been observed. This may indicate that various studies can produce completely different results and conclusions depending on the individual QOL instrument used. Thus, the selection of instruments for a single study should be thoroughly justified.

2. Dimensional structure of QOL: Our results of an integrative bottom-up approach by simultaneous analysis of seven different QOL measures (45 subscales) clearly indicates a consistent pattern of QOL dimensions. A domain of general life evaluations from depressed mood to psychological well-being seems to be the most relevant QOL factor (see next point no. 3). Dependent on the level of depression all other life areas are subjectively evaluated more or less negatively. Further relevant domains are physical health, vitality, spare time, material resources, intimate relationships, psychosocial attitudes, and affective conditions. These QOL domains are not represented by every single instrument, however, they establish something like a set union which is not totally closed across different instruments. When there are some QOL domains not represented by any of those instruments selected for the present analysis, the set may be extended by other domains.
3. The meaning of depression: Maybe the most relevant issue to solve for QOL research is the substantial relationship between QOL and depression. When viewed dimensionally and not just as a clinical category the degree of depression seems to have the most prominent impact on subjective QOL. Provocatively, one could ask if depression (versus psychological well-being) and subjective QOL are more or less the same construct. There is empirical evidence now supporting these assumptions. Depressive symptoms or similar clinical measures have been identified as one of the strongest predictors for subjective QOL in undifferentiated psychiatric patients[19–21], in patients with schizophrenia[22–28] or patients with depression[29,30]. Moreover, the high impact of depression can also be demonstrated outside the psychiatric field (e.g. in somatic patients[31,32] or multiple sclerosis[33]). These studies report common proportions of variance between 30% and 70%, which is close to the theoretical maximum (when error variance is also considered). Finally, results of the present studies further support this clear relationship: Depressive symptomatology was substantially related to nearly all QOL subscales. When improvement in depressive symptoms was controlled, the majority of QOL scales did not indicate improvements anymore. The center of the spatial configuration in Figure 1 is defined by a general QOL component from depressed mood to psychological well-being. The evaluation of all other life areas will be substantially distorted by the degree of depression the subject is experiencing, which may be the main reason for positive relationships between different QOL domains and questionnaires. This basic meaning of depression for subjective QOL can also explain the lower QOL levels in MD patients when compared to schizophrenia. Depression seems to be a candidate for the identification of the QOL G -factor.
4. The meaning of current mood: Usually QOL questionnaires ask for an evaluation of an interval of four (+/− two) weeks. They do neither ask for current states nor stable traits but something in-between. However, our results suggest that current mood during the self-report has a substantial impact on the evaluation of longer time intervals and should be more thoroughly controlled.

REFERENCES

1. Pukrop R, Möller HJ, Steinmeyer EM. Quality of life in psychiatry: a systematic contribution to construct validation and the development of the integrative assessment tool 'Modular System for Quality of Life'. Europ Arch Psychiatry Clin Neurosci 2000;250:120–132
2. Pukrop R, Schlaak V, Möller-Leimkühler A, et al. Reliability and validity of Quality of Life assessment in patients with depression and patients with schizophrenia. Psychiatry Res 2003;119:63–79
3. Oliver JPJ. The social care directive: Development of a quality of life profile for use in community services for the mentally ill. Soc Work Soc Sci Rev 1991;3:5–45
4. Lehman AF. A quality of life interview for the chronically mentally ill. Eval Prog Plan 1988;11:51–62
5. Stewart AL, Ware JE. Measuring functioning and well-being. Durham London: Duke University Press;1992
6. Bergner M, Bobbitt RA, Carter WB, et al. The Sickness Impact Profile: development and final revision of a health status measure. Med Care 1981;19:787–805
7. European Group for Quality of life and Health Measurement. European guide to the Nottingham Health Profile. Montpellier:Imprimerie Air Dauphin;1992
8. Andrews FM, Withey SB. Social indicators of well-being. Americans' perceptions of life quality. New York:Plenum;1976
9. Campbell A, Converse PE, Rodgers WL. The Quality of American Life. New York:Russell Sage Foundation;1976
10. Dupuy HJ. The psychological general well-being index. In: Wenger NK et al., eds. Assessment of quality of life in clinical trials of cardiovascular therapies. New York:Le Jacq,1984:170–183
11. Bullinger M, Kirchberger, I, Steinbüchel N. Der Fragebogen Alltagsleben - ein Verfahren zur Erfassung der gesundheitsbezogenen Lebensqualität [Questionnaire of everyday-living – a tool fort he assessment of health-related quality of life]. Z Med Psychol 1993;3:121–131
12. Guttman L. What is not what in statistics. Statistician 1977;26:81–107
13. Borg I, Lingoes JC. Multidimensional similarity structure analysis. New York:Springer;1987
14. Borg I, Shye S. Facet theory: form and content. Newbury Park:Sage;1995
15. Levy S. Lawful roles of facets in social theories. In: Canter D (ed) Facet theory - approaches to social research. New York:Springer;1985:59–96
16. Levy S. Louis Guttman on theory and methodology: selected writings. Aldershot:Dartmouth;1994
17. Atkinson M, Zibin S, Chuang H. Characterizing quality of life among patients with chronic mental illness: a critical examination of the self-report methodology. Am J Psychiatry 1997;154:99–105
18. Russo J, Roy-Byrne P, Reeder D, et al. Longitudinal assessment of quality of life in acute psychiatric inpatients: reliability and validity. J Nerv Ment Dis 1997;185, 166–175
19. Koivumaa-Honkanen HT, Viinamäki H, Honkanen R, et al. Correlates of life satisfaction among psychiatric patients. Acta Psychiatr Scand 1996;94:372–378
20. Corrigan PW, Buican B. The construct validity of subjective quality of life for the severely mentally ill. J Nerv Ment Dis 1995;183:281–285
21. Sullivan G, Wells KB, Leake B. Clinical factors associated with better quality of life in a seriously mentally ill population. Hosp Com Psychiatry 1992;43:794–798
22. Mechanic D, McAlpine D, Rosenfield S, et al. Effects of illness attribution and depression on the quality of life among persons with serious mental illness. Soc Sci Med 1994;39:155–164
23. Browne S, Roe M, Lane A, et al. Quality of life in schizophrenia: relationship to sociodemographic factors, symptomatology and tardive dyskinesia. Acta Psychiatr Scand 1996;94:118–124
24. Bechdolf A, Klosterkötter J, Hambrecht M et al. Determinants of subjective quality of life in postacute patients with schizophrenia. Europ Arch Psychiatry Clin Neurosci 2003;253:228–235
25. Law CW, Chen EY, Cheung EF, et al. Impact of untreated psychosis on quality of life in patients with first-episode schizophrenia. Qual Life Res 2005;14:1803–1811
26. Huppert JD, Smith TE. Anxiety and schizophrenia: the interaction of subtypes of anxiety and psychotic symptoms. CNS Spectr 2005;10:721–731

27. Hofer A, Baumgartner S, Edlinger M, et al. Patient outcomes in schizophrenia I: correlates with sociodemographic variables, psychopathology, and side effects. Eur Psychiatry 2005;20:386–394
28. Karow A, Moritz S, Lambert M, et al. PANSS syndromes and quality of life in schizophrenia. Psychpathol 2005;38:320–326
29. McKenna SP, Hunt SM. A new measure for quality of life in depression: testing the reliability and construct validity of the QLDS. Health Policy 1992;22:321–330
30. Gregoire J, de Leval N, Mesters P, et al. Validation of the quality of life in depression scale in a population of adult depressive patients aged 60 and above. Qual Life Res 1994;3:13–19
31. Waltz M. A longitudinal study on environmental and dispositional determinants of life quality: social support and coping with physical illness. Soc Indicat Res 1986;18:71–93
32. Abbey A, Andrews FM. Modeling the psychological determinants of life quality. Soc Indicat Res 1985;16:1–34
33. D'Alisa S, Miscio G, Baudo S, et al. Depression is the main determinant of quality of life in multiple sclerosis: a classification regression study. Disadbil Rehabil 2006;28:307–314

PART II

QUALITY OF LIFE IMPAIRMENT SYNDROME IN SEVERE MENTAL DISORDERS

CHAPTER 9

HEALTH RELATED QUALITY OF LIFE IN SUBJECTS AT RISK FOR A FIRST EPISODE OF PSYCHOSIS

ANDREAS BECHDOLF*, VERENA VEITH, RALF PUKROP AND JOACHIM KLOSTERKÖTTER

Department of Psychiatry and Psychotherapy, University of Cologne, Kerpenerstr. 6250937 Köln, Germany

Abstract: To date no data on health related quality of life (HRQL) in subjects at risk for a first episode of psychosis compared with first episode schizophrenia patients (FE) or healthy controls (HC) is available, although the concept of HRQL is of growing relevance in schizophrenia research. Therefore 45 subjects in a putatively early initial prodromal state (EIPS), 40 FE and 45 HC were assessed for demographics, psychopathology and HRQL as measured by the self-rating instrument Modular System for Quality of Life. Results indicated that on a descriptive level in most life areas HC experienced the highest HRQL scores followed in hierarchical order by EIPS and FE. EIPS and FE experienced significantly lower HRQL than HC in 5 and 6 of 7 HRQL domains. When comparing EIPS and FE, EIPS experienced a significantly lower HRQL level in affective HRQL. HRQL ratings in EIPS were negatively correlated with joining school or work and higher positive prodromal symptom scores. In FE HRQL scores were negatively correlated with age and depression. The major finding from the study is that HRQL in subjects at risk for a first episode of psychosis is substantially reduced when compared with HC. Thus, HRQL maybe already reduced prior to the onset of first positive schizophrenia symptoms. These data support the notion that subjects at risk for a first episode of psychosis constitute a clinical population in which further service and intervention research is indicated

Keywords: Prodrome, High risk research, First episode schizophrenia, Schizophrenia, Quality of life

INTRODUCTION

Schizophrenia is a chronic disease with devastating consequences for the affected individuals, their families and society[1]. Among psychiatric disorders, schizophrenia is most disabling and requires a disproportionate share of mental-health services.

* Cologne Early Recognition and Intervention Centre for mental crises - FETZ, Department of Psychiatry and Psychotherapy, University of Cologne, Kerpener Str. 62, 50924 Cologne, Germany, Tel.:(+49)-221-478-3869, Fax: (+49)-221-478-6030, email: andreas.bechdolf@uk-koeln.de

M.S. Ritsner and A.G. Awad (eds.), Quality of Life Impairment in Schizophrenia, Mood and Anxiety Disorders, 159–171.

For example, patients with schizophrenia occupy about 25% of all psychiatric hospital beds[2] and represent 50% of admissions to hospital[3]. The total costs of treating schizophrenia are high, estimated to be US $ 44.9 billion in the USA for the year 1994[4] and £ 2.6 billion in the UK for 1996[2]. The combined economic and social costs of schizophrenia place it among the world 's top ten causes of disability-adjusted life years[5] accounting for an estimated 2.3% of all burdens in developed countries, and 0.8% in developing countries[6].

A number of recent clinical findings and hypotheses have shifted research interests towards prevention of schizophrenia and have stimulated researchers to formulate the"early intervention hypothesis"[7] stating that the course of the illness may be improved when the disorder is detected and treated early. These findings and hypotheses are: (a) schizophrenia goes untreated for 5 years in 75% of the cases with non-psychotic prodromal symptoms (anxiety, depression, cognitive disturbances, negative symptoms[8], (b) social decline starts prior to the onset of positive symptoms[9](c) delayed treatment correlates with an unfavourable outcome[10,11], (d) psychotic episodes itself may have neurotoxic effects[12], (e) criteria have been developed, which enable to predict the onset of psychosis in prospective follow-along studies in 35-54% of the cases in a help-seeking population within 12 months [ultra-high risk (UHR), [13,14,15,16] and in 65% of the cases within 5.4 years on average [basic symptoms[17]].

Criteria have been established which enable clinicians to identify in a help-seeking population, those who have a risk to develop psychosis of up to 54% within 12 months[13,15,16] and of 70% within 5.4 years[17].

These findings and hypotheses stimulated early intervention efforts in subjects at risk for a first episode of psychosis around the world[18,19,20,21,22]. First descriptive data indicated that subjects at risk of developing psychosis are substantially compromised in psychosocial functioning (as measured by the Global Assessment of Functioning scale – GAF[23,21,14,22] and make an odyssey through the health system[24,25]. This population is significantly less psychopathologically symptomatic in terms of PANSS total score than patients with first episode schizophrenia (FE) and significantly less depressed than a sample suffering from adolescent depression[14]. Moreover, a number of neurobiological abnormalities have been described in this population.[26] However, to date no data of subjective quality of life (HRQL) in subjects at risk of developing a first episode of psychosis as compared to patients with schizophrenia and healthy controls (HC) is available, although the concept of HRQL is of growing relevance in schizophrenia research, and there is an increasing amount of literature on the subject[27,28,29].

Most studies of HRQL in patients with schizophrenia reported lower HRQL scores in patients than in healthy controls[28,30,31]. They found weak associations between subjectively reported levels of HRQL on one hand and sociodemographic characteristics, objective indicators, positive or negative psychotic symptoms on the other[32,33,34]. Like in the general population[35,36] depression scores and psychosocial concepts such as coping, social support and self-efficacy explained substantial portions of variance of overall HRQL[37,38,39].

The aim of the present study is to explore for the first time HRQL in subjects at risk for a first episode of psychosis as compared with FE and healthy controls (HC).

METHOD

Subjects

Subjects at risk for a first episode of psychosis were recruited from consecutive referrals to the Cologne Early Recognition and Intervention Centre for mental crisis (FETZ) at the Department of Psychiatry and Psychotherapy at the University of Cologne, Germany between July 2000 and February 2003. The centre is an outpatient department, specially designed for young people suffering from possible pre-psychotic symptoms, which aims to promote help-seeking and to avoid stigma. Besides self-referrals, referrals are made from health settings, primary health care, mental health professionals and other youth support services. In an extension of the international 'ultra-high risk' (UHR) criteria as defined by Yung and co-workers (1998). Broadly applied the criteria employ attenuated or transient psychotic symptoms or a combination of presence of a risk factor and recent significant decline in psychosocial functioning. A time-related distinction between an early (EIPS) and late initial prodromal state (LIPS) is made in the FETZ. Subjects in the EIPS are characterized by psychosis predictive basic symptoms and/or a risk factor combined with a recent decline in psychosocial functioning. Subjects meeting LIPS criteria are suffering from attenuated and/or transient psychotic symptoms[40,41].

All subjects at risk for a first episode of psychosis in this study met EIPS criteria, as defined by: (I) the presence of at least one self-experienced cognitive thought and perception deficit during the last 3 months prior to assessment [„basic symptoms", i.e. thought interferences, thought perseveration, thought pressure, thought blockages, disturbances of receptive language (heard or read), decreased ability to discriminate between ideas and perception, fantasy and true memories, unstable ideas of reference (subject-centrism), derealisation, visual perception disturbances (blurred vision, transitory blindness, partial sight, hypersensitivity to light, etc.), acoustic perception disturbances (e. g. hypersensitivity to sounds or noise, acoasms)], which were found to be predictive for transition to psychosis in 70% of the cases within 5.4 years[42] and/ or (II) the presence of clinical relevant decline of functioning (reduction in the Global Assessment of Functioning Score according to DSM-IV (APA 1994) of at least 30 points within the past year) in combination with well-established risk factors (first-degree relative with a lifetime-diagnosis of schizophrenia or a schizophrenic spectrum disorder or pre-/perinatal complications)[41].

Exclusion criteria were as follows: attenuated or transient positive symptoms; present or past diagnosis of a schizophrenic, schizophreniform, schizoaffective, delusional or bipolar disorder according to DSM-IV; present or past diagnosis of a brief psychotic disorder according to DSM-IV with a duration of more than one week or within the last 4 weeks regardless of its duration, diagnosis of delirium, dementia, amnestic or other cognitive disorder, mental retardation, psychiatric disorders due to a somatic actor or related to the consumption of psychotropic substances according to DSM-IV, alcohol- or drug abuse within the last three months prior to inclusion according to DSM-IV, diseases of the central nervous system and aged below 17 and above 35 years, treatment with antipsychotics. All EIPS gave informed consent to participate at a randomized controlled psychological intervention trial[37,18,19] and were assessed at intake of the trial.

Patients with FE were recruited from consecutive admissions to the inpatient unit of the Department of Psychiatry and Psychotherapy of the University of Cologne between July 1999 and December 2000. 26 patients gave informed consent to participate at a psychological group therapy as part of a randomized, controlled intervention trial[43], 14 gave informed consent to take part at a reliability and validation study of the HRQL measure used[44]. FE had to be aged between 18 and 40 years and met criteria for first episode of schizophrenia or a schizoaffective disorder (DSM IV 295.1-3, 295.5-9). At the time of the assessment most patients were partly in remission from acute psychosis. All FE-subjects were receiving psychotropic medication on an individual basis.

HC were recruited from the general population and matched for sex and age with the EIPS sample.

Measures

Objective information was assessed by a short demographic interview and was extracted from case notes.

HRQL was assessed using the „Modular System for Quality of Life“[44,43], a self-report instrument consisting of a demographic module and four HRQL modules (a core module, a partnership module, a family module and a professional occupation module). In this study we refer to the core module which comprises seven areas of HRQL (47 items on a 7-point-rating scale): Physical Health (e.g. limitations to perform moderate or vigorous activities), Vitality (e.g. feeling full of life; feel relaxed and comfortable; amount of time feeling tired), Psychosocial QoL (e.g. satisfaction with relation to other people; feeling self-confident), Affective HRQL (e.g. feeling under strain, stress or pressure; getting easily hurt), Material HRQL (e.g. with financial situation or living conditions), Spare Time HRQL (e.g. interest in hobbies at home or outdoor), General HRQL (e.g. having problems in general; enjoying life). Raw scores were rescaled to a range from 0 to 100 so that a higher score indicates higher HRQL. Median values for internal consistencies were 0.88 (range 0.73–0.92) in the general population and 0.83 (0.78–0.92) in patients with schizophrenia.

Predictive basic symptoms were observer rated with the symptom list of the Early Recognition Inventory and Interview for the Retrospective Assessment of the Onset of Schizophrenia-ERIraos[45]. This instrument was specially designed for the assessment of prodromal symptoms and consists of 105 items covering

12 categories. Kappas for “symptoms present in the year before interview” range between 0.41 and 0.87[46].

Psychopathology was observer rated using the „Positive and Negative Syndrome Scale – PANSS“[47]. For further analysis we used the components positive syndrome, negative syndrome and general psychopathology. A Depression score was derived from PANSS items anxiety (G2), depression (G6), guilty feelings (G3) and tension (G4) according to Wolthaus and co-workers (2000)[48]. This score was found to correlate with MADRS[49] depression-scores 0.87 (34) in patients with FES. In a reliability check intraclass correlation coefficients were 0.87 for the positive syndrome subscale, 0.73 for the negative syndrome subscale and 0.87 for the general psychopathology scale.

Data Analyses

Data were analyzed using the Statistical Package for Social Sciences[50]. Sample characteristics were compared using t-tests or χ^2-tests to check for demographic differences. Raw data distribution did not differ significantly from normal distribution indicated by Kolmogrow-Smirnow tests. Therefore, one-way ANOVAs with Bonferroni post-hoc tests were applied to investigate HRQL differences between EIPS, FE and HC. Results will be reported before and after a Bonferroni adjustment for multiple testing according to an overall alpha level of 5%. Exploratory bivariate correlations (pointbiserial or Spearman’s coefficients) with two-tailed tests of significance were calculated between HRQL core dimensions and sociodemographic parameters or psychopathology scores. For this purpose, the socio demographic variable employment status has been recoded to a dichotomous variable (occupation/training vs. no occupation/training).

RESULTS

Description of Sample

45 subjects at risk of developing a first episode of psychosis, 40 FE and 45 HC were included in the study (Table 1). EIPS and HC did not differ in age (25.7 vs. 25.0 years). Patients of the FE sample were 2 years older on average than the participants from the two other diagnostic groups. About 2/3 of each sub sample were male. Overall, there were no significant differences between sub samples in age and gender distribution. EIPS and HC were quite highly educated with 12 years of education. As could be expected, EIPS and HC were significantly higher educated than FE patients. Most participants in all samples were singles, but in the HC sample significantly fewer persons than in the FE group were singles. The employment status differed significantly between FE and HC. As could be expected, FE showed the highest unemployment rate followed by EIPS. Although in all groups most participants still attended schools, universities or other educational settings, the overall rate in FE was the smallest compared with the other groups. FE patients showed significantly higher positive,

Table 1. Sample characteristics for subjects in an early initial prodromal state of psychosis (EIPS), patients with first episode schizophrenia (FE) and healthy controls (HC)

	EIPS (n=45)	FE (n=40)	HC (n=45)	EIPS vs. FE (p-values)*	EIPS vs. HC (p-values)*	FE vs. HC (p-values)*
Age, years [mean (sd)]	25.7 (5.0)	27.8 (7.8)	25.0 (3.6)	0.230	1.00	0.067
Gender [n (%)]						
Female	14 (31.1)	16 (40.0)	16 (35.6)	0.496	0.823	0.823
Male	31 (68.9)	24 (60.0)	29 (64.4)			
Years of education [mean (sd)]	12.2 (1.5)	10.9 (1.5)	12.6 (0.8)	< 0.001	1.00	< 0.001
Marital status [n (%)]						
Married, Cohabitation	8 (17.8)	4 (10.0)	13 (28.9)			
Living alone, Divorced	37 (82.2)	36 (90.0)	32 (71.1)	0.362	0.319	0.034
Employment status [n (%)]						
Unemployed	9 (20.0)	15 (28.2)	4 (8.9)			
Full/part time	9 (20.0)	12 (30.0)	20 (44.4)	0.012	0.054	0.002
Training/ retraining	26 (57.8)	10 (25.0)	21 (46.7)			
Other/ unknown	1 (2.2)	3 (7.5)	0 (0)			
PANSS [mean (sd)]						
Positive	9.67 (2.54)	14.58 (4.99)		< 0.001	–	–
Negative	11.07 (3.55)	16.51 (6.42)	–	< 0.001	–	–
General	27.62 (5.34)	34.34 (10.31)	–	< 0.001	–	–
Depression	8.14 (5.34)	9.68 (3.21)	–	0.195	–	–

* values refer to Bonferroni or Fisher-exact post hoc tests

negative and general psychopathology scores than EIPS, whereas depression scores did not differ significantly between EIPS and FE samples.

HRQL

Descriptive statistics for HRQL in the three subsamples and tests for group comparisons are presented in Table 2. On a descriptive level HC experienced the highest HRQL scores followed in hierarchical order by EIPS and FE with two exceptions: vitality and affective HRQL levels were highest in HC, but lower in EIPS than in FE. Posthoc tests revealed that EIPS und FE both differed significantly from HC in 5 and 6 out of 7 HRQL areas, respectively. EIPS reported significantly lower HRQL scores than HC in: physical health, vitality, psychosocial, affective and general HRQL, wheras material and spare time HRQL

Table 2. Mean values (and standard deviations in parantheses) of health related Quality of Life (HRQL) and group differences (ANOVAS, p-values) between subjects in an early initialprodromal state of psychosis (EIPS), patients with first episode schizophrenia (FE) and healthy controls (HC)

	EIPS (n=45)	FE (n=40)	HC (n=45)	EIPS vs. FE (p-values)	HC vs. EIPS (p-values)	HC vs. FE (p-values)
Physical health	60.86 (17.73)	54.61 (19.06)	75.13 (15.33)	0.301	<0.000*	<0.000*
Vitality	47.39 (16.24)	50.78 (18.21)	61.57 (18.22)	1.000	0.001*	0.016
Psychosocial HRQL	54.89 (18.85)	49.81 (21.14)	67.57 (16.02)	0.642	0.005*	<0.000*
Material HRQL	62.42 (18.92)	56.67 (18.53)	70.28 (19.94)	0.508	0.162	0.004*
Spare time HRQL	57.88 (17.08)	50.21 (25.35)	66.67 (22.26)	0.318	0.171	0.002*
Affective HRQL	44.41 (19.47)	57.27 (16.65)	60.98 (17.98)	0.004*	<0.000*	1.000
General HRQL	55.74 (21.08)	53.02 (24.23)	72.27 (18.86)	1.000	0.001*	<0.000*

* remained significant after Bonferroni adjustment for multiple testing (adjusted level 0.007)

were comparable. FE differed significantly from HC in physical health, vitality, psychosocial, material, spare time and general HRQL. When comparing HRQL ratings of EIPS and FE no significant differences emerged except in the domain affective HRQL, in which the self-ratings were significantly lower in the EIPS than in the FE group. All differences remained significant after Bonferroni adjustment for multiple testing accept the difference in the vitality domain between HC and FE.

Bivariate Correlations

Exploratory bivariate correlations between sociodemographic parameters (age, gender, years of education, marital and employment status) and psychopathology on the one hand andHRQL domains on the other are presented for EIPS and FE in Table 3. In clients fulfilling EIPS criteria employment or being in training was significantly correlated with lower psychosocial and general HRQL scores. In addition in EIPS higher PANSS positive scores were significantly correlated with lower ratings constantly across HRQL domains psychosocial, spare time, affective and general HRQL. FE vitality and material HRQL were negatively correlated with age. With regard to psychopathology depression was negatively correlated with psychosocial, affective and general HRQL in FE. When group differences were retested by controlling for employment staus (dichotomous) and positive symptoms by ANCOVA all former significant group differences remained significant.

Table 3. Bivariate correlations between health related quality of Life (HRQL) in subjects in an early initial prodromal state of psychosis (EIPS) and patients with first episode schizophrenia (FE) and socio-demographic indicators or psychopathology (Pearson's or point serial coefficients)

	EIPS							FE						
	PH	V	PS	M	ST	A	G	PH	V	PS	M	ST	A	G
Sociodemographic indicators														
Age									−.341*		−.432**			
Gender														
Years of education														
Marital status[†]														
Employment status			−.331*				−.318*							
Psychopathology														
PANSS positive			−.405**		−.311*	−.398**	−.329*							
PANSS negative														
PANSS general														
Depression										−.482*			−.501*	−.507*

HRQL domains: PH, physical health; V, vitality; PS, psychosocial HRQL; M, material HRQL; ST, spare time HRQL; A, affective HRQL; G, general HRQL ; * $p \leq 0.05$; ** $p \leq 0.01$, non significant coefficients are not given,[†] because of the low proportion of unmarried FE no coefficient could be calculated

DISCUSSION

To our knowledge this is the first study of HRQL in subjects at risk for a first episode of psychosis compared with HRQL in FE and HC. On a descriptive level HC experienced the highest HRQL scores followed in hierarchical order by subjects at risk and FE in most life areas. Patients in a putative prodromal state and FE patients experienced significantly lower HRQL than HC in 5 and 6 out of 7 HRQL domains, respectively. Between subjects at risk and FE no significant differences could be observed except for affective HRQL, which was significantly lower in persons fulfilling at-risk criteria than FE. Not many correlations between sociodemographic variables or psychopatholgy on the one hand and HRQL measures on the other could be observed. However, lower HRQL ratings in persons at-risk were correlated with joining school or work and higher positive prodromal symptom scores. These relationships, however, could not account for the HRQL group differences reported above. In FE HRQL scores were negatively correlated with age and depression. Because depression scores did not differ significantly between EIPS and FE, HRQL group differences could not be due to different depression levels between the two diagnostic groups.

The major conclusion from the findings is that HRQL in subjects at risk for a first episode of psychosis is substantially reduced when compared with HC. Thus, HRQL maybe already reduced prior to the onset of first positive schizophrenia symptoms. Subjects in a putative prodromal state perceived themselves as primarily compromised (and also more compromised than FE) with regard to their affective condition and their vitality. They felt under strain, pressure, easily hurt, uncomfortable and tired. Moreover subjects at risk felt a lack of self-confidence, were unsatisfied with relations to other people and had difficulties in enjoying life in general (HRQL dimensions: psychosocial HRQL, general HRQL). As these HRQL scores were even more reduced when persons rated with higher positive symptom scores and still attended school, university or were having a job, one could state that these HRQL ratings may reflect the subjective experience of limited resources and coping strategies to overcome actual demands. This overwhelming situation may than result in social decline which usually takes place during prodromal states[18,8,21,14,22]. Once full schizophrenia had developed HRQL scores tend to further decrease (although not significantly).

With regard to the persons at risk of developing a first episode of psychosis, our findings are in accordance with the findings of Miller and co-workers (2003) indicating that subjects at risk for a first episode of psychosis show lower PANSS positive and PANSS negative scores than patients with FE. Earlier studies reporting lower HRQL values in patients with schizophrenia than in HC seem to supoort our data[28,30,29]. Moreover, relationships between high depression scores and low HRQL measures have been frequently reported for patients with schizophrenia[37,38,39].

Methodological Considerations

There are some methodological limitations of the findings: (I) The results for subjects at risk for a first episode of psychosis cannot be generalized beyond the help-seeking subset of persons at risk. It should be noted that despite intensive awareness-raising and educational efforts, the majority of young people who would have fulfilled at risk criteria, probably did not show up at the Early Recognition and Intervention Centre. (II) Because the future course of subjects at risk for a first episode of psychosis has still to be evaluated longitudinally, the respective sample may contain false positive cases, who will not develop schizophrenia. (III) Conclusions from our sample regarding subjects at risk for a first episode of psychosis meeting other at-risk-criteria than EIPS have to be interpreted with caution. (IV) Participants of the study were not randomly selected and all subjects at risk and most FE patients gave informed consent to participate at psychological intervention trials. Although we doubt that this is a significant source of bias the samples may not be representative and generalizations from the findings might be limited. (V) In contrast to the outpatient at-risk sample all FE patients were hospitalized, which was found to correlate with lower HRQL scores [32]. Therefore, lower HRQL scores in FE might be in part due to hospitalization. This limitation, however, stresses the level of subjective impairment in persons at risk. (VI) PANSS scores do not adequately describe prodromal psychopathology. Due to comparison purposes, however, all earlier studies used established scales in the field of psychopathology of schizophrenia to compare subjects at risk for a first episode of psychosis and patients with schizophrenia [18,39,20,21,14,22,41,15,16]. (VI) Future studies of HRQL in subjects at risk of psychosis should include psychosocial concepts like coping skills, self-constructs, stress which could explain relevant HRQL differences between the groups and contribute to distress and vulnerability which may constitute quality of life.

Clinical Consequences

Our data indicate that in addition to observer rated psychopathological and social functioning deficits or neurobiological deviances subjects at risk for a first episode of psychosis, who present at an Early Recognition and Intervention Centre, perceive themselves on a subjective level as substantially compromised, too. The relevant reduction of life satisfaction related with the syndrome support the notion that subjects at risk for a first episode of psychosis constitute a clinical population [18,19,39,17,20,21,22,41,15,16]. Therefore further research in specialized services and interventions for subjects at risk, who are help-seeking, worrying about their symptoms and wishing to receive treatment, seems justified. Drawing on the pattern of reduced HRQL, which we found in subjects at risk for a first episode of psychosis, interventions focusing on the affective condition and offering practical help with regard to the occupational situation seem to be most acceptable to this population. Interventions which have been recently developed and evaluated in persons at risk of developing psychosis, such as individual cognitive-behavioral therapy and/or atypical neuroleptic treatment both incorporating case management [18,19,39,20,21,22,41] are therefore likely to be acceptable to subjects at risk for a first episode of psychosis and to improve their HRQL [35,37,29].

ACKNOWLEDGEMENTS

This work is part of the German Research Network on Schizophrenia funded by the German Federal Ministry for Education and Research BMBF (grant 01 GI 9935). It was supported by grants from the German Research Foundation (Ste 523/2-2) and the Koln Fortune Program (191/1998)/Faculty of Medicine, University of Cologne, Germany.

REFERENCES

1. Mueser KT, McGurk SR, Schizophrenia. Lancet. 200; 363: 2063–72.
2. Terkelsen KG, Menikoff A. Measuring costs of schizophrenia: implications for the post-institutional era in the US. *Pharmacoeconomics* 1995; 8: 199–222.
3. Geller JL, An historical perspective on the role of state hospitals viewed form the "revolving door", *Am J Psychiatry* 1992; 149: 1526–33.
4. Rice DP. Economic burden of mental disorders in the United States. *Economics Neuroscience* 1999; 1: 40–44.
5. Murray CJL, Lopez AD, eds. The global burden of disease and injury series, vol. 1: a comprehensive assessment of mortality and disability from diseases, injuries, and risk factors in 1990 and projected to 2020. Cambridge MA: *Harvard University Press*, 1996.
6. US Institute of Medicine. Neurological, psychiatric, and developmental disorders: meeting the challenges in the developing world. Washington, DC: *National Academy of Sciences*, 2001.
7. McGlashan TH, Johannessen JO (1996) Early detection and intervention in schizophrenia. Rationale. Schizophr Bull 22: 201–222.
8. Häfner, H., Riecher-Rössler, A., Maurer, K. et al., 1992. First onset and early symptomatology of schizophrenia. Eur. Arch. Psychiatry. Clin. Neurosci. 242, 109–118.
9. Häfner H, Nowotny B et al. (1995) When and how does schizophrenia produce social deficits? Eur Arch Psychiatry Clin Neurosci 246: 17–28
10. Norman RM, Malla AK (2001) Duration of untreated psychosis: a critical examination of the concept and its importance. Psychol Med 31(3): 381–400.
11. Marshall M, Lewis S, Lockwood A, Drake R, Jones P, Croudace T (2005) Association between duration of untreated psychosis and outcome in cohorts of first-episode patients: a systematic review. Arch Gen Psychiatry 62(9): 975–983.
12. Pantelis C, Yücel M, Wood SJ, Velakoulis D, Sun D, Berger G, Stuart GW, Yung A, Phillips L, McGorry P (2005) Structural Brain Imaging Evidence for Multiple Pathological Processes at Different Stages of Brain Development in Schizophrenia. Schizophr Bull 31: 672–696.
13. Miller T.J., McGlashan T.H., Rosen J.L., Somjee L., Markovich P.J., Stein K., Woods S.W., 2002. Prospective diagnosis of the initial prodrome for schizophrenia based on the structured interview for prodromal syndromes: preliminary evidence of interrater reliability and predictive validity. Am. J. Psychiatry 159, 863–865.
14. Miller T.J., Zipursky R.B., Perkins D., Addington J., Woods S.W., Hawkins K.A., Hoffman R., Preda A., Epstein I., Addington D., Lindborg S., Marquez E., Tohen M., Breier A., McGlashan T.H., 2003. The PRIME North America randomized double-blind clinical trial of olanzapine versus placebo in patients at risk of being prodromally symptomatic for psychosis. II. Baseline characteristics of the"prodromal" sample. Schizophr. Res. 61(1), 19–30.
15. Yung, A.R., Phillips, L.J., Yuen, H.P., Francey, S.M., McFarlane, C.A., Hallgren, M., McGorry, P.D., 2003. Psychosis prediction: 12-month follow up of a high-risk ("prodromal") group. Schizophr. Res. 60(1), 21–23.
16. Yung, A.R., Phillips, L.J., Yuen, H.P., McGorry, P.D., 2004. Risk factors for psychosis in an ultra high-risk group: psychopathology and clinical features. Schizophr. Res. 67, 131–142.
17. Klosterkotter, J., Hellmich, M., Steinmeyer, E.M., Schultze-Lutter, F., 2001. Diagnosing schizophrenia in the initial prodromal phase. Arch. Gen. Psychiatry 58, 158–164.

18. Bechdolf A, Veith V, Berning J, Stamm E, Janssen B, Wagner M, Klosterkötter J (2005a) Cognitive-behavioural therapy in the pre-psychotic phase. An exploratory study. Psychiatr Res 136 (2–3): 251–55.
19. Bechdolf A, Ruhrmann S, Wagner M, Kuhn KU, Janssen B, Bottlender R, Wieneke A, Schultze-Lutter F, Maier W, Klosterkötter J (2005b) Interventions in the initial prodromal states of psychosis in Germany: Concept and recruitment. B J Psychiatry Suppl 48: S45–8.
20. McGlashan TH, Zipursky RB perkins D, Addington J, Miller T, Woods SW et al. (2006) A randomized, double-blind trial of olanzapine versus placebo in patients prodromally symptomatic for psychosis. Am J Psych 163(5): 790–99.
21. McGorry P.D., Yung A.R., Phillips L.J. et al., 2002. Can first episode psychosis be delayed or prevented? A randomized controlled trial of interventions during the prepsychotic phase of schizophrenia and related psychosis. Arch. Gen. Psychiatry 59(10), 921–928.
22. Morrison A.P., French P., Walford L., Lewis S.W., Kilcommons A., Green J., Parker S., Bentall R. P., 2004. A randomised controlled trial of early detection and cognitive therapy for the prevention of psychosis in people at ultra-high risk. Br. J. Psychiatry 185, 291–7.
23. American Psychiatric Association, 1994. Diagnostic and Statistical Manual of Mental Disorders (Fourth Edition). Washington, D.C.
24. Kohn, D., Niedersteberg, A., Wieneke, A., Bechdolf, A., Pukrop, R., Ruhrmann, S., Schultze-Lutter, F., Maier, W., Klosterkötter, J., 2004. Early course of illness in first episode schizophrenia with long duration of untreated illness - a comparative study. Fortschr. Neurol. Psychiatr. 72(2): 88–92.
25. Preda, A., Miller, T.J., Rosen, J.L., Somjee, L., McGlashan,T.H., Woods, S.W., 2002. Treatment histories of patients with a syndrome putatively prodromal to schizophrenia. Psychiatr. Serv. 53(3), 342–4.
26. Brockhaus-Dumke, A., Tendolka, I., Pukrop, R., Schultze-Lutter, F., Klosterkötter, J., Ruhrmann, S., 2005. Impaired mismatch negativity generation in prodromal subjects and patients with schizophrenia. Schizophr. Res. 73(2–3), 297–310.
27. Baker, F., Intagliata, J., 1982. Quality of life in the evaluation of community support systems. Evaluation and Program Planning 5, 69–79.
28. Lehman A.F., Ward N.C., Linn L.S., 1982. Chronic metal patients: the quality of life issue. Am. J. Psychiatry, 139(10), 1271–76.
29. Ritsner, M. (2006) The distress/protection vulnerability model of quality of life impairment in mental disorders. Current evidence and future research directions (chapter one, this book).
30. Pukrop, R., Schlaak, V., Möller-Leimkühler, A.M., Albus, M., Czernik, A., Klosterkötter, J., Möller, H.J., 2003. Reliability and validity of quality of life assessed by the short-form 36 and the Modular System for Quality of Life in patients with schizophrenia and patients with depression. Psychiatr. Res. 119, 63–79.
31. Ritsner, M., Modaj, I., Endicott, J., Rivkin, O., Nechamkin, Y., Barak, P., Goldin, V., Ponizovsky, A., 2000. Differences in quality of life domains and psychopathologic and psychosocial factors in psychiatric patients. J. Clin. Psychiatry. 6(11), 880–89.
32. Atkinson, M., Zibin, S., Chuang, H., 1997. Characterizing quality of life among patients with chronic mental illness: a critical examination of the self-report methodology. Am J Psychiatry. 154, 99–105.
33. Browne, S., Roe, M., Lane, A., Gervin, M., Morris, M., Kinsella, A., Larkin, C., O'Callagahn, E., 1996. Quality of life in schizophrenia: relationship to sociodemographic factors, symptomatology and tardive dyskinesia. Acta. Psychiatr. Scand. 94, 118–124.
34. Kaiser, W., Priebe, S., Hoffmann, K., Isermann, M., 2000. Profiles of subjective quality of life in schizophrenic in- and out-patient samples. Psychiatry Res. 66, 153–166.
35. Andrews, F.M., Whitey, S.B., 1978. Social indicators of well-being. Americans' perceptions of life quality. Plenum, New York.
36. Campbell A., Converse, P.E., Rodgers, W.L., 1976. The quality of American life. Russel Sage Foundation, New York.

37. Bechdolf, A., Klosterkötter, J., Hambrecht, M., Knost, B., Kuntermann, C., Schiller, C., Pukrop, R., 2003. Determinants of subjective quality of life in post acute patients with schizophrenia. Eur. Arch. Psychiatry Clin. Neurosci. 253, 228–235.
38. Carpiniello B, Lai G, Pariante C M, et al., 1997. Symptoms, standards of living and subjective quality of life: a comparative study of schizophrenic and depressed out-patients. Acta Psychiatr. Scand. 96, 235–241.
39. Reine, G., Lancon, C., Di Tucci, S., Sapin, C., Auquier, P., 2003. Depression and subjective quality of life in chronic phase schizophrenic patients. Acta Psychiatr. Scand. 108, 297–303.
40. Hafner, H., Maurer, K., Ruhrmann, S., Bechdolf A., Klosterkotter, J., Wagner, M., Maier, W., Bottlender, R., Möller, H.J., Gaebel, W., Wölwer, W., 2004. Are early detection and secondary prevention feasible? Facts and visions. Eur. Arch. Psychiatry Neurosci. 254, 117–128.
41. Ruhrmann, S., Schultze-Lutter, F., Klosterkötter, J., 2003. Early detection and intervention in the initial prodromal phase of schizophrenia. Pharmacopsychiatry 36 (suppl 3), 162–167.
42. Yung, A.R., Phillips, L.J., McGorry, P.D. et al., 1998. Prediction of psychosis. Br. J. Psychiatry 172 (suppl. 33), 14–20.
43. Bechdolf, A., Knost, B., Kuntermann, C., Schiller, S., Hambrecht, M., Klosterkötter, J., Pukrop, R., 2004. A randomized comparison of group cognitive-behavioural therapy and group psychoeducation in patients with schizophrenia. Acta Psychiatr. Scand. 110, 21–28.
44. Pukrop, R., Möller, H.J., Steinmeyer E. M., 2000. Quality of life in psychiatry. A systematic contribution to construct validation and the development of the integrative assessment tool"Modular system of quality of life." Eur. Arch. Psychiatry Clin. Neurosci. 250, 120–132.
45. Maurer, K., Könnecke, R., Schultze-Lutter, F. Häfner, H., 2000. Early Recognition Inventory. Unpublished Manual and Instrument. Mannheim, Germany, 2000.
46. Maurer, K., Horrmann, F., Schmidt, M., Trendler, G., Hafner, H. 2004. The early recognition inventory: structure, reliability and initial results. Schizophr. Res. 67 (suppl), 34.
47. Kay S.R., 1987. The Positive and Negative Symptom Scale (PANSS) of schizophrenia. Schizophr. Bull. 13, 261–276.
48. Wolthaus, J.E.D., Dingemans, P.M.A.J., Schene, A.H., Linzen, D.H., Knegtering, H., Holthausen, E.A.E., Cahn, W., Hijman, R., 2000. Component structure of the PANSS in patients with recent-onset schizophrenia and spectrum disorders. Psychopharmacology 150, 399–403.
49. Montgomery S, Asberg M., 1979. A new depression scale designed to be sensitive to change. Br. J. Psychiatry 134, 382–389.
50. SPSS Statistical package for the social sciences, 2001. SPSS Release 11.0, SPSS, Chicago.

CHAPTER 10

QUALITY OF LIFE IMPAIRMENT SYNDROME IN SCHIZOPHRENIA

MICHAEL S. RITSNER AND ANATOLY GIBEL

Department of Psychiatry, The Rappaport Faculty of Medicine, Technion - Israel Institute of Technology and Acute Department, Sha'ar Menashe Mental Health Center, Israel

Abstract: Patients with schizophrenia exhibit an exceedingly wide range of symptoms, and a broad spectrum of cognitive impairments. In addition, it has become increasingly apparent that the disorder is, to variable degrees, accompanied by quality of life impairments. This chapter addresses the question of whether the health-related quality of life (HRQL) impairment or deficit is a syndrome in schizophrenia. Therefore, first, we discuss what the general and domain-specific HRQL impairments are. Then, we address distressing and protective factors, and a factor structure of HRQL impairment. The literature, as well as new and previously published findings from the Shaar Menashe Longitudinal Study of Quality of Life will be presented in detail.

We argue that HRQL deficit is highly prevalent and fairly marked in schizophrenia patients: 49% of the patients are clinically severely impaired regarding general life quality, 42% - in general activities, 39% - in subjective feelings, 30% - in both leisure time activities and social relationships. The HRQL impairment has been observed before individuals exhibit the signs and psychotic symptoms of schizophrenia; it is relatively stable throughout the course of the illness. HRQL impairment syndrome appears to be relatively independent of symptomatology and neurocognitive deficit. Finally, the authors suggest that impairment in general and the domain-specific quality of life in particular is sufficiently reliable, stable, and specific enough syndromes to warrant inclusion in the diagnostic criteria for schizophrenia. Limitations in the current knowledge in this area are identified, and suggestions for future research are provided

Keywords: Schizophrenia, Quality of life, Impairment, Factors, Predictors, Model

INTRODUCTION

Historically, interest in measuring the quality of life (QOL) and health-related quality of life (HRQL) was stimulated by the plight of deinstitutionalized individuals with mental illness and by a parallel interest in assessing dimensions of daily life such as personal safety, isolation, poverty, and transience of accommodation. This was evident after the deinstitutionalization process that took place in the 1960s and 1970s in western countries. Although existing HRQL studies are difficult to

M.S. Ritsner and A.G. Awad (eds.), Quality of Life Impairment in Schizophrenia, Mood and Anxiety Disorders, 173–226.

compare, in the last decades the assessment of HRQL has gained importance as a measure of treatment, rehabilitation, and social intervention in schizophrenia. Empirical findings from these studies have been summarized in several comprehensive reviews[1–6]. The prevalent conception underlying previous studies viewed poor HRQL as an 'outcome' of symptomatology or treatment interventions that disagrees with the new findings (see Chapter 1).

Well-known from an objective standpoint, the quality of life of schizophrenia patients is low in most life areas: they are generally unemployed, unmarried, have limited social and family relations, and rely on disability funds for financial support. For assessment of HRQL use, both self-report and observer-rated (proxy) instruments provide distinct types of information. However, they also substantially differ in terms of the underlying quality of life concepts and the data collection procedures. In particular, an observer-rated instrument, such as the Quality of Life Scale (QLS)[7], includes items for rating the schizophrenic-deficit syndrome. On the other hand, the issue of self-report instruments in the psychiatric population may be influenced by affective bias, poor insight, and recent life events[1,8]. However, Voruganti et al.[9] found that the patients' self-reports were highly consistent and their HRQL scores correlated significantly with clinician's estimates. Based on comparisons of self-reports and observer-rated measures, it was concluded that self-report instruments are valid and should form part of the overall assessment of the quality of life among patients with psychotic disorders[10]. There is also a general consensus regarding the importance of using both self-report and observer-rated measures of HRQL.

This chapter focuses mainly on discussing a number of specific issues regarding the general and domain-specific HRQL impairments among schizophrenia patients by a comparison with healthy subjects through the course of illness using cross-sectional and longitudinal observations. Our aim was also to clarify the role of various distressing and protective factors that probably influence HRQL impairment and are linked to the stress process, illness manifestations, and treatment interventions using relevant literature, and the database of the Shaar Menashe Longitudinal Study of Quality of Life (Shaar Menashe study). A detailed description of the design, data collection, and measures used in the Shaar Menashe study are presented in an Appendix; demographic and clinical characteristics of patients' samples are presented in Table 1. This chapter presents both previously reported[11–28] and new findings from this project.

FACTS ABOUT HRQL IMPAIRMENT SYNDROME

Schizophrenia compared to healthy subjects. Differences in the quality of life levels between schizophrenia patients and healthy subjects have emerged as the most powerful findings regarding schizophrenia patients, according to many studies made in the last two decades[11–32]. Moreover, today there is compelling evidence that schizophrenia patients, compared to healthy subjects, demonstrate significant impairment or deficits regarding satisfaction with their quality of life. As depicted in Figure 1, schizophrenia patients from the database of the Shaar Menashe study

Table 1. Characteristics of schizophrenia patients: Shaar Menashe study

Variables	Initial sample (n = 237)		Follow up sample (n = 148)		Community based sample (n = 133)		Total sample (initial and community based samples, n = 370)	
	Mean	SD	Mean	SD	Mean	SD	Mean	SD
Age, yr.	37.9	9.8	38.2	9.5	39.6	9.3	38.5	9.7
Education, yr.	10.2	2.8	10.2	2.7	10.6	2.6	10.4	2.8
Age of onset, yr.	23.4	7.8	22.9	7.1	23.5	7.5	23.5	7.7
Number of admissions	7.6	4.5	7.8	4.5	6.6	6.3	7.3	5.3
Illness duration, yr.	14.3	9.4	15.0	9.1	23.5	7.5	17.6	9.8
General quality of life ($Q\text{-}LES\text{-}Q_{index}$)	3.4	0.8	3.5	0.8	3.4	0.9	3.4	0.7
Physical Health	43.6	12.3	45.3	10.8	45.8	11.1	44.4	11.9
Subjective Feeling	48.8	13.1	52.0	13.2	50.1	11.9	49.2	12.7
Leisure Time Activities	19.6	6.4	21.1	6.7	19.5	5.7	19.5	6.2
Social Relationships	36.8	9.7	37.6	10.7	34.8	8.8	36.1	9.4
General Activity	46.6	12.0	47.7	11.8	47.0	10.7	46.8	11.5
Life Satisfaction	3.3	1.2	3.5	1.1	3.4	0.9	3.4	1.1
Satisfaction with medicine	3.5	1.2	3.7	1.1	3.6	0.9	3.5	1.1
Illness severity (CGI scale)	4.5	0.9	4.1	1.2	3.0	0.8	4.0	1.1
Symptom severity (PANSS)	84.4	19.5	82.3	20.8	57.5	18.5	74.7	23.1
Emotional distress (TBDI)	1.3	0.8	1.1	0.8	0.8	0.6	1.1	0.8
Somatization (BSI)	0.8	0.8	0.7	0.8	0.3	0.5	0.6	0.7
Side effects (DSAS)	0.4	0.3	0.2	0.2	0.2	0.2	0.3	0.3
General functioning (GAF)	47.4	11.5	53.6	12.9	63.4	11.3	53.2	14.0

are significantly impaired across general and all domain-specific life qualities compared with healthy subjects measured with the Q-LES-Q (MANOVA, $F=17.2$, $df=14, 1056$, $p < 0.001$).

Schizophrenia compared to other mental disorders. Since different subtypes of mental disorders may be accompanied by various clinical and psychosocial factors, it is of interest to establish whether they could also be distinguished on the basis of quality of life measures. Unfortunately, the comparative literature on various mental disorders is limited.

Some cross-sectional comparisons failed to find considerable differences in the general quality of life levels between *schizophrenia and schizoaffective disorders*[11,33,34]. However, schizophrenia inpatients were significantly less satisfied than schizoaffective patients in two specific domains: subjective feelings ($F=3.1$, $p < 0.05$), and social relationships ($F=5.8$, $p < 0.05$)[11]. Differences between schizophrenia and schizoaffective patients regarding satisfaction with social relationships remained significant after controlling for the confounding effects of symptom severity, emotional distress, side effects, suicide risk, coping styles, self-variables and illness duration[3].

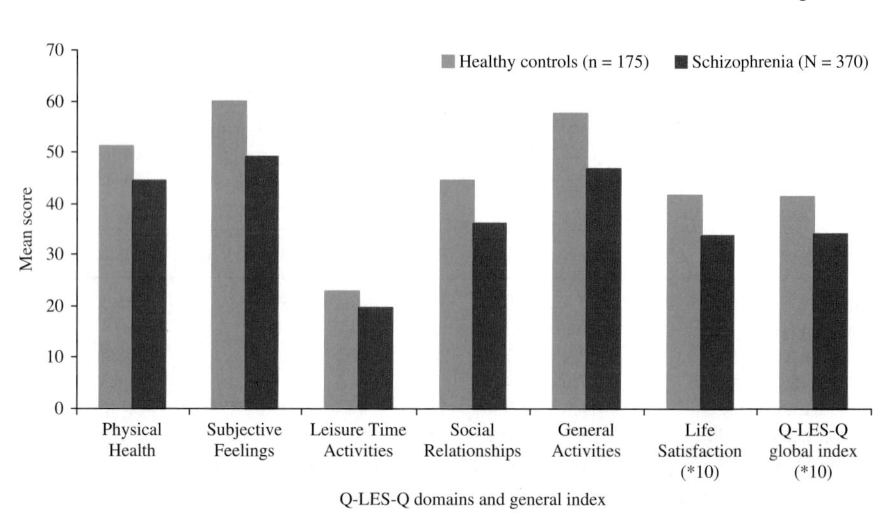

Figure 1. Quality of life profile of 370 patients with schizophrenia and 175 healthy subjects

Depressed patients reported greater dissatisfaction with the quality of life than schizophrenia patients. Rudolf and Priebe[35] demonstrated that depressed women expressed dissatisfaction with 4 out of 8 life domains and with life as a whole (LQOLP), and had a lower quality of life than women with schizophrenia. The differences remain statistically significant when controlling for the influence of age and anxiety/depression, and therefore, changes in depressive symptoms do not fully explain HRQL fluctuations. At the same time, mood and schizoaffective disorder patients had quite similar ratings on all life quality domains measured with Q-LES-Q[3]. When schizophrenia (n = 403), depression (n = 349), and anxiety (n = 139) were concurrently investigated, schizophrenia patients again recorded better life satisfaction than patients with major depression and anxiety disorders[36].

Patients with *anxiety disorders* reported the lowest levels of HRQL among the psychiatric disorders. For example, respondents with pure generalized anxiety disorder showed significantly lower quality of life scores (SF-36) compared with respondents with pure major depressive disorder[37]. Obsessive-compulsive disorder patients reported significantly lower HRQL scores than schizophrenia outpatients[38]. However, these findings were not replicated[39].

Patients with severe quality of life impairment syndrome. Clinically severe quality of life deficit was operationally defined when Q-LES-Q scores were two or more standard deviations below the healthy subjects[3,40]. The proportion of patients with severe impairment in general HRQL varied with different diagnoses: major depressive disorder (63%), double depression (85%), PTSD (59%), dysthymic disorder (56%), premenstrual dysphoric disorder (31%), obsessive-compulsive disorder (26%), social phobia (21%), and panic disorder (20%)[40]. Regression analyses conducted for each disorder suggested that illness-specific symptom scales were significantly associated with the baseline quality of life but explained only a small to modest proportion of the variance in Q-LES-Q scores.

We found clinically severe impairment in the quality of life among 27-49% of schizophrenia inpatients (initial sample, n = 237). Specifically, 26.6% had a severe impairment regarding satisfaction with physical health, 30% with leisure-time activities and social relationships, 38.8% with subjective feelings, 42.2% with general activities, and 49.4% with general life quality ($Q\text{-}LES\text{-}Q_{index}$). When 148 of 237 patients were re-assessed about 16 months later, severe impairment in Q-LES-Q domains was observed in 18.2%, 19%, 28.4%, 34.5%, 41.2%, and 39.2% of the patients, respectively. Furthermore, 38.4% of the schizophrenia inpatients showed severe impairment regarding 1-3 domains, 53.6% of the inpatients felt severe deficits regarding 4-5 domains, and only 8% of the inpatients had no severe impairment in life quality (30.4%, 59.4%, and 10.2% of the patients, respectively; follow-up sample, n = 148). Thus, clinically severe impairment in life quality is a prevalent disturbance in schizophrenia.

Quality of life impairment syndrome through the course of illness. Several studies provide information about the occurrence of HRQL deficit in the early phases of schizophrenia, before the onset of psychotic symptomatology, and regarding HRQL deficit remaining throughout the onset of schizophrenia psychotic symptomatology. Stone et al.[41] reported that subjects who met the criteria for *vulnerability to schizophrenia* ("schizotaxia") received significantly lower QLS total ratings, including the interpersonal relations subscale. They also showed significantly higher emotional distress scores (Hopkins Symptoms Checklist), and demonstrated particular elevations on the obsessive-compulsive, anxiety, and hostility subscales. Several studies have suggested that there may be an association of poor premorbid adjustment with poor quality of life in schizophrenia[42,43]. In particular, Czernikiewicz et al.[44] reported that poor premorbid adjustment significantly correlated with poorer life quality (QLS) among 120 schizophrenia patients.

Erickson et al.[45] reported that *before the first episode of illness*, schizophrenia subjects had fewer and less satisfactory social relationships than a matched healthy comparison group, and subjects with affective psychosis. Recently, Bechdolf and associates[46] demonstrated that in subjects in a putative *prodromal state* (at risk for a first episode of psychosis) subjective HRQL is substantially reduced when compared with healthy controls. These subjects felt primarily compromised with regard to their mood and their vitality; they felt stressed and pressured, easily hurt, and often tired; moreover, they reported a lack of self-confidence, and dissatisfaction with their social relationships and expressed discontent with life in general. Despite the limitations, these important findings suggest that the subjective quality of life is already compromised *prior to the onset of schizophrenia symptoms.* Furthermore, Browne et al.[42] found a diminished quality of life in the sample of first-episode psychosis patients presenting to a catchment area psychiatric service.

Does quality of life impairment syndrome significantly change over time, and if so, over what period? If quality of life impairment is an outcome of progression of the schizophrenia process, long-term patients should be reporting lower satisfaction with their life quality than first-admitted patients. However, this assumption disagrees with empirical findings. For example, Priebe et al.[47] reported that HRQL appears to be lower in 86 first-admitted patients than in groups with long-term illness. Whereas some researchers did not find a statistically significant improvement in the subjective HRQL of schizophrenia patients during a 9-month follow-up[47] or a 7-10-year period[48,49], others reported positive changes in the domain-specific quality of life after a one year follow-up[50,51]. For example, Malla et al.[52] found a significant improvement in most dimensions of the Wisconsin Quality of Life Index after a one year follow-up of 41 patients, with a first episode of psychosis, who were generally independent of changes in symptoms. Another study underscores the relatively stable quality of life deficit, as documented in a one-year follow-up after hospitalization[31]. Ruggeri and associates[53], in a three-year study, did not find changes in the quality of life of 107 schizophrenia patients attending the South Verona community-based mental health center; however, the mean symptom severity and some types of needs for care actually worsened.

In the framework of the Shaar Menashe study, 148 schizophrenia patients were followed about 16 months after hospital admission[15,17]. The general quality of life appears to be slightly, but statistically significantly increased during the follow-up period (from 3.4 ± 0.7 to 3.5 ± 0.7, paired $t = 2.3$, $p < 0.05$). Likewise, paired *t*-test revealed that schizophrenia patients significantly improved regarding satisfaction with subjective feelings (from 48.5 ± 12.8 to 52.0 ± 13.2; $t = 2.9$, $p < 0.01$) and leisure-time activities (from 19.3 ± 6.4 to 21.2 ± 6.6; $t = 3.0$, $p < 0.01$). Changes in physical health, social relationships, general activities, and satisfaction with medicine did not reach a statistically significant level. Changes in most Q-LES-Q domains did not relate to changes in the treatment setting in the follow-up period ($F = 0.60$–1.93, $df = 2,148$, $p > 0.05$), except for social relationships, which improved among discharged patients and worsened among patients who remained hospitalized ($F = 4.01$, $df = 2, 148$, $p = 0.020$; Bonferroni multiple comparison test, $p < 0.05$).

Thus, taken together, the presented findings underscore the relatively stable quality of life impairment syndrome that began before the manifestation of psychotic symptomatology with slight fluctuations in general, and domain-specific HRQL scores through the course of schizophrenia.

OVERVIEW OF STRESS PROCESS-RELATED FACTORS

Since the vast majority of the literature focuses mainly on clinical correlates, the assessment of *stress process-related or psychosocial factors* such as emotional distress, somatization (or somatic distress), expressed emotion, personality traits, self-constructs, coping styles, and perceived social support influencing HRQL impairment was relatively neglected. These issues in the quality of life field have received more attention in recent years after the findings of Swedish researchers[54–58], and those from the Shaar Menashe study[11–28] were published.

Cognitive-relational theory defines stress as a particular relationship between the person and the environment, which is appraised by the person as taxing or exceeding his or her resources, and which endangers his or her well-being[59]. The stress process must be analyzed as an active, unfolding process that includes stressors (daily hassles, life events), the stress response (emotional and somatic distress, anxiety), and determinants or mediators of the stress response (self-constructs, coping, social support, etc.). Table 2 presents correlation coefficients between stress-process-related variables among schizophrenia patients. As shown, emotional and somatic distress are highly positively correlated ($r = 0.57$, $p < 0.001$), and together with the expressed emotion, they were negatively associated with self-esteem, self-efficacy, task- and avoidance-coping styles, and social support. Factor analysis for these variables revealed two factors: the first factor with a protective effect accounted for 52% of the total variance of the variables used (eigenvalue=2.72) and was composed from self-efficacy, task- and avoidance-coping styles, and social support scores, and the second factor, with a distressing effect, accounted for 47.8% of the total variance of variables used (eigenvalue=2.51); it includes emotional distress, somatization, expressed emotion, emotion-related coping style, and low self-esteem scores.

Emotional distress. This is an unspecific reaction of an individual to external and internal stressors and is characterized by a mixture of sub-threshold distress symptoms, such as obssesiveness, depression, hostility, sensitivity, anxiety, hopelessness, helplessness, dread, confused thinking, sadness, and paranoid ideation[60,61]. Although self-reported distress highly correlates with subjective measures of depression, conceptually and empirically, it is distinct from depressive symptoms. Apart from depressive components, emotional distress incorporates other nonspecific psychological manifestations, has stronger relations with common psychosocial factors, and tends to be milder and more transient than depression[62]. A theoretical model for the underlying distress mechanisms suggests that two different realities coexist within a mentally ill individual: one clinical (external) and another psychological (internal)[63]. The clinical reality is presented with psychopathological symptoms that can be ascertained by a clinician directly observing behavior and interviewing using observer-rated scales, whereas the psychological reality expressing the subjective experience of disease or suffering may be accessible only through the patient's self-report.

Accumulating evidence suggests that schizophrenia patients are sensitive to life events[64], daily hassles[65], and report more emotional, and somatic distress (somatization) compared with healthy controls[66]. Emotional distress experienced by schizophrenia patients is positively associated with symptom expression[63,67], side effects of antipsychotic agents[12,68], with temperament types, emotion-oriented coping, and weak self-constructs[22], with the burden associated with daily-living and the satisfaction of services[69].

Emotional distress is a primary importance factor of poor quality of life in schizophrenia[11,63,70,71]. From the correlation coefficients shown in Table 3, it can be seen that general emotional distress ($TBDI_{index}$) is negatively associated with the Q-LES-Q scores (r ranged from -0.53 to -0.28, $p < 0.001$). The negative relationship remains significant when the effect of both severity of symptoms

Table 2. Pearson correlations between stress process-related factors (follow-up sample, n=148)

Variables	Emotional distress	Somatization	Self-esteem	Self-efficacy	Task coping	Emotion coping	Avoidance coping	Expressed emotion
Emotional distress								
Somatization	0.57***							
Self-esteem	−0.69***	−0.37***						
Self-efficacy	−0.63***	−0.40***	0.62***					
Task coping	−0.47***	−0.23**	0.50***	0.65***				
Emotion coping	0.39***	0.25***	−0.32***	−0.12	0.19*			
Avoidance coping	−0.30***	−0.12	0.35***	0.52***	0.81***	0.29***		
Expressed emotion	0.36***	0.40***	−0.33***	−0.26***	−0.23**	0.25***	−0.18*	
Social support	−0.48***	−0.27***	0.45***	0.51***	0.60***	0.00	0.56***	−0.42***

*p < 0.05, **p < 0.01, ***p < 0.001

Table 3. Pearson correlations between quality of life and stress process-related factors in schizophrenia (Shaar Menashe study)

Variables	Q-LES-Q index	Physical health	Subjective feeling	Leisure time	Social relationships	General activities	Life satisfaction	Satisfaction with medicine
Emotional distress index[1]	−0.50***	−0.40***	−0.48***	−0.28***	−0.26***	−0.53***	−0.49***	−0.29***
(partial correlations)[1a]	(−0.39)***	(−0.25)***	(−0.39)***	(−0.23)***	(−0.20)**	(−0.42)***	(−0.42)***	(−0.20)**
Obssesiveness[1]	−0.46***	−0.37***	−0.48***	−0.30***	−0.27***	−0.42***	−0.37***	−0.20**
Hostility[1]	−0.20**	−0.12	−0.15*	−0.11	−0.16*	−0.26***	−0.24***	−0.07
Sensitivity[1]	−0.40***	−0.30***	−0.39***	−0.22**	−0.22**	−0.41***	−0.40***	−0.22**
Depression[1]	−0.52***	−0.42***	−0.51***	−0.28***	−0.24***	−0.55***	−0.52***	−0.37***
Anxiety[1]	−0.41***	−0.34***	−0.40***	−0.23***	−0.17*	−0.45***	−0.38***	−0.29***
Paranoid ideation[1]	−0.30***	−0.25***	−0.26***	−0.12	−0.17*	−0.34***	−0.29***	−0.09
Somatisation[1]	−0.38***	−0.36***	−0.35***	−0.20**	−0.40***	0.34***	−0.35***	−0.21**
(partial correlations)[1a]	(−0.23)***	(−0.19)***	(−0.22)**	(−0.10)	(−0.05)	(−0.29)***	(−0.27)***	(−0.09)
Expressed emotion[4]	−0.43***	−0.39***	−0.40***	−0.30***	−0.36***	−0.31***	−0.33***	−0.31***
Emotion oriented coping[3]	−0.21**	−0.17*	−0.21**	−0.10	−0.04	−0.25***	−0.23***	−0.08
Novelty seeking[2]	−0.17	−0.08	−0.16	−0.10	−0.07	−0.28**	−0.21*	−0.04
Harm avoidance[2]	−0.62***	−0.55***	−0.60***	−0.38***	−0.43***	−0.60***	−0.48***	−0.32**
Reward dependence[2]	0.08	0.10	0.01	0.10	0.23*	0.02	0.01	0.01
Self-esteem[1]	0.58***	0.43***	0.57***	0.41***	0.34***	0.56***	0.49***	0.28**
Self-efficacy[1]	0.53***	0.39***	0.51***	0.48***	0.40***	0.49***	0.36***	0.20**
Task oriented coping[3]	0.35***	0.25***	0.32***	0.29***	0.34***	0.33***	0.23***	0.14
Avoidance coping[3]	0.34***	0.23***	0.27***	0.32***	0.36***	0.33***	0.18**	0.15*
Social support[1]	0.45***	0.28***	0.35***	0.33***	0.41***	0.49***	0.37***	0.21**

*$p < 0.05$, **$p < 0.01$, ***$p < 0.001$

[1]Total sample = 370 patients, [1a]controlling for symptoms (PANSS), and size effects (DSAS);

[2]Temperament, TPQ (n = 90); [3]CISS (n=237); [4]LEE (follow-up sample, n = 148)

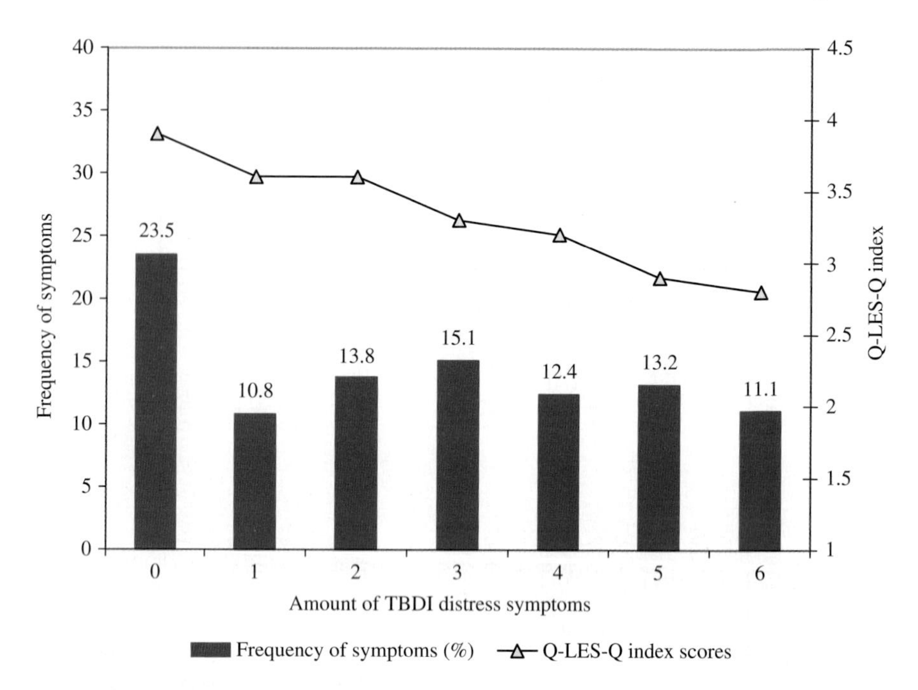

Figure 2. General quality of life levels across distress symptoms among 370 schizophrenia patients

(PANSS) and side effects (DSAS) was removed from the correlation matrix. Since the correlation between the severity of emotional distress and the depression factor (PANSS items: G1, G2, G3, G6) was $r=0.48(n = 370, p < 0.001)$, we tested the concurrent effects of these variables on Q-LES-Q_{index} by partial correlation analysis. The findings indicate that the correlation between Q-LES-Q_{index} and the depression factor decreased from $r=-0.32(p < 0.01)$ to $r = -0.10(p > 0.05)$, after adjusting for $TBDI_{index}$ scores, whereas the correlation between Q-LES-Q_{index} and $TBDI_{index}$ before and after adjusting for the depression factor ratings remained highly significant ($r=-0.50$, and $r=-0.42$, respectively, both $p < 0.001$). Furthermore, the distribution of 370 schizophrenia patients according to the number of distress symptoms and corresponding values of Q-LES-Q_{index} is shown in Figure 2. A total of 23.5% respondents reported no symptoms at all, whereas 11.1% had all six-distress symptoms during the last month. As the number of symptoms grew, the Q-LES-Q_{index} scores gradually decreased from 3.9 (SD=0.6) for those with any distress symptoms to 2.8 (SD=0.7) for those having six symptoms (ANOVA, $F=17.2$, $df=6, 370$, $p < 0.001$).

Thus, emotional distress uniquely contributed to quality of life impairment in schizophrenia. The relationship between emotional distress and poor HRQL appears, at least partly, to be independent of symptom severity (including depression) and the side effects of antipsychotic medications.

Somatization. At least 30% of the schizophrenia patients reported some somatization symptoms: faintness or dizziness (56%), nausea (48%), and heart/chest pain (38%) are the most common somatic presentations[66]. Higher somatization scores among schizophrenia patients are correlated with elevated emotional distress and with the side effects of antipsychotic agents.

General and domain-specific quality of life dimensions are negatively correlated with somatization scores: r ranged from -0.40 ($p < 0.001$) to -0.20 ($p < 0.01$) (Table 3). After controlling for symptoms and side effects (PANSS and DSAS, respectively), the correlation coefficients between Q-LES-Q and the somatization scores lose their significance with three domains (satisfaction with leisure time, social relationships, and with medicine). Based on these findings, we suggest that somatization perhaps mediates the impact of symptoms and side effects on leisure time, social relationships, and satisfaction with medicine. The relationship of somatization with other domains (satisfaction with general activity, life satisfaction, physical health, subjective feelings) did not relate to the severity of the symptoms and side effects.

Expressed emotion. Highly expressed emotion (EE) is a measure of hostile, critical, and emotionally over involved attitudes expressed by a family member about a mentally ill relative. In some studies a relationship was shown between high EE and relapse in schizophrenia[72,73]. Hooley and Campbell[73] have shown that high-EE relatives of schizophrenia patients attribute more personal control to the patient and are more behaviorally controlled than low-EE relatives. High EE is considered a marker of a dysfunctional family interaction in patients with schizophrenia.

Mubarak and Barber[74] studied the association between the emotional expressiveness of the primary caregivers and the HRQL of 174 schizophrenia patients. Emotional involvement of key caregivers was found to have a significant association with the HRQL deficit of patients.

Likewise, poor HRQL of schizophrenia patients from the Shaar Menashe study is correlated with expressed emotion scores (r ranged from -0.43, to -0.30, $p < 0.001$; Table 3). The negative contribution of expressed emotion to the general Q-LES-Q_{index} of schizophrenia patients was 17.1% ($R^2 = 0.17$, $F = 28.9$, $df = 1, 148$; $p < 0.001$; in a follow-up sample, $n = 148$). This value became lower until reaching 4.5% ($\beta = -0.12, t = 2.5, p = 0.013$, partial $R^2 = 4.5\%$) after controlling the confounding effect of the following factors: emotional distress ($\beta = -0.49, t = 7.3, p < 0.001$, partial $R^2 = 28.4\%$), self-esteem ($\beta = 0.28, t = 4.3, p < 0.001$, partial $R^2 = 11.9\%$), self-efficacy ($\beta = 0.20, t = 3.2, p = 0.001$, partial $R^2 = 7.2\%$), and emotion-related coping ($\beta = 0.17, t = 3.5, p < 0.001$, partial $R^2 = 8.2\%$), which accounted for 74% of the general Q-LES-Q_{index} scores (model's properties: $R^2 = 0.74, F = 78.8, df = 5, 148, p < 0.0001$). Age, education, age of onset, illness duration, and symptom severity did not reach a significant level. Further research is needed to understand the role of EE in the context of its relationship with quality of life impairment.

Personality traits. Personality is a broad concept involving genetically determined traits (i.e. temperament) and environment, the organized nature of personality, as well as its direct impact on adjustment and coping and their subsequent effects on various degrees of psychopathology. A considerable body of recent research shows that stable personality differences exist among schizophrenia patients, and that they most likely predate the onset of illness[33,74].

Several studies sought to find a relationship between HRQL impairment and personality traits of schizophrenia patients. Kentros et al.[33] noted that global life quality (LQOLP) positively correlated with extroversion and agreeability, and negatively correlated with the domain of neuroticism measured with the NEO Personality Inventory. This study has some methodological limitations including its small sample size ($n = 21$) and a lack of a control for additional HRQL-related factors. Hansson and associates[55] showed that lower levels of harm avoidance and higher levels of self-directedness significantly correlated with better subjective HRQL (LQOLP). Self-directedness accounted for 12.3% of the variance in overall subjective HRQL scores compared with 26.8% for psychopathology. Importantly, harm avoidance and self-directedness measured with the Tridimensional Personality Questionnaire (TPQ) was not associated with subtypes of schizophrenia or the demographic characteristics of the respondents[57]. More recently, Eklund et al.[56] examined 117 outpatients with schizophrenia, schizophreniform, and schizoaffective disorders and found that trait-like properties, such as the self-factor and personality, explained most of the variation in the LQOLP scores.

The findings from the Shaar Menashe study also indicate that lower levels of harm avoidance (TPQ) significantly correlated with better life quality across all Q-LES-Q domains (r ranged from -0.62, to -0.32, $p < 0.001$) of 90 stable schizophrenia patients[18]. Novelty seeking is negatively associated with general activities ($r = -0.28$, $p < 0.01$) and life satisfaction ($r = 0.21$, $p < 0.05$), whereas reward dependence is positively correlated with social relationships ($r = 0.23$, $p < 0.05$; Table 3). After controlling for emotional distress, social support, self-esteem, avoidance coping, side effects, age, and depression scores, the best-fitted regression models for the Q-LES-Q domains (except leisure-time activities) included temperament factors. In particular, novelty seeking ratings markedly contributed (16.4%) to the prediction of Q-LES-Q_{index}, subjective feelings (12.7%), and physical health (6.2%). Regression analysis revealed a positive contribution of reward dependence (15.6%) to patients' satisfaction from their social relationships. Higher levels of harm avoidance, associated with less satisfaction with general activities, accounted for 10.4% of the variance, and medication accounted for 6.2% of the variance. Thus, harm avoidance, novelty seeking, and reward dependence are associated with different domains of HRQL impairment of schizophrenia patients. These TPQ temperament factors explain from 6% to 16% of the variability in the Q-LES-Q domain scores.

In addition, Kurs and associates[23] used a triplet design, comparing 47 outpatients with 47 of their non-affected siblings and 56 healthy control subjects. When data were adjusted for gender, age at examination, and education, ANOVA indicated that TPQ temperament factors significantly contributed to impairment in physical health, social relationships, and subjective feelings in schizophrenia patients; harm avoidance was associated with a general Q-LES-Q_{index} independent of psychopathology. We suggest that harm avoidance may act as a moderator or mediator between the severity of psychopathology and specific domains of Q-LES-Q. The association of reward dependence with social relationships is not linked to illness because of the lack of correlation between symptom severity and reward dependence. This may be indicative of the deficit in interactive social relationships typical of schizophrenia patients, resulting from their lack of motivation to seek reward. Interestingly, just as Q-LES-Q scores were similar for non-affected siblings, and healthy controls, there were no differences between the two groups in terms of TPQ temperament factors. Thus, temperament factors apparently do not affect the life quality of non-affected siblings.

Thus, temperament traits, which are not necessarily part of the deterioration process of the illness, are associated with HRQL impairment in schizophrenia. Based on a significant association between HRQL impairment and harm avoidance, which might serve as a marker for underlying genetic vulnerability to schizophrenia[76], we hypothesize that HRQL impairment is a syndrome caused by vulnerability to schizophrenia. Longitudinal observations that may potentially provide an answer to this hypothesis are lacking.

Self-esteem. Conscious self-concept or self-esteem refers to some subjective entity closely related to the personal sense of identity, as the factor of importance in evaluating the patient's quality of life. It was found that self-esteem and life satisfaction are clearly discriminable constructs[77]. A few studies found lower levels of self-esteem among patients with psychological distress, depression, and psychotic states compared with a control group[54,78,79].

Self-esteem positively correlated with self-efficacy ($r=0.62$, $p < 0.001$), task- and avoidance-related coping styles ($r=0.50$ and 0.35, $p < 0.001$, respectively), and with social support ($r=0.45$, $p < 0.001$) among schizophrenia patients (Table 2). As a positive factor, self-esteem has been incorporated into a subjective HRQL model of severe mental disorders[11,54,80]. Cross-sectional analysis showed a positive association of self-esteem with Q-LES-Q domains (r ranged from 0.58, to 0.34, $p < 0.001$; for satisfaction with medicine $r=0.28$, $p < 0.01$; Table 3); self-esteem contributes 6-7% of the variance in the appraisal of subjective HRQL among schizophrenia patients[54,81]. Longitudinal data indicate a positive correlation between elevation in self-esteem scores and improvement in Q-LES-Q domains over time, which account for 7–9% of the variance in the change domain-specific quality of life[17].

Self-efficacy. General self-efficacy is the belief in one's competence to cope with a broad range of stressful or challenging demands[82]. General self-efficacy may explain a broader range of human behaviors and coping outcomes when the context is less specific. It might be useful when studying the well-being or behavior of patients who have to adjust their lives to multiple demands owing to illness.

Schizophrenia patients showed a high positive correlation of self-efficacy with task- and avoidance-related coping styles ($r=0.65$ and 0.52, $p < 0.001$, respectively), and with perceived social support ($r=0.51$, $p < 0.001$), whereas a negative association with emotional distress ($r=-0.63$, $p < 0.001$), and somatization ($r=-0.40$, $p < 0.001$; Table 2). Changes in the experience of self-efficacy appear to be associated with variability and improvement in the coping strategies scores over time[83].

There is evidence of a close relationship between general self-efficacy and subjective life quality[11,15,54]. Indeed, self-efficacy and Q-LES-Q scores positively correlated across all domains: r ranged from 0.53 to 0.36 ($p < 0.001$), for satisfaction with medicine $r=0.20$($p < 0.01$; Table 3). Self-efficacy accounts for 16%, 21%, and 22% of the variance in physical health, leisure-time activities and general activity scores, respectively[11]. However, individual changes in specific quality of life domains over 16 months are influenced by changes in self-efficacy that accounted only for 4% of the variance in general Q-LES-Q_{index}, 4.5% in subjective feeling, 5.8% in general activity, and 8.1% in leisure-time activities[17].

Coping behavior. Coping styles or strategies are defined as the individual's ability to overcome and adapt to stresses caused by negative life events. According to the cognitive-transactional theory of stress[59], coping has been defined as one's cognitive and behavioral effort to manage the internal and external demands of a person-environment transaction that is considered taxing or exceeding one's resources. There are two approaches to coping behavior: one that emphasizes coping as a trait or personality characteristic, and another that emphasizes coping as a process, that is, efforts to manage stress that changes over time and in accordance with the situational contexts in which it occurs[84,85]. Various coping strategies may be categorized into three main styles: task-, emotion-, and avoidance-oriented[86]. Task-oriented coping is used to actively solve an underlying problem, cognitively reconceptualize it, and potentially minimize its adverse effects. Emotion-oriented coping strategies are person-oriented, and include emotional responses, e.g., self-preoccupation, self-blame, and fantasizing reactions. Avoidance-oriented coping involves both task and person orientations: one may avoid a stressful situation either by using social diversion, i.e., choosing to be with other people and seeking emotional support, or via self-distraction from a stressful situation, e.g., "giving up", denial, or engaging in a substitute task. Research has indicated that schizophrenia patients are inflexible in their use of coping strategies[87], tend to use maladaptive or emotion-oriented coping styles[88], and rely more on passive avoidant strategies and less on active problem solving[89].

As can be seen from the Shaar Menashe findings, quality of life impairment correlated positively with task- and avoidance-oriented coping styles and highly negatively with emotion-oriented coping (Table 3). Regression analysis revealed the various contributions of avoidance-oriented (12-18%), task-oriented (10-12%), and emotion-oriented (2-4%) coping styles in the structure of domain-specific quality of life in schizophrenia inpatients[11]. Different coping strategies may reduce the negative influence of specific symptoms and emotional distress on the subjective life quality in schizophrenia. Specifically, emotion-oriented coping mediated the relationship between the severity of activation, anxiety/depression symptoms, and Q-LES-Q scores, whereas avoidance-oriented coping mediated between the quality of life and paranoid symptoms[16]. Given these results, coping abilities appear to be an important factor in the structure of HRQL impairment in schizophrenia.

Perceived social support. In spite of different theoretical and methodological approaches, there is a general consensus among researchers that social support plays a positive role in promoting good mental health. Several studies have shown a significant positive relationship between perceived social support and HRQL impairment in schizophrenia[11,55,90,91]. Inspection of Table 3 shows that family support, support by friends, and other significant support were positively associated with both general and domain-specific scores. Staff support also correlated with general HRQL scores (r=0.32, p=0.002)[91]. Perceived social support plays an important role in improving the life quality of patients with schizophrenia: social support from significant others considered important accounted for 9.9% of the improvement of the general quality of life, and for 3.5% of the improvement in leisure-time activities[17,27].

ILLNESS-RELATED VARIABLES

Duration of untreated psychosis (DUP) is the time from the manifestation of the first psychotic symptom to the initiation of adequate treatment. In the period of untreated psychosis, 117 young patients with first-episode schizophrenia had poorer HRQL (SF-36, and WHOQOL-BREF) than their counterparts in the community[32]. It has been hypothesized that a longer DUP leads to a poorer HRQL outcome. Indeed, a number of studies confirmed this assumption[42,43,92], whereas others failed to find such evidence. For instance, after controlling for the effects of age at onset, the DUP initial psychosis did not significantly impair the subsequent quality of life[93]. However, findings from a naturalistic study of 318 first-episode patients followed up 8 years after the initial treatment suggest that, after controlling for the effects of other factors, a shorter DUP correlated moderately with enhanced HRQL[94]. Overall, the findings from longitudinal studies of first-episode schizophrenia suggest that poor HRQL is associated with long delays in treatment[95].

PANSS items. The step-wise multiple regression analysis, used to predict the general quality of life, revealed that only 10 of 30 PANSS items significantly contributed to the structure of the global quality of life of 370 schizophrenia patients (Shaar Menashe study, total sample). The best-fitted model explains 22% of the variance in Q-LES-Q_{index} scores ($F=11.2$, $df=10, 369$, $p < 0.001$). Specifically, active social avoidance accounts for 4.5% (G_{16}; $t=4.1$, $p < 0.001$), hostility for 3.8% (P_7; $t=3.8$, $p < 0.001$), depression −3.6% (G_6; $t=3.6$, $p < 0001$), grandiosity −2.3% (P_5; $t=2.9$, $p=0.003$), difficulty in abstract thinking – 2% (N_5; $t=2.7$, $p=0.006$), mannerisms and posturing – 1.7% (G_5; $t=2.5$, $p=0.013$), lack of judgment and insight (G_{12}; $t=2.3$, $p=0.020$), poor impulse control (G_{14}; $t=2.3$, $p=0.024$), somatic concern (G_1; $t=2.2$, $p=0.028$), and hallucinatory behavior (P_3; $t=2.1$, $p=0.038$) – between 1.1 and 1.5% of the variability in the Q-LES-Q_{index} scores. Note that any stress process-related variables with possible confounding effects on Q-LES-Q_{index} scores were not entered into this analysis.

Depressive symptoms. Persistence of depression occurs in 30-40% of patients during the longitudinal course of schizophrenia[96]. Expression of depressive symptoms has been well documented as a poor HRQL factor in cross-sectional

Table 4. Pearson correlations quality of life scores with severity of symptoms in schizophrenia (total sample, n = 370)

Symptom severity	Q-LES-Q index	Physical health	Subjective feeling	Leisure time	Social relationships	General activities	Life satisfaction	Satisfaction with medicine
PANSS, total score	−0.20**	−0.18**	−0.21**	−0.10	−0.07	−0.21**	−0.18*	−0.16*
Negative syndrome[1]	−0.15*	−0.15*	−0.16*	−0.08	−0.07	−0.15*	−0.09	−0.09
Positive syndrome[1]	−0.05	−0.06	−0.05	−0.01	0.00	−0.09	−0.14	−0.05
General psychopathology[1]	−0.25***	−0.22***	−0.27***	−0.14	−0.10	−0.26***	−0.22***	−0.21**
Negative factor[2]	−0.22***	−0.21**	−0.24***	−0.14	−0.11	−0.22***	−0.14	−0.14
Positive factor[2]	−0.08	−0.08	−0.07	−0.03	−0.01	−0.13	−0.14	−0.09
Activation factor[2]	−0.19*	−0.17*	−0.19**	−0.12	−0.10	−0.19**	−0.16*	−0.13
Dysphoric mood[2]	−0.31***	−0.26***	−0.34***	−0.19**	−0.14	−0.30***	−0.24***	−0.29***
Autistic preoccupations[2]	−0.14	−0.13	−0.16*	−0.06	−0.04	−0.17*	−0.13	−0.13
Anxiety/Depression[3]	−0.30***	−0.24***	−0.33***	−0.18**	−0.15*	−0.30***	−0.24***	−0.30***
Depressive symptoms[4]	−0.32***	−0.27***	−0.36***	−0.20**	−0.15*	−0.31***	−0.24***	−0.29***
Mood factor[5]	−0.31***	−0.26***	−0.34***	−0.19**	−0.14	−0.30***	−0.24***	−0.28***

[1]Kay et al. (1978); [2]White et al. (1997); [3]Lindenmayer et al. (1994); [4]Lancon et al. (1998); [5]Mohr et al. (2004). *) $p < 0.05$, **) $p < 0.01$, ***) $p < 0.001$

[11,97,98], and longitudinal studies[27,99,100,101]. However, Malla et al.[52] showed that a significant improvement in most dimensions of HRQL within one year were generally independent of changes in the depressive symptoms.

Findings from the Shaar Menashe study have indicated that the severity of the depressive symptoms is inversely correlated with the quality of life deficit (Table 4). Reduction in the severity of depressive symptoms is associated with improvements in general life quality[27], which does not appear to predict variability of the domain-specific quality of life over time[15,17]. Significant correlations between changes in the severity of depression symptoms and changes in the general Q-LES-Q_{index} lost their significance when emotional distress, coping styles, self-constructs, and side effect scores were controlled in an ANCOVA model. In addition, a correlation between a depressive subscale of emotional distress (DEP_{TBDI}) and Q-LES-Q_{index} $r = -0.50 (p < 0.001)$ remains significant $r = -0.43$ $(p < 0.001)$ after controlling for depressive symptoms (PANSS, G_6; DEP_{PANSS}), whereas a correlation between DEP_{PANSS} and Q-LES-Q_{index} $r = -0.33$ $(p < 0.001)$ decreased and lost significance $r = -0.12$ $(p > 0.05)$ after controlling for DEP_{TBDI} (total sample, n=370). The correlation coefficient between DEP_{TBDI} and DEP_{PANSS} was $r = 0.46$ $(p < 0.001)$. Thus, the contribution of depressive symptoms to the variability of HRQL impairment in previous studies was overestimated because it did not account for the impact of confounding factors.

Anxiety symptoms. Anxiety has been a well-documented negative factor influencing impairment in the quality of life[102–105]. Moreover, Huppert and associates [100,101] reported that anxiety but not depressive symptoms predicted satisfaction with daily activities, family contacts, and quality of health among 53 patients with either schizophrenia or schizoaffective disorder. They found that correlations between anxiety symptoms and QOLI remained significant even after controlling for depressive, positive, and negative symptoms. Some serious limitations of this study are (1) both anxiety and depression were assessed using a single BPRS item, (2) the sample size was small, and (3) this sample included patients with two disorders.

Findings from the Shaar Menashe study, based on 370 schizophrenia patients, disagree with the conclusion of Huppert and associates[100,101]. Indeed, correlation of Q-LES-Q_{index} with DEP_{PANSS} was found significant both before and after controlling for ANX_{PANSS} scores: $r = -0.33$ and -0.27, respectively (both $p < 0.001$). On the contrary, the significant correlation of Q-LES-Q_{index} with ANX_{PANSS} lost its significance after controlling for DEP_{PANSS} symptoms: $r = -0.21$ $(p < 0.01)$ and $r = -0.07$ $(p > 0.05)$, respectively.

Yet anxiety as a subscale of emotional distress (ANX_{TBDI}) and anxiety as an observer-rated symptom (ANX_{BPRS}) were moderately correlated ($r = 0.39, p < 0.001$). When both ANX_{TBDI} and ANX_{BPRS} scores were entered into a regression model predicting the Q-LES-Q_{index} scores (model properties: $R^2 = 0.26, F = 40.8, df = 2, 236, p < 0.001$), the contribution of ANX_{TBDI} (14.9%) was three times higher than ANX_{BPRS}(4.7%). Similarly, Wetherell et al.[106] found that anxiety as a subscale of emotional distress (ANX_{BSI}) has a significant negative impact on the QWB and SF-36 values in a sample of middle-aged and older patients with schizophrenia and schizoaffective disorder. These findings underscore that anxiety as an expression of emotional distress rather than anxiety as an observer-rated symptom should be taken into account for interpretation of the quality of life data in schizophrenia.

Negative symptoms. The association of a poor quality of life with the severity of negative symptoms has been well-documented in cross-sectional studies [11,98,107]. However, correlation of a negative factor with HRQL measures proved to be weak ($r = -0.15 - 0.24$; Table 4). PANSS negative syndrome, considered to be a cross-sectional indicator, does not appear to predict variability of domain-specific Q-LES-Q scores over time [17]. Improvement in the Q-LES-Q_{index} over time is associated with a reduction of the anergia factor (N_1, N_2, G_7, G_{10}) together with additional variables [27].

Positive symptoms. The influence of positive symptoms on HRQL deficit in schizophrenia is controversial [98,108]. Although the correlation of the severity of positive factors with HRQL scores did not reach a significant level (Table 4), hallucinatory behavior (P_3), grandiosity (P_5), and hostility (P_7) together accounted for 7.2% of the Q-LES-Q_{index} scores (see PANSS items section). In addition, we found an association between a reduction in the paranoid factor (P_6, P_7, G_8; $\beta = -0.30$, $p < 0.01$, partial $R^2 = 12.7\%$) and an improvement in the Q-LES-Q_{index} of 59 schizophrenia patients over time, controlling for age, education, follow-up duration, and other variables [27].

Cognitive impairment. Deficit in attention, memory, and executive functions is present in at least 70% of patients with schizophrenia [109]. Some studies revealed contradictory findings, with weak-to-moderate correlations between different measures of cognition functioning and HRQL, including span of apprehension and iconic memory [110], verbal memory [111], and executive ability with the Wisconsin Card Sorting Test (WCST) [112]. McDermid and Heinrichs [113] demonstrated that memory-impaired schizophrenia patients experience significantly poorer HRQL than their memory-unimpaired counterparts. Sota and Heinrichs [114] reported that cognitive data such as general intellectual ability, executive ability (WCST), verbal memory, and manual dexterity predicted subjective HRQL (Sickness Impact Profile) in a sample of 55 atypical neuroleptic-naive schizophrenia patients assessed at index and 3 years later. Changes in memory over time rather than performance levels were related to life quality, as revealed at the follow-up exam. Alptekin et al. [115] found that deficits in executive function and working memory appeared to have a direct impact on patients' perceived HRQL, especially in the social domain. Wegener et al. [116] have shown that cognitive flexibility, verbal fluency, verbal ability, and sustained attention explained up to 28% of the variance in the four domains of WHOQOL-BREF in a sample of 51 young people who experienced a first episode of psychosis.

In order to further investigate the different relations between HRQL and cognitive functions, we performed a series of stepwise regression analyses in the group of sixty-two medicated schizophrenia patients, examined with the Cambridge Neuropsychological Test Automated Battery (CANTAB), Q-LES-Q, QLS, PANSS,

TBDI, and the Extrapyramidal Symptom Rating Scale (ESRS)[117]. Table 5 presents the obtained cognitive predictors of the Q-LES-Q and QLS domain scores. Specifically, impairment of performance in executive functions (IED, SWM, and SOC), attention and memory (SSP, PRM, and DMS), sustained attention (RVP), and motor skills (RTI, BLC) appear to significantly contribute to predicting HRQL impairment syndrome in schizophrenia. The best-fitted regression models show that cognitive functions explain from 15% (general activities) to 32% (subjective feelings and social relationships) of the variability in the scores of self-report Q-LES-Q domains, and from 13% (interpersonal relations) to 35% (intrapsychic foundations) of the variability in observer-rated QLS scores. Regression models, based on cognitive dysfunctions, accounted for 29% and 23% of the variance Q-LES-Q_{index} and QLS total scores, respectively.

In the second step of the analysis, cognitive variables and additional well-known factors influencing the quality of life (PANSS, TBDI, ESRS, age, education, and illness duration) were entered into a backward selection procedure. Six best-fitted regression models for predicting self-reported quality of life accounted for 26–69% of the total variance in the Q-LES-Q domain scores. According to the first model, significant indicators of the total variance in the Q-LES-Q_{index} scores were performance in executive functions (IED, $R^2=20.3\%$, and SWM, $R^2=8.3\%$) and motor skills (RTI, $R^2=7.2\%$). Variability in physical health scores was associated with impairment in executive functions (SOC, $R^2=18.3\%$), sustained attention ($R^2=9.9\%$), and memory (PRM, $R^2=11.9\%$; and SSP, $R^2=10.2\%$; second model). The next three models revealed that 22.4% of the variance in subjective feelings was explained by impairment in executive functions (IED, $R^2=12.9\%$) and memory (PRM, $R^2=9.5\%$); 16.5% of the social relationships – by impairment in executive functions (IED), and 24.9% of the general activities – by deficit in motor skills (RTI). However, cognitive variables did not reach significant levels for prediction of the leisure time activities (model 6). In addition, these regression models included dysphoric mood ($R^2=9-41\%$), emotional distress ($R^2=3-31\%$), autistic preoccupations ($R^2=30\%$), activation ($R^2=13-16\%$), and positive ($R^2=28\%$) symptoms, side effects ($R^2=2.5\%$), age at examination ($R^2=13-14\%$), education ($R^2=12-14\%$), and illness duration ($R^2=9\%$).

Four of the five best-fitted regression models for predicting the QLS domain ratings included cognitive dysfunctions excluding common objects and activities. In particular, performance in executive functions appears to significantly contribute to predicting the general quality of life or QLS, total score (IED, $R^2=8.5\%$), instrumental role functioning (IED, $R^2=12.3\%$, and SOC, $R^2=10.5\%$), and interpersonal relations and social network (IED, $R^2=15\%$). Deficits in memory functions (SRM) accounted for 13.5% of the total variance in instrumental role functioning, and for 7.7% of the variability in intrapsychic foundations. In addition, sustained attention (RVP) and motor skills (RTI) markedly contributed to the prediction of the total variance in the instrumental role functioning ratings (14.1%, and 11.8%, respectively). As expected, the severity of autistic preoccupations ($R^2=24-37\%$),

Table 5. Predicting quality of life domain scores from neurocognitive functions in schizophrenia patients

CANTAB tasks	Metrics	Predicting Q-LES-Q domains						Predicting QLS domains				
		Index[a]	Health	Feel.	Leis.	Soc.	Act.	Total[b]	Int.	Instr.	Psych.	Obj.
Big/Little Circle[1]	BLC									+		+
Reaction Time[1]	RTI2	+	+	++	+						+	+?
Delayed Matching to Sample[2]	DMS1						+?					
	DMS2					+?						++
	DMS3									+++	+?	
Pattern Recognition Memory[2]	PRM			+?		+?						
Spatial Span[2]	SSP		+									
Paired Associates Learning[3]	PAL1					++						
Rapid visual information processing[4]	RVP A'							+				
Spatial working memory[5]	SWM2	+?		++							+?	
Intra/Extra Dimensional Set Shift[5]	IED1	+++	++	+++	++	++	+	++	++		++	
Stockings of Cambridge[5]	SOC2										+	+

Cognitive functions: [1]motor skills, [2]attention and memory, [3]learning, [4]sustained attention, [5]executive function

Models' properties: **Q-LES-Q$_{index}$** ($R^2=0.29$, F=6.6, df=3, 62, $p < 0.001$); **Health** - physical health ($R^2=0.22$, F=4.3, df=3,62, p=0.009); **Feel.** - subjective feelings ($R^2=0.32$, F=5.6, df=4,61, $p < 0.01$); **Leis.** - leisure time activities ($R^2=0.17$, F=9, df=2,61, p=0.012); **Soc.** - social relationships ($R^2=0.32$, F=5.8, df=4, 62, $p < 0.001$); **Act.** - general activities ($R^2=0.15$, F=4.6, df=2, 61, p=0.014); **QLS, total** ($R^2=0.23$, F=7.4, df=2, 62, p=0.002); **Int.** - interpersonal relations ($R^2=0.13$, F=8.0, df=1,61, p=0.006); **Instr.** - instrumental role ($R^2=0.20$, F=7.5, df=2,62, p=0.001); **Psych.** - intrapsychic foundations ($R^2=0.35$, F=4.7, df=5,62, p=0.002); **Obj.** - common objects and activities ($R^2=0.33$, F=5.5, df=4,61, p=0.001).

Significance: (+?) p=0.061–0.097; (+) $p < 0.05$; (++) $p < 0.01$; (+++) $p < 0.001$

Metrics: BLC =percent correct; RTI2=five choice reaction time; DMS1=Percent correct (all delays); DMS2=percent correct (simultaneous); DMS3=prob. error given error; PRM=percent correct; SSP=span length; PAL1=total errors (adjusted); RVP A'=visual sustained attention; SWM2=strategy; IED1=stages completed; SOC2=subsequent thinking time.

Model's properties: [a]$R^2=0.29$, F=6.6, df=3, 62, $p < 0.001$; [b]$R^2=0.23$, F=7.4, df=2, 62, p=0.002

emotional distress ($R^2 = 12 - 30\%$), positive symptoms ($R^2 = 16.2\%$), and dysphoric mood ($R^2 = 11\%$) also predicted poor QLS ratings.

Overall, deficit in frontal lobe/executive functions tested with the IED tasks was a significant predictor of impairment in the general quality of life (Q-LES-Q_{index}, QLS_{total}), and in five domains (subjective feelings, social relationships, interpersonal relations, instrumental role functioning, and intrapsychic foundations), with the SWM tasks – in two dimensions (Q-LES-Q_{index}, instrumental role functioning), and in the SOC tasks – in two (physical health and instrumental role functioning). The inability to sustain attention (RVP) was found to be associated with a deficit in physical health (Q-LES-Q) and instrumental role functioning (QLS). Among memory tests, performance on the PRM and SSP tasks were found to be valid predictors of the following self-report HRQL domains: physical health (PRM, SSP) and subjective feelings (PRM), whereas a deficit in motor skills (RTI) was found to be a valid predictor of the variability in the quality of Q-LES-Q_{index}, general activities, and instrumental role functioning.

Thus, the cognitive dysfunctions show a significant association with the quality of life impairment in schizophrenia. Although patients' cognitive dysfunctions may be embedded in their overall appraisal of quality of life, in the light of these results, it is reasonable to assume that the relationship between cognitive deficit and quality of life impairment may be affected by those factors related to the vulnerability to illness influencing both the cognitive functioning and the quality of life of schizophrenia patients. Further studies investigating the role of cognitive dysfunction in HRQL impairment syndrome within the context of large-scale studies should be encouraged.

Insight into illness. Insight into schizophrenia has been defined as a complex phenomenon that includes the patient's awareness of the disorder, its social consequences, and the need for treatment[118]. Since the vast majority of the literature in this area views insight as a product of the disease process and therefore focuses on the neurocognitive and clinical correlates, the relationship of insight with the quality of life is less clear[11,119,120]. Lysaker et al.[120] have shown that patients with poor insight had significantly poorer observer-rated QLS than unimpaired insight subjects, despite having equivalent deficit symptoms. No significant relationships between subjective life quality and insight among clinically stable schizophrenia outpatients have been reported by Browne et al.[119].

The Shaar Menashe study revealed a significant inverse association between the Q-LES-Q domains and the self-report insight (IS) scores among schizophrenia inpatients (r ranged from -0.21 to -0.30, $p < 0.001$, $n = 237$), and after a 16-month follow up (r ranged from -0.28 to -0.38, $p < 0.001$, $n = 148$). Using an observer-rated ITAQ scale, we did not find significant associations with the quality of life domains among inpatients. A follow up examination indicated that better insight is associated with poorer satisfaction with physical health ($r = -0.21$, $p < 0.05$), subjective feeling ($r = -0.20$, $p < 0.05$), general activities ($r = -0.27$, $p < 0.001$), and general life quality ($r = -0.23$, $p < 0.01$). Satisfaction with medicine did not correlate with insight measured with both instruments. The correlation between IS and ITAQ scores was 0.46 ($p < 0.001$). When changes over time in insight and Q-LES-Q scores were analyzed, slight negative associations between changes in awareness and fluctuations in social relationships, general activities, and satisfaction with medicine domains were confirmed[17]. Thus, the better self-report insight, the poorer perceived quality in schizophrenia. Longitudinal findings support the notion that precise analysis of insight is important in clarifying the factors involved in HRQL construct in schizophrenia.

Sleep quality. Poor sleep quality has been recognized as an important construct associated with poor quality of life of individuals of the general population[121,122] and patients with various somatic pathologies. Findings from the literature suggest that poor sleep quality was a more frequent problem among schizophrenia patients (45%) than in the general population (20-30%)[123]. We hypothesized that the HRQL deficit and poor sleep quality among schizophrenia patients would also be associated. To test this assumption, (a) Q-LES-Q domain scores were compared among 145 schizophrenia patients with poor and good sleep quality, as measured by the Pittsburgh Sleep Quality Index; (b) the relationship between the quality of life and the quality of sleep ratings before and after partialling out the effects of emotional distress, side effects, depression and other symptoms was evaluated; and (c) the contribution of sleep quality components for prediction of quality of life ratings was determined[19]. We found that the poor quality of life and poor sleep quality are substantially associated among schizophrenia patients.

Furthermore, poor sleep quality might be negatively associated with HRQL deficit acting synergistically with depression, emotional distress and side effects, and/or independently. Our findings indicated that schizophrenia patients with prominent sleep complaints were more distressed and depressed, had a greater number of adverse events concerning medications and accompanying mental and somatic discomfort than those who had no sleep complaints. Overall quality of sleep was moderately correlated with current levels of depression and side effects, and highly correlated with the distress index. There are several explanations for the confounding effects of distress, depressive symptoms, and the side effects of medication on the sleep quality/life quality relationship. First, stress and depression are the leading causes of insomnia[124] and thus may substantially contribute to the association between sleep and life quality. Second, impaired sleep, resulting in daytime dysfunction, may be an important additional source of stress for many individuals, setting up a vicious, self-perpetuating cycle[125]. Third, adverse effects of medication, including sleep disturbances, may induce additional distress influencing this association. Finally, all of the above factors have been shown to be important determinants of the quality of life in chronic schizophrenia patients.

To test the hypothesis that HRQL deficit syndrome and poor sleep quality of schizophrenia patients may be independently associated, depression, distress and side effect ratings were partialled from the correlation matrix. According to partial correlations, depression and adverse effect scores did not reduce the sleep-life quality relationship, but controlling exclusively for emotional distress in this manner, significantly reduced the association between sleep quality and satisfaction with subjective feelings, leisure time activities, and social relationships. Finally, after adjusting for potential covariate effects (the Montgomery and Äsberg Depression Rating Scale, DSAS, and TBDI), satisfaction with the overall quality of life (Q-LES-Q_{index}) and the general activities' domain still remained significantly associated with sleep quality scores. Thus, the association between poor sleep quality and HRQL impairment syndrome appears both independently and synergistically with depression, distress, and the side effects of medications. We speculated that HRQL impairment syndrome might be associated with a slow-wave sleep deficit and shortened rapid eye movement sleep stage that have been commonly observed in schizophrenia[126]. Future research is expected to clarify the nature of the sleep-life quality relationship.

Suicidal behavior. Suicide risk in patients with schizophrenia is alarmingly high, between 10% and 13% of all people with this disorder die by suicide[127], and as much as half of all patients were reported to have made suicide attempts at some time during the course of the disorder[128]. Poor quality of life has been recognized as an important predictor of suicide behavior in the general population[129]. In order to examine the relationship between HRQL deficit and suicidal ideations, the Suicidal Ideation Scale (SIS) was composed as an average of the four following items: in the past month 'Did you feel that you would be better off dead or wish you were dead?' 'Did you want to harm yourself?' 'Did you think about suicide?' 'Did you have an active suicide plan?' Responses are scored on a 1- to 5-point scale (1-never, 2-rarely, 3-occasionally, 4-fairly often, and 5-very often) with higher scores indicating a higher level of suicidal ideations and consequently a higher risk for a suicidal attempt (Cronbach's $\alpha = 0.82$). Schizophrenia inpatients had lower SIS scores (mean = 1.2, SD = 0.5, n = 237) compared with 70 schizoaffective (mean = 1.4, SD = 0.6, $p = 0.022$) and 32 mood disorder patients (mean = 1.4, SD = 0.6, $p = 0.035$). Increased suicidal ideations among schizophrenia patients were slightly, but significantly associated with poorer subjective feeling ($r = -0.20$, $p < 0.01$), general activities ($r = -0.19$, $p < 0.01$), and general life quality (Q-LES-Q_{index}, $r = -0.18$, $p < 0.01$).

Ponizovsky et al.[14] explored the relationship between the quality of life and suicidal attempts among 227 schizophrenia patients from the database of the Shaar Menashe study, which was divided into two subgroups: with and without a lifetime history of suicide attempts. The study tested the hypothesis that Q-LES-Q scores are negatively associated with a history of suicidality. Indeed, the patients who had attempted suicide multiple times were less satisfied with regard to a larger number of life domains than the non-attempters and the single attempters. The differences remained significant after controlling the age of onset, the number and length of hospitalizations, and the severity of positive, negative, and depressive symptoms. Thus, dissatisfaction with the quality of life in general and with reference to four specific domains was associated with repeated suicide attempts. The evaluation of schizophrenia patients that are suspected to be suicidal should include measuring HRQL.

TREATMENT-RELATED FACTORS

Since the role of these factors will be discussed in Part III 'Quality of life outcomes' of this monograph, only a few factors will be briefly mentioned here.

Unmet needs. The needs of individuals with schizophrenia are varied and extensive. The relationship of needs and HRQL in schizophrenia has been repeatedly investigated for the last two decades [130,131]. The most frequently occurring unmet needs among schizophrenia patients in five European countries were daytime activities, company and intimate relationships, psychotic symptoms, psychological distress, and information [132]. The association between high numbers of unmet needs and poor subjective HRQL appears increasingly robust across several studies [131,133]. Future research should investigate whether changes in needs precede changes in the quality of life deficit.

General functioning and working. Quality of life and general functioning are quite separate constructs that rely on different sources of information and which have distinct contributions to make to the evaluation process [134]. Brekke et al. [135] examined whether neurocognitive measures of executive functioning (WCST) moderated the relationship between psychosocial functioning and HRQL levels among 40 schizophrenia patients. Multiple regression and correlation analyses show that executive functioning was a strong moderator. Specifically, individuals with schizophrenia with impaired executive functioning displayed a positive and statistically significant association between psychosocial functioning and HRQL. However, among schizophrenia patients with intact executive performance, psychosocial functioning was negatively associated with HRQL. These findings indicate that executive functioning plays a major role in moderating the relationship between HRQL and psychosocial functioning in schizophrenia. Recently, Becker et al. [136] concluded that scores of functioning level are related to HRQL in a complex way, and types of unmet need impinge on the relationship.

Using findings from the Shaar Menashe study and multiple regression analysis, we examined the contribution of general functioning (GAF) to variability of the general quality of life impairment (Q-LES-Q_{index}). The relative contribution of general functioning to the quality of life structure was 2.4% among relapsed schizophrenia patients (1st model, $n=237$ inpatients) and 7.9% among stabilized patients (2nd model, follow up sample, $n=148$). Note that (a) GAF ratings were inversely associated with the Q-LES-Q_{index} scores ($\beta=-0.13, p=0.029$ and $\beta=-0.28, p=0.007$, 1st and 2nd models, respectively), and (b) the 1st regression model also included emotional distress, PANSS total scores, side effects, and social support, whereas the 2nd model included the same variables excluding the PANSS ratings. Thus, although a weak contribution, general functioning appears to be a distressing factor regarding the HRQL level in schizophrenia.

On the other hand, there is evidence that a daily work occupation is associated with a markedly better HRQL for people with schizophrenia [137,138]. Bond et al. [139] examined the cumulative effects of work on HRQL of 149 unemployed patients with severe mental illness receiving vocational rehabilitation throughout the 18-month study period. The authors found that the competitive work group showed higher rates of improvement in satisfaction with vocational services, leisure, and finances, in symptoms and self-esteem than did participants in a combined minimal work-no work group. Thus, working is positively correlated with a better HRQL, but a clear causal relationship has not been established[140].

Treatment settings. There are controversial comparative findings regarding the quality of life among inpatients vs. outpatients with schizophrenia. It would be reasonable to assume that inpatients would report a poorer HRQL than outpatients because of higher levels of illness symptoms. However, patients who are hospitalized in a psychiatric hospital for a longer time seem to achieve a stabilization of the level and structure of the subjective quality of life. Kaiser et al.[141], for instance, found similarities in LQOLP levels among schizophrenia inpatients with a longer stay ($\geq$ 2 years) and outpatients, whereas several significant differences could be shown between inpatients with a shorter present stay and outpatients. Moreover, Franz et al.[142] showed that schizophrenia patients who lived under hospitalized conditions for a long time are significantly more satisfied with important life domains (health and the family situation in the present study) than patients who have been hospitalized for less than 3 months and expect to leave this setting in the near future. Results indicate that the experiences of restricted and deprived living conditions induce accommodation processes and response-shifts that should be taken into account in the interpretation of quality-of-life data.

Although some studies examined HRQL changes following hospital discharge and found contradictory results[50,143], global well-being is generally higher among community-based patients than among hospital residents[144,145]. When the impact of community resettlement on the life quality (QOLI) of people with long-term psychiatric disorders was evaluated in a longitudinal study, no differences were found apart from a higher satisfaction with the living situation after discharge[134]. At the same time, significant changes in the objective quality of life indices include improved living conditions, higher levels of social contact and increased leisure activities. It is probably not the place itself that influences the HRQL, but apart from personal, sociodemographic and illness-related factors, no doubt a major contributing factor is the amount of social support that is provided in different settings[146].

Antipsychotic agents. Traditionally, HRQL is conceptualized as the outcome of illness, and, therefore, the pharmaceutical industry is increasingly incorporating HRQL measures for assessment of efficacy in the drug development process. This approach, used to distinguish the effects of competing treatments, indicates trade-offs between survival, functional status, and well being, and provides comprehensive information on the effects of treatment on the patient outcomes. Comparative HRQL findings concerning specific antipsychotic agents were recently broadly reviewed [1,4,5] (see Chapter 16 in this book). However, when HRQL impairment is conceptualized as the core feature of an illness, the goal in treating schizophrenia is to find agents for treatment that will promote an improvement in the HRQL impairment. Although there are reports that improvement over time in the overall HRQL is not influenced by the type of medication prescribed[147], here we will review olanzapine and complementary medicines possibly relevant to treating HRQL impairment.

Various post-marketing studies report controversial findings regarding the effects of olanzapine (OL) on HRQL impairment. There is evidence-supporting an advantage for OL versus haloperidol in improving the quality of life over time[148]. A few studies reported that OL improved the HRQL deficit after treatment for 12 months[149–151]. However, others failed to find any differences between OL and other antipsychotic agents regarding HRQL impairment[152,153]. When HRQL levels were compared between risperidone (RP) and OL-treated patients, OL resulted in superior[154] or comparable efficacy[20,155].

Some possible limitations in these studies regarding the effectiveness of antipsychotic agents on the quality of life deficit should be addressed. First, these studies are difficult to compare because different self-report or observer-rated scales are used. Second, comparative studies are usually designed without adjustment for confounding factors. Third, short-term randomized, double-blind, placebo-controlled trials may not accurately reflect typical clinical practice. Furthermore, no study to date has examined the influence of antidepressants, mood stabilizers, anxiolytics and anti-parkinson adjunctive drugs on HRQL impairment among patients treated with antipsychotic agents. Finally, despite their own limitations, long-term naturalistic studies combining random assignment with comprehensive global evaluation may add information that is more easily generalized to various routine practice profiles.

In order to address these limitations, we examined specific predictors of the efficacy of OL, risperidone (RP), and first-generation antipsychotic agents (FGAs), the role of confounding factors, and concomitant agents such as antidepressants, anxiolytics, and mood stabilizers in the treatment of quality of life impairment of schizophrenia patients[28]. This was a community-based, open label, parallel group naturalistic study of 124 schizophrenia outpatients who received OL, RP, FGA, or combined agents (CA). Evaluations were performed at baseline and 12 months later. It was found that OL was superior to RP, FGAs, and CA in terms of quality of life. FGAs revealed greater therapeutic benefit than RP, which was more beneficial than combined therapy. Improvement in Q-LES-Q scores was revealed in patients who received antidepressants and anxiolytics, but not mood stabilizers, or anti-parkinson drugs. This effect was independent of treatment groups and gender. Regression models revealed that changes in emotional distress and side effects were common predictors for HRQL changes across treatment groups. Specific predictors of HRQL efficacy included self-efficacy for OL, negative and positive symptoms for RP, dysphoric mood and positive symptoms, and daily doses and self-efficacy for FGA-treated patients. Thus, these findings suggest that OL is beneficial in the treatment of HRQL impairment in schizophrenia, compared with RP, FGAs, and CA.

Complementary medicines. Although antipsychotic drugs are an indispensable component of the management of schizophrenia, clinical response of psychopathological symptoms, cognitive impairment, and quality of life deficit remains incomplete in chronic schizophrenia. Consequently the development of more effective treatments is an important research goal. One promising direction is the use of complementary medicines such as ginkgo and hydergine as cognitive enhancers, passion flower and valerian as sedatives, St John's wort and s-sadenosylmethionine as antidepressants, adenosylmethionine, and seleniumand folate to complement antidepressants (see review[156]). The evidence of neural degeneration in the pathophysiology of schizophrenia suggests that there may be treatment opportunities through neural protection. Potentially useful substances include rivastigmine, omega-3 fatty acids and antioxidants, modafinil, and neurosteroids as HRQL enhancers in schizophrenia. However, their effectiveness regarding HRQL deficit is often not established.

Rivastigmine. Facilitation of central cholinergic activity may form a potential treatment strategy for cognitive and HRQL impairments in schizophrenia. Recently, it was reported that rivastigmine, a CNS-selective cholinesterase inhibitor, increased cerebellar activity and influenced attentional processes[157]. Lenzi et al.[158] attempted to determine whether rivastigmine would improve HRQL, assessed using the Satisfaction with Life Domains Scale (a self-report scale containing 10 "satisfaction" items) in 16 clinically stable schizophrenia patients in the residual phase. Study subjects began rivastigmine treatment at a dose of 1.5 mg bid (a maximum of 6 mg bid). All subjects were monitored for 12 months. Rivastigmine treatment resulted in significant improvements in HRQL, which were paralleled by significant improvements in cognitive function, learning, and memory, with some improvement in attention. The BPRS factor "anergia" showed significant improvement. The authors concluded that rivastigmine significantly improved HRQL in schizophrenia due to the drug's effects on cognitive deficits and negative symptoms. Further investigation of the effects of cholinergic agents (choline, lecithin, physostigmine, tacrine, 7-methoxyacridine, ipidacrine, galantamine, donepezil, rivastigmine, eptastigmine, metrifonate, arecoline, RS 86, xanomeline, cevimeline, deanol, and meclofenoxate) on HRQL impairment is warranted.

Omega-3 fatty acids and antioxidants. The importance of omega-3 fatty acids for physical health is now well recognized and there is increasing evidence that omega-3 fatty acids may also be important for maintaining adequate mental health. The two main omega-3 fatty acids in fish oil, eicosapentaenoic acid (EPA) and docosahexaenoic acid (DHA) have important biological functions in the central nervous system[159]. Both EPA and DHA can be linked with many aspects of neural function, including neurotransmission, membrane fluidity, ion channel and enzyme regulation, as well as gene expression. Arvindakshan et al.[160] reported the supplementation with a mixture of EPA/DHA and antioxidants (vitamin E/C) to 33 schizophrenia patients for 4 months. They found a significant improvement in HRQL deficit (QLS). Future studies need be done in placebo-controlled trials and also with a comparison group supplemented with fatty acids alone in a larger number of patients, and for a longer duration of treatment while the dietary intake is monitored. This may establish the efficacy of omega-3 fatty acid supplementation in terms of HRQL deficit management.

Modafinil, a novel wake-promoting cognitive enhancer works through the sleep-wake centers of the brain to activate the cortex and selectively improves neuropsychological task performance in healthy volunteers and adult patients with attention deficit hyperactivity disorder. It has been argued that persistent cognitive deficits in patients with schizophrenia are responsible for the failure of many of them to rehabilitate socially even when psychotic symptoms are in remission. One report suggests that modafinil may be an effective and well-tolerated adjunct treatment that improves global functioning and clinical condition, and reduces fatigue in patients with schizophrenia or schizoaffective disorder[161]. Moreover, in narcolepsy patients, modafinil significantly improved HRQL (SF-36) measures of mental and physical component summary scores from weeks 1 through 6[162].

Dehydroepiandrosterone (DHEA). Both clinical and biological data indicate that schizophrenia patients are impaired in their biological response to stress; this is associated with a dysregulated hypothalamic-pituitary-adrenal (HPA) axis. DHEA is a neurosteroid and serves as a precursor for both androgenic and estrogenic steroids. Its sulfated form (DHEAS) is the most abundant steroid found in the body. These neurosteroids exhibit a variety of properties including anti-stress and neuroprotective properties. Some findings indicated alterations in the DHEA metabolism in schizophrenia[163–165] that are attributed to emotional distress, anxiety, severity of dysphoric mood, and activation symptoms influencing HRQL impairment[27].

DHEA administration to patients receiving ongoing antipsychotic medication was investigated in a multicentered, 12-week double-blind, randomized, placebo-controlled, crossover trial[166]. Schizophrenia patients were randomly allocated to 2 treatment groups receiving either DHEA (200 mg/day) or a placebo for 6 weeks with the crossover between DHEA and placebo occurring after 6 weeks. Patients continued to receive their regular antipsychotic medication for the duration of the study. Compared to a placebo, DHEA administration did not produce a significant improvement in clinical symptoms, side effects, and HRQL scores (Q-LES-Q and QLS). However, six weeks of DHEA administration (but not a placebo) was associated with a significant improvement in PANSS ratings compared with the baseline. Furthermore, six weeks of DHEA treatment was associated with a significant improvement in the cognitive functions of visual sustained attention, as well as visual and movement skills compared to placebo conditions. DHEA augmentation was associated with elevations of both DHEA and DHEAS serum concentrations. DHEA treatment was well-tolerated without any serious adverse effects. Thus, this short-term study does not support DHEA's value as an effective adjunct in the treatment of symptoms, side effects, and quality of life impairment in schizophrenia; however, it suggests that DHEA improves sustained attention, visual, and movement skills. During this study no significant change in two quality of life measures was observed under DHEA treatment. However, taking into account the complexity of the factor structure of HRQL impairment in schizophrenia, and the short-term study time, additional analysis with relevant confounding variables is warranted. Systematic clinical trials are needed to test the ability of other neurosteroids and promising substances that may be an integral part of the neuroprotective system ameliorating HRQL in schizophrenia.

Side effects. Antipsychotic medications have a wide range of side effects that may physically and psychologically affect schizophrenia patients. Moreover, side effects are considered a major source of subjective discomfort (distress), and noncompliance with treatment. Most cross-sectional studies of the effects of antipsychotic agents on HRQL reported that HRQL measures inversely correlate with side effects [12,167,168]. In this context, HRQL has become an important measure of the safety of antipsychotic agents in the treatment of schizophrenia patients.

Decreasing severity of adverse events is related to improvement in physical health and leisure time activities during the follow-up period, as supported by findings from cross-sectional investigations [12,110,169]. Browne et al.[119] assessed the relationship between the subjective response to neuroleptics and observer-rated QLS in a selected group of clinically stable schizophrenia outpatients. A significant portion of the variance in the HRQL of these patients was explained by a combination of a dysphoric response to antipsychotic agents and protracted illness duration. Gerlach and Larsen[170] studied 53 chronic schizophrenic out-patients receiving maintenance depot antipsychotic treatment, and no correlation was found between subjective HRQL and the degree of side effects. Dernovsek et al.[168] studied the influences on QLS with 200 chronic schizophrenia outpatients treated with depot antipsychotics. They found that the parkinsonism score was inversely correlated with the QLS ratings. Allison and associates[171] presented evidence that weight gain is directly associated with reduced HRQL (Psychological Well-Being Index) in 286 schizophrenia patients who were taking antipsychotic medications. When gender and use of antipsychotics were controlled, gain of weight was related to a poorer quality of life and reduced well-being and vitality. However, in this study confounding factors that predisposed people to both gain of weight and poorer HRQL were not addressed. Olfson et al.[172] reported that sexual dysfunction occurred in 45.3% of 139 adult males with schizophrenia and is associated with diminished HRQL, decreased occurrence of romantic relationships, and reduced intimacy when relationships are established. However, a high frequency of sexual dysfunction was reported by both drug-free and neuroleptic-treated schizophrenia patients, who are associated with both FGAs and SGAs[173].

According to findings from the Shaar Menashe study, 161 schizophrenia patients with side effects reported less satisfaction with subjective feelings and general activities than patients without side effects[12]. General Q-LES-Q_{index} was inversely associated with distress attributed to the following adverse events: sleep disturbances and fatigue (both $r = -0.31$, $p < 0.001$), tachycardia ($r = -0.30$, $p < 0.001$), tremor and sexual dysfunction ($r = -0.28$, $p < 0.001$), headache ($r = -0.25$, $p < 0.001$), polyuria/disuria and dizziness (both $r = -0.20$, $p < 0.01$), hypertension or hypertension ($r = -0.18$, $p < 0.05$), dyskinetic movements, and constipation or diarrhea (both $r = -0.16$, $p < 0.05$). However, patients' subjective responses to adverse events, and not the number of events themselves appear to be more predictive of HRQL.

Two studies attempted to assess the contribution of side effects to the structure of HRQL scores in schizophrenia. The first study, showing that subjective distress is caused by akathisia (11%) and neuroleptic dysphoria (6%), accounted for 17% of the variance of subjective HRQL[70]. In the second study, clinical and stress process-related variables were used for regression analysis[12]. When the number of adverse events and attributed distress were simultaneously entered, the initial regression model accounted for 13% of the variance of general Q-LES-Q_{index}, which is comparable with Awad et al.'s data[70]. After controlling for clinical and stress process-related variables, side effects contributed only 3.2% to the variance of the HRQL scores. Thus, side effects of antipsychotic drugs influence subjective HRQL of schizophrenia patients in a significantly lesser degree than other stress process-related variables and illness-associated factors. Moreover, the patient's subjective response to the side effects of medications is more predictive of HRQL than the number of those events.

Demographic and background variables. Empirical findings from a rather large body of studies have shown that the relationships of demographic and background characteristics have a controversial, or a weak, or no relationship to subjective HRQL levels and are therefore still not considered to significantly influence HRQL impairment[11,27,90]. For instance, Skantze et al.[174] (1992) found that the quality of life was independent of gender, marital status, and the standard of living. Browne et al.[42] found HRQL to be independent of gender and age at the onset among first-episode patients. Findings from Canada, Cuba, and the United States (102 outpatients) suggest no discernable differences in the perception of HRQL (LQLI) for men and women with schizophrenia[175]. Another study reported that the HRQL (SF-36) of female and male patients did not differ at baseline, but the improvement after treatment was greater in female patients treated in a day hospital[176]. Young[91] reported that sex, age, education, marital status, religion, and place of birth were not related to subjective general HRQL. Duration of illness, and the cumulative length of previous hospitalization were inversely related to HRQL deficit of schizophrenia patients who were attending a catchment area rehabilitation center[169].

We did not find a significant association of general HRQL scores with gender differences ($F = 0.41, df = 1, 370, p = 0.52$), and marital status ($F = 0.43, df = 2, 370, p = 0.65$) of long-term schizophrenia patients, when controlling for age and education. Likewise, domain-specific Q-LES-Q scores were not related to gender and marital status (all $p > 0.05$). Furthermore, the correlation of Q-LES-Q dimensions with age, education, age of onset, illness duration, and number of admissions did not reach a significant level (all $p > 0.05$). Yet correlation coefficients between changes in the Q-LES-Q measures and age, length of education, age of illness onset, number of admissions, duration of the disorder, length of follow-up period, and of last hospitalization were also not significant[17,27].

FACTOR STRUCTURE OF HRQL DEFICIT SYNDROME

Variability in estimations of specific factors (determinants, indicators, and predictors) that constituted the structure of HRQL deficit syndrome in schizophrenia is dependent on the study design (cross-sectional or longitudinal), the stage of the illness (first episode or chronic patients; exacerbation or stabilization phase), treatment settings (hospital, outpatient clinic, rehabilitation facilities, etc.), used HRQL instruments (self-report or observer-rated), and from a list of illness-related and the patient's characteristics entered into multivariate analysis.

Factor analysis. For this analysis we used data from 148 schizophrenia patients who were examined twice: at the exacerbation of the illness and after 16 months. Variables with an absolute loading greater than the amount set in the minimum loading option (> 0.4) were selected (age, education, age of onset, and illness duration were removed to avoid augmenting scores; Table 6). The first and second factors with a distressing (a negative or harmful) effect on the general HRQL included severity of emotional distress, somatization, illness symptoms and side effects, emotionally oriented coping, and general functioning. The third factor (termed 'protective') was constructed using self-esteem, self-efficacy, task and avoidance coping styles, and social support. The general quality of life scores were significantly associated with the 1st and 3rd factors, but not with the 2nd (symptoms). Insight, expression emotion, and side effects (AIMS) scores did not reach the minimum loading option. Two distressing and protective factors accounted for 68.4% and 27.9% of the total variance of used variables, respectively. Factor analysis of data obtained from the same patients 16 months later revealed a similar two-arm factor structure. In addition, insight scores were found to be inversely associated with a "protective factor" (negative loading), whereas higher expression emotion scores and lack of other significant support were identified among the distressing factors. Thus, in a framework of a relatively stable 'distress-protective' factor structure of HRQL, an impairment in schizophrenia was observed, creating changes in the role of some variables.

Cross-sectional indicators. As expected, regression analysis for HRQL levels in schizophrenia, based on clinical variables, indicated that the severity of the symptoms and side effects explain about 32% and 17% of their variance, respectively[70]. For older patients with schizophrenia and other psychoses, path analysis revealed that the general quality of the well-being (QWB) scores is influenced by symptoms, psychosocial factors, and social maladjustment[177]. Among 137 middle-aged and elderly outpatients with schizophrenia or schizoaffective disorder, age at the onset of illness, depressive symptoms, and cognitive functioning predicted 39% of the variance in the SF-36[30]. Findings from the EPSILON study suggest that poor HRQL is associated with anxiety, depression, psychotic symptoms, more previous psychiatric admissions, alcohol abuse, having no reliable friends nor daily contact with family, being unemployed, and having few leisure activities[132]. Hansson et al.[54] reported that no clinical characteristics, but subjective factors were significantly associated with subjective HRQL (satisfaction with health, 36.3%, and self-esteem, 7.3% of the variance), together with having a close friend. This model

Table 6. Factor loadings after Varimax rotation of variable values of schizophrenia patients examined twice with an interval of about 16 months (follow-up sample, n=148)[1]

Variables		Initial examination			Follow up examination		
		Factor 1 $(4.76)^2$ *'distress'*	Factor 2 $(3.18)^2$ *'symptoms'*	Factor 3 $(3.19)^2$ *'protective'*	Factor 1 $(5.75)^2$ *'distress'*	Factor 2 $(4.44)^2$ *'protective'*	Factor 3 $(3.78)^2$ *'symptoms'*
General quality of life (Q-LES-Q_{index})		**0.4104**	−0.1064	**0.5695**	**−0.6382**	**0.5243**	0.1532
General functioning (GAF)		0.0669	**−0.4173**	0.0309	−0.0375	−0.0043	**0.5846**
Symptoms:	Negative symptoms	−0.0314	**0.6117**	−0.1880	−0.0318	−0.0416	**−0.8012**
(PANSS)	Positive symptoms	−0.1233	**0.7333**	0.1006	0.2028	0.1170	**−0.6926**
	Activation symptoms	−0.0647	**0.7934**	−0.1027	0.0511	0.1234	**−0.8236**
	Dysphoric mood	−0.3509	0.3193	−0.0871	**0.4683**	−0.1621	−0.4201
	Negative symptoms	−0.1728	**0.7363**	−0.0808	−0.0206	−0.1432	**−0.8912**
Insight:	Observer-rated (ITAQ)	−0.2638	−0.3322	−0.1468	0.0955	**−0.4535**	0.2828
	Awareness (IS)	−0.3549	−0.1607	−0.3728	0.2544	**−0.4524**	0.0998
	Needs (IS)	−0.0863	−0.2198	−0.2386	0.0515	**−0.4137**	0.1480
Side effects:	AIMS	−0.0646	0.3249	−0.0272	0.0159	0.0126	−0.3351
	DSAS	**−0.4012**	0.1264	−0.0892	**0.5580**	−0.1207	−0.1404
Emotional distress:	Obssesiveness	**−0.6619**	−0.0368	−0.2256	**0.5506**	−0.5110	−0.1871
(TBDI)	Hostility	**−0.4555**	0.2916	0.0314	**0.3865**	−0.2942	−0.2371
	Sensitivity	**−0.7521**	0.0359	−0.1736	**0.6732**	−0.3428	−0.1726
	Depression	**−0.7961**	−0.0153	−0.2129	**0.7085**	−0.4096	−0.2190
	Anxiety	**−0.6818**	0.1952	−0.1035	**0.5892**	−0.3934	−0.1915
	Paranoid ideation	**−0.6930**	0.0670	−0.0334	**0.4922**	−0.4854	−0.2314

Somatization (BSI)		**−0.6180**	0.1377	0.0289	**0.6155**	−0.1244	−0.0690
Expressed emotion:	Irritability	−0.1187	0.1500	−0.0896	**0.6108**	0.0768	0.0989
(LEE)	Criticism	−0.1818	0.1177	−0.1954	**0.5801**	−0.0660	0.1023
	Lack of emotion support	0.0239	0.3543	−0.0930	**0.6516**	0.0413	0.0926
Self-esteem (RSES)		0.5142	0.0913	**0.5235**	**−0.5682**	0.4474	0.1295
Self-efficacy (GSES)		0.2484	−0.0682	**0.5911**	−0.4109	**0.6342**	0.1895
Coping styles:	Task oriented coping	−0.0330	0.0495	**0.7758**	−0.1366	**0.7902**	0.1723
(CISS)	Emotion oriented coping	**−0.6286**	0.0533	0.3250	**0.5246**	0.1676	−0.0888
	Avoidance oriented coping	−0.0710	−0.0988	**0.7305**	0.0056	**0.7292**	−0.0211
Social support:	Family support	−0.0407	−0.2877	**0.4361**	−0.3136	**0.4063**	0.0367
(MSPSS)	Friend support	0.1849	−0.1442	**0.4700**	−0.1369	**0.6187**	−0.0098
	Other significant support	0.0409	−0.2482	**0.4037**	**−0.4657**	0.219884	−0.1083
Factors' contribution (%)		**41.0**	**27.5**	**27.4**	**39.7**	**30.7**	**26.1**

[1]Variables with an absolute loading greater than the amount set in the minimum loading option (≥0.40) were selected

[2]Eigenvalue

explained 52.3% of the variance in global subjective HRQL for a sample of 418 schizophrenia patients from 10 sites.

The UK700 Group[178] reported that unmet needs accounted for 20% of the variance for subjective LQOLP, compared with 19% for depression and positive symptoms, and 7% for social variables of 708 patients with severe mental illness. The number of hospitalizations, the perceived sense of freedom and social support could explain about one third of the variance in overall life satisfaction of 88 patients with long-term mental illness[91].

Analysis of the initial data from the Shaar Menashe study revealed various patterns of predictors with the following estimations of their contribution to variability for general and domain-specific Q-LES-Q dimensions of 161 schizophrenia inpatients: emotional distress (10-37%), somatization (11-20%), depressive and negative symptoms (8-15%), side effects (4-5%), self-efficacy (16-21%), coping styles (11-22%), perceived social support (8-29%), and insight (1%)[11].

Logistic regression analysis with stress process-related variables, symptom severity, demographic and background data of schizophrenia patients at initial examination (1st model) and at follow-up examination (2nd model) was applied to search for predictors of severe HRQL impairment. Table 7 presents a summary of the regression analysis for two groups of patients: being severely impaired versus those with mild or no impairment. The 1st model indicates 6 significant predictors: three negative (emotional distress, somatization, and age), and three positive (self-esteem, self-efficacy, and social support). The 2nd revealed only three significant predictors of severe HRQL impairment syndrome (emotional distress, self-esteem, and self-efficacy). Both models correctly classified about 80% of the patients as being severely impaired versus those with mild or no impairment. Symptom severity did not reach a significant level in these prediction models.

Three research groups that explored other samples of schizophrenia patients confirmed our conclusions. For instance, Bechdolf and associates[179] reported that depressive symptoms (PANSS), negative coping, perceived social support, and self-efficacy were the strongest significant HRQL determinants among 66 post-acute schizophrenia patients. Furthermore, Eklund et al.[56–58], using a structural equation modeling approach, concluded that the self-factor and personality explained most of the variation in self-rated HRQL among 117 individuals with schizophrenia. However, Chan et al.[180] found that symptoms of distress accounted for the most variance (33%) in the subjective quality of life, followed by psychological integration (3%) and physical integration (2%) for 154 patients with severe mental illness living in the community.

Longitudinal predictors. The database of the Shaar Menashe study presents an opportunity to establish two different kinds of longitudinal predictors: first, based on changes in the values of factors (independent variables) over time (between T_1 and T_2) for predicting the changes in the HRQL scores (dependent variable) over the same time, and second, a selected set of factors at the initial examination (T_1) for prediction of HRQL levels at the follow-up examination (T_2).

Predicting changes in general Q-LES-Q$_{index}$. The first set of predictors accounted for 36% of the total variance of the individual changes in the Q-LES-Q$_{index}$ scores

Table 7. Summary of logistic regression analysis for discrimination between severe and mild quality of life impairment in schizophrenia patients (follow-up sample, n = 148)

Regression model	β[a]	χ^2	p	Last R^{2b}
Initial examination (1st model)				
Intercept	−3.89	8.3	0.004	0.0351
Emotional distress	−0.75	5.8	0.016	0.0247
Somatization	−0.81	6.2	0.012	0.0263
Self-esteem	0.13	6.1	0.013	0.0258
Self-efficacy	0.12	16.1	0.001	0.0658
Social support	0.02	4.4	0.035	0.0189
Age at examination	−0.04	4.9	0.027	0.0208
Severe HRQL impairment=117 patients, mild – 120 patients.				
Model's properties: $R^2=0.36$, $df=6$, $\chi^2=130.9$, $p < 0.001$; *correctly classified*=79.2%				
Follow-up examination (2nd model)				
Intercept	−3.97	4.6	0.032	0.0310
Emotional distress	−1.22	9.4	0.002	0.0613
Self-esteem	0.16	3.9	0.049	0.0262
Self-efficacy	0.11	7.2	0.007	0.0478
Severe HRQL impairment=58 patients, mild – 90 patients.				
Model's properties: $R^2=0.37$, $df=3$, $\chi^2=85.7$, $p < 0.001$; *correctly classified*=81.8%				

[a]β is the estimated value of regression coefficient that was calculated using the Newton-Raphson method to solve the nonlinear, maximum likelihood equations.
[b]Last R^2 reflects the amount that this variable adds to the overall R^2 when it is added to the logistic regression equation.

of 148 schizophrenia patients during a 16-month follow-up period ($R^2=0.36$, adj. $R^2=0.33$, $F=15.7$, $df=5$, $p < 0.001$). Improved general life quality is associated with a reduced paranoid factor (PANSS), emotional distress, somatization, and increased self-efficacy and self-esteem ratings. Changes in distress and somatization intensity account for 12.2%, in self-constructs – 11.7%, and in paranoid factor – for 2.8% of the fluctuations in the Q-LES-Q_{index} scores[15]. More meticulous results were obtained when these patients where divided into three subgroups by place at the follow-up examination: reassessed during the same hospitalization ('hospitalized'), at discharge from an additional admission ('discharged'), and in an outpatient clinic ('outpatients'). Regression analysis for each subgroup revealed that fluctuations in Q-LES-Q_{index} scores were influenced by different patterns of predictors, which accounted for 30-67% of the change in variance[27]. Established models suggest that a reduction in paranoid severity and emotional distress, together with increasing self-esteem and social support ratings, are associated with improvement in the general life quality among hospitalized patients during the follow-up period. A decrease in the anergia factor and emotional distress scores accounts for 8.5% and 25%, respectively, of improved perception of the subjective life quality of patients discharged at the follow-up assessment. For outpatients, subgroup regression analysis revealed four negative (anergia, emotional distress, side effects, and emotionally related coping style) and two positive (avoidance coping and expressed emotion) predictors associated with improvement of the general quality of life over a studied time.

Predicting changes in domain-specific HRQL. Fluctuations in satisfaction with specific Q-LES-Q domains were influenced by different patterns of predictors that accounted for 12-36% of the change in variance [17]. In particular, improvement in *physical health* ($R^2 = 0.12$) appeared to be related to a reduction of somatization and side effects, which accounted for 7.6% and 3.3%, respectively. Improved subjective feelings ($R^2 = 0.33$) were associated with decreased emotional distress (13.5%), and somatization (3.4%), and with elevated self-esteem (8.9%) and self-efficacy (4.5%) levels. Positive changes in leisure time activities ($R^2 = 0.16$) may arise due to a reduction in the number of adverse symptoms (3.2%), together with increased self-efficacy (8.5%) and support from others (3.5%). Improved satisfaction with social relationships ($R^2 = 0.19$) is associated with lower emotional distress (obssesiveness, 11.9%), and a higher awareness of illness (5.8%). Changes over time in the severity of somatization (7.7%) and paranoid symptoms (PANSS, 5.3%), improvement in self-esteem (6.9%) self-efficacy (5.8%), and awareness of illness (3.3%) may significantly contribute to higher satisfaction with general activities ($R^2 = 0.36$). In contrast, changes in the awareness of the illness and the need for treatment together account for 14.8% of the fluctuation in satisfaction with medicine scores over time.

Predicting general life quality levels at the follow-up examination from factors at the initial examination [27]. Baseline values of the PANSS activation factor, emotional distress, task-oriented coping style and self-esteem accounted for 41% of the Q-LES-Q_{index} scores at the follow-up examination ($F = 10.7$, $df = 9, 148$; $p < 0.001$). In particular, emotional distress (accounting for 11.6 % of the variance) was inversely associated with improvement of the general life quality over time, whereas the impact of the depression factor did not reach a significant level ($\beta = -0.14$, $t = 1.9$, $p > 0.05$, partial $R^2 = 2.7\%$). Higher activation factors, task-oriented coping, self-esteem, and support from friends predicted better satisfaction with life quality after 16 months (they account for 4.1%, 2.8%, 2.7%, and 2.9% of the Q-LES-Q_{index} scores at the follow-up examination, respectively). Age, education and follow-up duration did not influence changes in general HRQL.

Two recent publications mentioned similar variables predicting subjective HRQL for schizophrenia patients. Ruggeri and associates [181] reported that psychological status, self-esteem, and satisfaction with service were the most important predictors of changes at 2 and 6 years in subjective HRQL domains of 261 individuals attending a community mental health service. Caron et al. [69] used repeated HRQL measures on the same subjects (after a 6-month interval) and a regression model that accounts for 50% of the variance in HRQL at T_1 and 43% at T_2, which indicated that the best predictors of HRQL were social support, the severity of daily hassles, the coping strategy of changing the situation, the level of education and the life-time hospitalization length.

Thus, longitudinal predictors such as cross-sectional indicators, show that changes in stress process-related factors, rather than the severity of the symptoms is associated with fluctuations in HRQL impairment syndrome among schizophrenia patients.

IS THE HRQL IMPAIRMENT A PARTICULAR SYNDROME IN SCHIZOPHRENIA?

The present findings and conclusions should be interpreted in light of potential limitations of the HRQL studies in schizophrenia.

- Collected samples for HRQL studies have concluded diagnostically various types of patients: "schizophrenia", "schizophrenia, and schizoaffective and schizophreniform disorders", "psychiatric patients", and "schizophrenia spectrum".
- Samples consist of patients that differed by the stage of illness (first episode or chronic patients), current mental health status (exacerbation or stabilization phase), treatment settings (hospital, outpatient clinic, rehabilitation facilities, etc.) and interventions.
- As a rule, researchers focused on assessment of illness-related variables, whereas the most distressing and protective factors were neglected. Lacking a patient's characteristics when entering data in the multiple regression analysis may have introduced bias in the assessment of the contribution of various factors to the HRQL structure.
- The diversity of the study design (cross-sectional or longitudinal; naturalistic study or clinical trial), and the instruments used (self-report or observer-rated) may also be factors.
- Another limitation of HRQL studies is the sample size, which is frequently rather small, so the statistical analysis may not have been sufficiently sensitive to show the relationships between the variables.
- Assessment of satisfaction in such domains as religion, family, and social relations are influenced by culture, the characteristics of the surroundings, and the local style of living[182].

In spite of these difficulties, we suggest that HRQL impairment is a particular syndrome in schizophrenia, which may be summarized as follows:

1. Schizophrenia patients are significantly impaired in both the general and the all domain-specific quality of life compared with healthy subjects.
2. The deficit in social relationships is more notable for schizophrenia, whereas the deficit in subjective feelings - is more observed in mood disorder patients. HRQL impairment is reportedly the lowest for anxiety and mood disorder patients, and somewhat higher for schizophrenia patients.
3. The quality of life deficit occurs in the early phases of schizophrenia, before the onset of psychotic symptomatology, and remains throughout its course.
4. The course of HRQL impairment is relatively stable with slight fluctuations in general and domain-specific HRQL scores over time.
5. There is a distress-protective factor structure underlying HRQL impairment in schizophrenia. Although this factor structure is relatively stable, changes in the illness phase and treatment settings over time might be associated with changes in the contribution of some factors to the structure of HRQL impairment.
6. About 40–50% of chronic schizophrenia patients have a severe impairment in general life quality; 38% of the patients showed severe impairment in 1-3

domains, and 54% of the patients felt severe deficits in 4-5 domains of life quality, and only 8% of the patients reported satisfaction with all five domains.

7. Logistic regression models, which correctly classified about 80% of the patients as being severely impaired versus those who were mild or non-impaired, revealed six significant predictors: three distressing (emotional distress, somatization, and age), and three protective (self-esteem, self-efficacy, and social support).
8. Stress process-related factors, rather than the severity of symptoms and treatment-related factors contribute to the levels of HRQL impairment in schizophrenia. The contribution for general and domain-specific quality of life varies: emotional distress (10-37%), somatization (11-20%), self-efficacy (16-21%), coping styles (11-22%), perceived social support (8-29%), temperament factors (6-16%), depressive and negative symptoms (8-15%), side effects (4-5%), and insight (1%).
9. A significant association exists between impairment in HRQL and various cognitive functions across domains, using cognitive tests, QOL instruments, and confounding factors.
10. The Distress/Protection Vulnerability Model hypothesizes that the level of HRQL impairment is a transaction due to the interaction of an array of distressing and protective factors: HRQL impairment decreases if the distressing factors outweigh the protective factors, and vice versa (see Chapter 1). Illness and treatment-related variables (secondary factors) influence HRQL impairment via mainly personal characteristics (primary factors)[11,27]. Table 8 summarizes those factors influencing HRQL impairment according to the Distress/Protection Vulnerability model, which merits further validation and development.

Table 8. Factors influencing quality of life impairment in schizophrenia, according to the Distress/Protection Vulnerability model

	Distressing or negative factors	Protective or positive factors
Primary Factors	• Emotional distress uniquely contributed to HRQL impairment, the relationship between emotional distress and poor HRQL appears, at least partly, independent from symptom severity (included depression) and side effects of medications. General and domain-specific HRQL impairments are negatively correlated with somatization scores. Somatization may be mediates the impact of symptoms and side effects on leisure time, social relationships, and satisfaction with medicine. Harm avoidance as temperament factor (TPQ) negatively associated with HRQL impairment. Emotion oriented coping style shows negative relationship with HRQL impairment.	• Novelty seeking, and reward dependence as temperament factors. • Self-directedness, extroversion and agreeability as personality traits. • Self-esteem • Self-efficacy • Task oriented and avoidance oriented coping styles • Perceived social support.

Secondary Factors	• Poor premorbid adjustment • A longer duration of untreated psychosis leads to a poorer HRQL. • Severity of depressive, anxiety, and negative symptoms. • Cognitive deficit in executive functions, attention and memory, sustained attention, visual and movement skills appear associated with HRQL impairment. This relationship various across domains, using cognitive tests, quality of life instruments, and confounding factors. • Better insight to illness is associated with severe HRQL impairment in physical health, subjective feeling, general activities, and general life quality. • Poor sleep quality is negatively associated with HRQL impairment ratings. Expressed emotion inversely correlated with HRQL impairment. • Repeated suicide attempts are associated with dissatisfaction with general HRQL and with four specific domains. • The high numbers of unmet needs the lower HRQL impairment. • General functioning inversely associated with HRQL impairment levels. • Side effects of antipsychotic drugs. The patient's subjective response to side effects of medication is more predictive of HRQL impairment than the number of those effects. • Sexual dysfunctions and weight gain are associated with diminished HRQL.	• Employment or working positively correlated with better HRQL scores. • Being community based resident. • Amount of social support provided by treatment settings. • Being in a psychiatric hospital for a longer time. • Olanzapine is beneficial in the treatment of HRQL impairment. • Concomitant treatment with antidepressants, anxiolytics and complementary medicines also may improve HRQL scores.

Finally, in mentioning the above evidence, the following question arises: *Should HRQL impairment be a diagnostic criterion for schizophrenia?* We believe that impairments in general and domain-specific quality of life may be viewed as an independent syndrome influenced by various hosts' vulnerability and resistance factors, and it is sufficiently reliable, stable, and relatively specific to warrant inclusion in the diagnostic criteria for schizophrenia.

CONCLUSIONS

In this chapter we concluded the following:

- The literature provides evidence that HRQL impairment syndrome is highly prevalent and fairly marked in adult patients with schizophrenia who are suffering from HRQL impairment-like symptoms of severity and cognitive deficit. There is generally more severity in affective disorders than in schizophrenia and schizophrenia may have a relatively specific pattern.

- The HRQL impairment syndrome has been found in adolescents, and in young adults before they exhibit the signs and psychotic symptoms of schizophrenia.
- This impairment may be caused by distressing and protective factors related to the vulnerability of the illness.
- In many instances, the HRQL impairment appears to be relatively stable across the course of the illness.
- The empirical findings also suggest that the HRQL impairment syndrome appears to be relatively independent from symptomatology and cognitive deficit.
- Finally, it is concluded that the evidence that the HRQL impairment is a prominent feature of schizophrenia is sufficiently compelling to warrant inclusion of this syndrome in the diagnostic criteria for schizophrenia.

APPENDIX

Shaar Menashe Longitudinal Study of Quality of Life

This is a large ongoing naturalistic prospective investigation that started in1998 and whose aim is to examine HRQL impairment and related factors among patients with severe mental disorders. A detailed description of the study design, data collection, and measures was reported elsewhere [11,15]. In brief, the participants met DSM-IV criteria for schizophrenia, schizoaffective, major depression or bipolar disorder; they were age 18-65, inpatients, and able to provide written informed consent for participation in the study. Patients with comorbid mental retardation, organic brain diseases, severe physical disorders, drug/alcohol abuse, and those with low comprehension skills were not enrolled. The Internal Review Board and the Israel Ministry of Health approved the study, and all participants gave written informed consent after receiving a detailed explanation of the study procedures.

Participants

The database of the project included the following samples of schizophrenia patients and healthy subjects who will be mentioned in this chapter.

1) *The initial sample* includes data about 237 schizophrenia patients that were examined at the exacerbation phase of illness. The initial data concerning all adult patients consecutively admitted to closed, open, and rehabilitation hospital settings of the Shaar Menashe Mental Health Center were collected since August 1998. The sample was 188 (79.3%) male and 49 (20.7%) female. Among 176 presented with paranoid type, 38 were with residual type, 11 with disorganized type, 1 with catatonic type, and 11 with undifferentiated type.
2) *A follow-up sample* represented 148 of 237 patients of the initial sample (62.4%), followed for 16 months (SD=4.6): The sample was 121 (83.1%) male and 27 (18.2%) female (see more details [15,17,27]).

3) *The community based sample* includes data about 133 schizophrenia outpatients examined at the stabilization phase of the illness. Among them were 102 (76.7%) males and 31 (23.3%) females (see more details[20,21,28]).
4) In order to reach a more representative sample of a schizophrenia population, the data of the initial (n=237) and community-based (n=133) samples were combined into a *total sample* (n=370). The total sample was 290 (78.4%) male and 80 (21.6%) female. Marital status: never married=227 patients (61.4%), married=74 patients (20%), other=69 patients (18.6%).
5) The *control sample* included 175 hospital staff members excluding physicians. Inclusion was based on the availability of respondents for the interview. Controls had no psychiatric history and did not fulfill DSM-IV criteria for any mental disorder. This sample was 37.1% male, with a mean age of 38.4 years (SD 9.9; range 20 - 61 years), and a mean length of education 13.6 years (SD=2.2). A total of 132 subjects (75.9%) were married, 24 (13.8%) were single, and 18 (10.3%) were widowed or divorced. The control subjects were comparable to the patients with regard to age, but female, married and more educated subjects were overrepresented.

Demographic, background, clinical, and HRQL characteristics of patients' samples are presented in Table 1.

Measures

All respondents participated in the initial interview; diagnoses were according to DSM-IV criteria. The Schedule for Assessment of Mental Disorder[11], a semi-structured interview, was used for collecting data covering background and demographic characteristics, family psychiatric history, personal psychiatric history, details of the present illness and medication, general medical history, and current laboratory tests. Information from a patient's relative, close companion, or file records supplemented the SAMD. The Checklist for Patients not Entered into Database (SAMD-0) was used to register non-enrolled patients.

The *Quality of Life Enjoyment and Life Satisfaction Questionnaire* (Q-LES-Q) consisted of 93-items grouped into ten summary scales as follows: Physical Health, Subjective Feelings, Leisure-Time Activities, Social Relationships, General Activities, Work, Household Duties, Medication Satisfaction, School/Course Work, Life Satisfaction, and Enjoyment[183]. Responses are scored on a 1- to 5-point scale, with higher scores indicating better HRQL. In the current analysis, we used seven of the Q-LES-Q domains, excluding the following three domains (Household Duties, School/Courses, and Work) as irrelevant for hospitalized patients. Internal consistency of the seven summary scales of the instrument, as measured by Cronbach's α coefficient, ranged from 0.85 to 0.93. We added the general Q-LES-Q_{index}, which was an average of the scores of the 60 items of the seven Q-LES-Q domains (Cronbach's α=0.95). Quality of life was operationally defined as *severely impaired* when Q-LES-Q scores decreased more than two standard deviations below the healthy subjects (control sample, n=175): < 2.72 for physical health, < 3.28 for subjective feelings, < 2.68 for leisure-time activities, < 2.96 for social relationships, < 3.28 for general activities, and < 3.41 for Q-LES-Q_{index} scores[3].

The observer-rated *Quality of Life Scale* includes 21 items rated by the clinician on a 7-point scale (0-1 severe impairment to 5-6 normal or unimpaired functioning) and includes four domains: interpersonal relations, instrumental role, intrapsychic foundations, common objects, and activities[7].

The overall level of functioning was assessed with the *Global Assessment of Functioning Scale*[184].

Severity of psychopathology was assessed using 30 items of the *Positive and Negative Syndromes Scale* (PANSS), which were analyzed by 5-factor models of Kay et al.[185]: anergia (N_1, N_2, G_7, G_{10}), thought (P_2, P_3, P_5, G_9), activation (P_4, G_4, G_5), paranoid (P_6, P_7, G_8), and depression factor (G_1, G_2, G_3, G_6), and White et al.[186]: negative factor (N_1, N_2, N_3, N_4, N_6, G_5, G_7, G_8, G_{13}, G_{14}), positive factor (P1, P_3, P_5, G_1, G_9), activation (P_4, P_7, N_3, G_4, G_8, G_{14}), dysphoric mood (G_1, G2, G_3, G_4, G_6) and autistic preoccupations (P_3, N_5, N_7, G_{11}, G_{13}, G_{15}). In addition, depressive factor (G_1, G_2, G_3, G_6)[187], anxiety-depression factor (G_1, G_2, G_3, G_6, G_{15})[188], and mood factor (G_1, G_2, G_3, G_4, G_6)[189] were evaluated.

Neurocognitive functions were assessed using tests from the computerized *Cambridge Automated Neuropsychological Test Battery* (CANTAB). For a description of the nature of these tests, the performance measures used and how the test scores are derived, see (*http://www.cantab.com/cantab/site/home.acds*). Briefly, these tests were run on an IBM-compatible personal computer with a touch-sensitive screen. Neuropsychological testing lasted approximately 2 h. Subjects completed the tests in a fixed order with a break half-way through. The nonverbal nature of the CANTAB tests makes them largely language-independent and culture-free. Overall, 13 neuropsychological tests were grouped into five cognitive domains: visual and movement skills, attention and memory, learning, sustained attention, and executive function: Motor Screening (MOT), Big/Little Circle (BLC), Reaction Time (RTI), Matching to Sample Visual Search (MTS), Delayed Matching to Sample (DMS), Pattern Recognition Memory (PRM), Spatial Recognition Memory (SRM), Spatial Span (SSP), Paired Associates Learning (PAL), Rapid visual information processing (RVP), Spatial working memory (SWM), Intra/Extra Dimensional Set Shift (IED), and Stockings of Cambridge (SOC). Neurocognitive functions of schizophrenia patients are presented by standardized z-scores that are given as the number of standard deviations from the mean performance computed relative to an extensive database of raw scores for normal, healthy adult subjects matched by age and sex. The average value of the z-scores of the CANTAB neurocognitive tasks was used to determine the cognitive indices in specific domains. Negative values of the z-scores indicate poorer than average performance.

For assessment of insight for illness, the *Insight and Treatment Attitudes Questionnaire* (ITAQ)[190] was employed. Responses are scored on a 3-point scale (0 – no, 1 – questionable, and 2 – good insight). Participants also completed the *Insight Self-report Scale* (IS)[191] with Cronbach's $\alpha=0.86$.

The presence and severity of adverse effects of medication as well as psychological responses to them were measured with the *Distress Scale for Adverse Symptoms* (DSAS)[11,12]. The DSAS is a 22-item checklist covering mental, neurological, somatic, and autonomic dysfunctions caused by current medication. Adverse symptoms are rated in a face-to-face interview on a 5-point intensity scale (0-none or questionable symptom to 4-extreme expression of the symptom). The patient is then asked "How much discomfort has each of these symptoms caused you during the previous week?" Responses are scored in the same way, with higher mean scores indicating greater intensity of associated distress. Three DSAS indices related to adverse events were computed: Number of Adverse Symptoms (NAS), Mental Distress Index (MDI, Cronbach's $\alpha = 0.89$), Somatic Distress Index (SDI, Cronbach's $\alpha = 0.81$), and the DSAS index that covers both observer-rated and self-report items of the DSAS. Higher index scores indicate a greater number of adverse events (NAS) and that higher distress levels are attributed to a given side effect (MDI, SDI).

Intrarater reliability scores for these rating scales were established by calculating interclass correlation coefficients (ICCs) in 22 patients who were assessed by two raters. All ICCs for QLS, PANSS, ITAQ, GAF, and DSAS (NAS) dimensions were significant ($p < 0.001$) and varied between 0.81 and 0.93.

Assessment of emotional distress was done using the *Talbieh Brief Distress Inventory* (TBDI). Construction, properties of the TBDI, its internal consistency, and validity are reported in detail elsewhere[63,192]. The TBDI is a 24-item questionnaire covering the six psychological symptoms: obssesiveness, hostility, anxiety, and paranoid ideation (each with 3 items), sensitivity (4 items), and depression (7 items). Responses are scored on a 0 to 4-point scale, with higher scores indicating greater intensity of distress, and a particular symptom severity. A general TBDI index, as an average of 24 items, is computed (range=0-4). To analyze specific psychological symptoms, differential criterial thresholds for each of TBDI subscales were made and validated elsewhere[193]. The threshold magnitudes of the mean scores for the following symptom subscales were obssesiveness and hostility (both> 1.3), sensitivity (> 0.8), anxiety (> 1.5), paranoid ideation (> 0.9), and depression (> 1.4). Thus, subjects who scored above these thresholds were considered to have the symptom, and those scoring lower than the threshold were asymptomatic. For each respondent, we calculated the TBDI mean number of symptoms independent of modality. The rationale for the symptom count is the observation that the number of symptoms increases with the intensity of psychological distress.

The *Somatization Scale* is derived from the *Brief Symptom Inventory* (BSI)[194]. The BSI-somatization scale reflects distress arising from perceptions of bodily dysfunction. Task-, emotion-, and avoidance-oriented coping styles were evaluated with the *Coping Inventory for Stressful Situations* (CISS)[86]. The *Rosenberg Self-Esteem scale* is a well-known 10-item self-report questionnaire for measuring self-esteem and self-regard (RSES)[195]. The *General Self-Efficacy Scale* is a 10-item standard abridged version of the GSES for evaluating a sense of personal competence in stressful situations (GSES)[82]. The *Multidimensional Scale of*

Perceived Social Support (MSPSS)[196] was used as a measure of social support. The MSPSS is a psychometric instrument used to assess emotional support and the degree of satisfaction with perceived social support from family, friends, and others considered important. The *Level of Expressed Emotion* (LEE) scale was developed to provide an index of the perceived emotional climate in a person's influential relationships[197]. In addition to providing an overall score, the 60-item scale assesses the following four characteristic attitudes or response styles of others considered important: intrusiveness, emotional response, attitude toward illness, and tolerance/expectations. The LEE scale has sound psychometric properties of internal consistency; reliability; independence from sex, age, and the number of contacts; and construct validity[198].

Personality traits were measured with the *Tridimensional Personality Questionnaire* (TPQ)[199], a 100-question self-report instrument that discriminated between different major temperament traits using three dimensions: novelty seeking (NS), harm avoidance (HA), and reward dependence (RD). Novelty seeking is viewed as a heritable bias in the activation or initiation of behaviors such as frequent exploratory activity in response to novelty, impulsive decision-making, extravagance in the approach to cues of reward, and quick loss of temper and active avoidance of frustration. Harm avoidance is considered a heritable bias in the inhibition or cessation of behaviors such as pessimistic worry in anticipation of future problems, passive avoidant behaviors such as fear of uncertainty and shyness of strangers, and rapid fatigability. Reward dependence is viewed as a heritable bias in the maintenance or continuation of ongoing behaviors, and is manifested as sentimentality, social attachment, and dependence on the approval of others.

Data Analysis

The NCSS-2000 PC program was used for all analyses (*http://www.ncss.com*)[200]. Differences between groups regarding continuous variables were evaluated with two-tailed t-tests. Mean values with standard deviation (SD) are presented. Differences in frequency of categorical variables were examined with a χ^2 test. The confounding effect of key variables on HRQL impairment was tested with analysis of covariance (ANCOVA). Pearson correlation coefficients were calculated between the key variables. Partial correlation analysis was also applied. The principle axis method of factor analysis with the varimax rotated factor matrix was performed for each group of patients with SMD. The eigenvalues are used to determine how many factors to retain. One rule-of-thumb is to retain those factors whose eigenvalues are greater than one. Multivariate regression analysis was used for predicting the HRQL scores from clinical and psychosocial independent variables. Before testing regression models, we used a step-wise backward selection procedure in order to reduce the number of independent variables to a much smaller number of predictors.

ABBREVIATIONS

AIMS	Abnormal Involuntary Movement Scale
BPRS	Brief Psychiatric Rating Scale [101,102]
BSI	Brief Symptom Inventory [194]
CANTAB	Cambridge Automated Neuropsychological Test Battery
CGI	Clinical Global Impression Scale
CISS	Coping Inventory for Stressful Situations [86]
CSES	General Self-Efficacy Scale [82]
DSAS	Distress Scale for Adverse Symptoms [11,12]
DSM-IV	Diagnostic and Statistical Manual of Mental Disorders
EPSILON	European Psychiatric Services: Inputs Linked to Outcome Domains and Needs study [132]
GAF	Global Assessment of Functioning Scale [184]
Q-LES-Q	Quality of Life Enjoyment and Life Satisfaction Questionnaire [183]
QLS	Heinrichs-Carpenter Quality of Life Scale [7]
QOLI	Lehman's Quality of Life Interview [29,144]
QWB	Quality of Well-Being scale [106,177]
IS	Insight Self-report Scale [191]
ITAQ	Insight and Treatment Attitudes Questionnaire [190]
LEE	Level of Expressed Emotion [197]
LQOLP	Lancashire Quality of Life Profile [33,35,55,56,141,178]
MSPSS	Multidimensional Scale of Perceived Social Support [196]
PANSS	Positive and Negative Syndromes Scale [185]
RSES	Rosenberg Self-Esteem Scale [195]
SF-36	Short Form 36-Item Health Questionnaire [30,32,37,106,162,176]
TBDI	Talbieh Brief Distress Inventory [63,192]
TPQ	Tridimensional Personality Questionnaire [199]
WCST	Wisconsin Card Sorting Test [112,114,135]
WHOQOL-BREF	World Health Organization Quality of Life-Bref Scale [32,116]

ACKNOWLEDGEMENTS

The authors would like to thank the following Drs.: O. Rivkin, MD, E. Bistrov, MD, H. Farkash, MD, H. Ilan, MD, G. Perelroyzen, MD, Y. Ratner, MD, and E. Shinkarenko, MD for their assistance in compiling data. We acknowledge the considerable support from our departmental staff, and research assistants Ms. Michal Z'ada, and Ms. Moria Dayan. We wish to thank our patients who participated in this project. We also with to express gratitude to Professor Ilan Modai, MD, M.H.A. (Director, Shaar Menashe Mental Health Center), Professor Jean Endicott, Ph.D. (Columbia University, New York, USA), and to Ms. Rena Kurs, B.A. (Lev-hasharon Mental Health Center, Netanya, Israel) for their fruitful support. The authors acknowledge the dedicated editorial assistance of Steve Manch.

REFERENCES

1. Ritsner M, Kurs R. Impact of antipsychotic agents and their side effects on the quality of life in schizophrenia. Expert Review and Pharmacoeconomic Outcomes Research 2002; 2:89–98.
2. Ritsner M, Kurs R. Quality of life outcomes in mental illness: schizophrenia, mood and anxiety disorders. Expert Review and Pharmacoeconomic Outcomes Research 2003; 3:189–199.
3. Ritsner, M., Kurs, R. Quality-of-Life Impairment in Severe Mental Illness: Focus on Schizoaffective Disorders. In: William H. Murray, ed. Schizoaffective Disorder: New Research. NOVA Publishers, NY, 2006, pp. 69–107.
4. Awad AG, Voruganti LN. Impact of atypical antipsychotics on quality of life in patients with schizophrenia. CNS Drugs. 2004, 18:877–893.
5. Lambert M, Naber D. Current issues in schizophrenia: overview of patient acceptability, functioning capacity and quality of life. CNS Drugs. 2004; (18 Suppl) 2:5–17.
6. Hansson L.Determinants of quality of life in people with severe mental illness. Acta Psychiatr Scand 2006; (429 suppl):46–50.
7. Heinrichs DW, Hanlon TE, Carpenter WT. The Quality of Life Scale: an instrument for rating the schizophrenic deficit scale. Schizophrenia Bull. 1984; 10:388–398.
8. Atkinson M, Zibin S, Chuang H. Characterizing quality of life among patients with chronic mental illness: a critical examination of the self-report methodology. *Am J Psychiatry* 1997; 154:99–105.
9. Voruganti L, Heslegrave R, Awad AG, Seeman MV. Quality of life measurement in schizophrenia: reconciling the quest for subjectivity with the question of reliability. Psychol Med 1998; 28:165–172.
10. Whitty P, Browne S, Clarke M, McTigue O, Waddington J, Kinsella T, Larkin C, O'Callaghan E. Systematic comparison of subjective and objective measures of quality of life at 4-year follow-up subsequent to a first episode of psychosis. J Nerv Ment Dis. 2004; 192:805–809.
11. Ritsner M, Modai I, Endicott J, et al. Differences in quality of life domains and psychopathologic and psychosocial factors in psychiatric patients. J Clin Psychiatry 2000; 61:880–889.
12. Ritsner M, Ponizovsky A, Endicott J, et al. The impact of side-effects of antipsychotic agents on life satisfaction of schizophrenia patients: a naturalistic study. Eur Neuropsychopharmacol 2002; 12:31–38.
13. Ritsner M, Modai I, Kurs R, et al. Subjective Quality of Life Measurements in Severe Mental Health Patients: Measuring Quality of Life of Psychiatric Patients: Comparison Two Questionnaires. Quality Life Research 2002; 11:553–561.
14. Ponizovsky AM, Grinshpoon A, Levav I, Ritsner MS. Life satisfaction and suicidal attempts among persons with schizophrenia. Comprehensive Psychiatry 2003; 44:442–447.
15. Ritsner M, Kurs R, Gibel A, et al. Predictors of quality of life in major psychoses: a naturalistic follow-up study. Journal of Clinical Psychiatry 2003; 64:308–315.
16. Ritsner M, Ben-Avi I, Ponizovsky A, et al. Quality of life and coping with schizophrenia symptoms. Quality of Life Research 2003; 12:1–9.
17. Ritsner M. Predicting changes in domain-specific quality of life of schizophrenia patients. J Nerv Ment Dis 2003; 191:287–294.
18. Ritsner, M., Farkas, H., Gibel, A. Satisfaction with quality of life varies with temperament types of patients with schizophrenia. Journal of Nervous and Mental Disease 2003; 191:668–674.
19. Ritsner M, Kurs R, Ponizovsky A, Hadjez J. Perceived quality of life in schizophrenia: relationships to sleep quality. Quality of Life Research 2004; 13:783–791.
20. Ritsner, M., Perelroyzen, G., Kurs, R., et al. Quality of life outcomes in schizophrenia patients treated with atypical and typical antipsychotic agents: A naturalistic comparative study. International Clinical Psychopharmacology 2004; 24:582–591.
21. Ritsner M, Perelroyzen G, Ilan H, Gibel A. Subjective response to antipsychotics of schizophrenia patients treated in routine clinical practice: a naturalistic comparative study. J Clin Psychopharmacol. 2004; 24:245–54.

22. Ritsner M, Susser E. Temperament types are associated with weak self-construct, elevated distress and emotion-oriented coping in schizophrenia: evidence for a complex vulnerability marker? Psychiatry Res. 2004; 128:219–28.
23. Kurs R, Farkas H, Ritsner M. Quality of life and temperament factors in schizophrenia: comparative study of patients, their siblings and controls. Quality of Life Research 2005; 14:433–440.
24. Ritsner M, Kurs R, Gibel A, Ratner Y, Endicott J. Validity of an abbreviated quality of life enjoyment and satisfaction questionnaire (Q-LES-Q-18) for schizophrenia, schizoaffective, and mood disorder patients. Qual Life Res. 2005; 14:1693–703.
25. Ritsner M, Kurs R, Ratner Y, Gibel A. Condensed version of the Quality of Life Scale for schizophrenia for use in outcome studies. Psychiatry Res. 2005; 135:65–75.
26. Ritsner MS, Ratner Y. The long-term changes in coping strategies in schizophrenia: temporal coping types. J Nerv Ment Dis. 2006; 194:261–267.
27. Ritsner M, Gibel A, Ratner Y. Determinants of Changes in Perceived Quality of Life in the Course of Schizophrenia. Quality of Life Research 2006; 15:515–526.
28. Ritsner MS, Gibel, A. The effectiveness and predictors of response to antipsychotic agents to treat impaired quality of life in schizophrenia: a 12-month naturalistic follow-up study with implications for confounding factors, antidepressants, anxiolytics, and mood stabilizers. Progress in Neuro-Psychopharmacology & Biological Psychiatry 2006;30:1442–1452 .
29. Lehman, A. F. The well-being of chronic mental patients: assessing their quality of life. Archives of General Psychiatry 1983; 40:369–373.
30. Sciolla A, Patterson TL, Wetherell JL, et al. Functioning and well-being of middle-aged and older patients with schizophrenia: measurement with the 36-item short-form (SF-36) health survey. Am J Geriatr Psychiatry. 2003; 11:629–637.
31. Gorna K, Jaracz K, Rybakowski F. Objective and subjective quality of life in schizophrenic patients after a first hospitalization. Rocz Akad Med Bialymst. 2005; (50 Suppl) 1:225–7.
32. Law CW, Chen EY, Cheung EF, Chan RC, Wong JG, Lam CL, Leung KF, Lo MS. Impact of untreated psychosis on quality of life in patients with first-episode schizophrenia. Qual Life Res. 2005;14:1803–1811.
33. Kentros MK, Terkelsen K, Hull J, et al.The relationship between personality and quality of life in persons with schizoaffective disorder and schizophrenia. Quality of Life Research, 1997, 6,118–22.
34. Tollefson GD, Andersen SW. Should we consider mood disturbance in schizophrenia as an important determinant of quality of life? Journal of Clinical Psychiatry 1999; (60 Suppl)5:23–29.
35. Rudolf H, Priebe S. Subjective quality of life in female in-patients with depression: a longitudinal study. International Journal of Social Psychiatry 1999; 45:238–246.
36. Koivumaa-Honkanen HT, Honkanen R, Antikainen R,et al. Self-reported life satisfaction and treatment factors in patients with schizophrenia, major depression and anxiety disorder. Acta Psychiatrica Scandinavica 1999; 99:377–384.
37. Wittchen HU, Carter RM, Pfister H, et al. Disabilities and quality of life in pure and comorbid generalized anxiety disorder and major depression in a national survey. *Int Clin Psychopharmacol.* 2000; 15:319–328.
38. Bystritsky A, Liberman RP, Hwang S, et al. Social functioning and quality of life comparisons between obsessive-compulsive and schizophrenic disorders. *Depress Anxiety.* 2001; 14:214–218.
39. Bobes J, Gonzalez MP, Bascaran MT, et al. Quality of life and disability in patients with obsessive-compulsive disorder. *Eur Psychiatry* 2001;16:239–245.
40. Rapaport MH, Clary C, Fayyad R, Endicott J. Quality-of-life impairment in depressive and anxiety disorders. American Journal of Psychiatry 2005; 162:1171–1178.
41. Stone WS, Faraone SV, Seidman LJ, et al. Concurrent validation of schizotaxia: a pilot study. Biol Psychiatry 2001; 50:434–440.
42. Browne S, Clarke M, Gervin M, et al. Determinants of quality of life at first presentation with schizophrenia. Br J Psychiatry 2000; 176:173–176.
43. Melle I, Haahr U, Friis S, et al. Reducing the duration of untreated first-episode psychosis - effects on baseline social functioning and quality of life. Acta Psychiatr Scand. 2005; 112:469–473.

44. Czernikiewicz A, Gorecka J, Kozak-Sykala A. [Premorbid adjustment and quality of life in schizophrenia]. Pol Merkuriusz Lek. 2005; 19:659–662.
45. Erickson DH, Beiser M, Iacono WG, et al. The role of social relationships in the course of first-episode schizophrenia and affective psychosis. Am J Psychiatry. 1989; 146:1456–1461.
46. Bechdolf A, Pukrop R, Kohn D, et al. Subjective quality of life in subjects at risk for a first episode of psychosis: a comparison with first episode schizophrenia patients and healthy controls. Schizophr Res. 2005; 79:137–143.
47. Priebe S, Roeder-Wanner UU, Kaiser W. Quality of life in first-admitted schizophrenia patients: a follow-up study. Psychological Medicine 2000; 30:225–230.
48. Tempier R, Mercier C, Leouffre P, Caron J. Quality of life and social integration of severely mentally ill patients: a longitudinal study. Journal of Psychiatry and Neuroscience 1997; 22:249–255.
49. Skantze K. Subjective quality of life and standard of living: a 10-year follow-up of out-patients with schizophrenia. Acta Psychiatrica Scandinavica 1998; 98:390–399.
50. Hoffmann K, Kaiser W, Isermann M, Priebe S. [How does the quality of life of long-term hospitalized psychiatric patients change after their discharge into the community?] *Gesundheitswesen* 1998; 60:232–238.
51. Henkel H, Schmitz M, Berghofer G. [Quality of life of the mentally ill] Wiener medizinische Wochenschrift 2000; 150:32–36.
52. Malla AK, Norman RM, McLean TS, McIntosh E. Impact of phase-specific treatment of first episode of psychosis on Wisconsin Quality of Life Index (client version). Acta Psychiatrica Scandinavica 2001;103:355–361.
53. Ruggeri M, Lasalvia A, Tansella M, et al. Heterogeneity of outcomes in schizophrenia. 3-year follow-up of treated prevalent cases. Br J Psychiatry 2004; 184:48–57.
54. Hansson L, Middelboe T, Merinder L, et al. Predictors of subjective quality of life in schizophrenic patients living in the community. A Nordic multicentre study. : International Journal of Social Psychiatry 1999; 45:247–258.
55. Hansson L, Eklund M, Bengtsson-Tops A. The relationship of personality dimensions as measured by the temperament and character inventory and quality of life in individuals with schizophrenia or schizoaffective disorder living in the community. Quality of Life Research 2001; 10:133–139.
56. Eklund M, Backstrom M, Hansson L. Personality and self-variables: important determinants of subjective quality of life in schizophrenia out-patients. Acta Psychiatrica Scandinavica 2003; 108:134–143.
57. Eklund M, Hansson L, Bengtsson-Tops A. The influence of temperament and character on functioning and aspects of psychological health among people with schizophrenia. Eur Psychiatry 2004; 19:34–41.
58. Eklund M, Backstrom M. A model of subjective quality of life for outpatients with schizophrenia and other psychoses. Quality of Life Research 2005; 14:1157–1168.
59. Lazarus, R. S., & Folkman, S. Stress, appraisal, and coping. New York: Springer. 1984.
60. Dohrenwend BP, Shrout PE, Egri G, Mendelsohn FS. Nonspecific psychological distress and other dimensions of psychopathology. Measures for use in the general population. Arch Gen Psychiatry 1980; 37:1229–1236.
61. Derogatis LR, Coons HL. Self-report measures of stress. In: Goldberger L and Breznittz S, ed. Handbook of Stress: Theoretical and Clinical Aspects. Second edition. The Free Press, New York,pp, 200–233 1993; pp, 200–233. 1993.
62. Coyne JC, Schwenk TL. The realtionship of distress to mood disturbance in primary care and psychiatric populations. J Consult Clin Psychol 1997; 65:161–168.
63. Ritsner M, Modai I, Ponizovsky A. The Talbieh Brief Distress Inventory: validation for use with psychiatric patients. Compr Psychiatry 2002; 2:144–163.
64. Norman RMG, Malla AK: Stressful life events and schizophrenia, I: a review of the research. Br J Psychiatry 1993; 162:161–166.
65. Myin-Germeys, I., Jim van, O.S., Schwartz, J.E., Stone, A.A., Delespaul, P.A. Emotional reactivity to daily life stress in psychosis. Archives of General Psychiatry 2001; 58:1093–1208.

66. Ritsner M. The attribution of somatization in schizophrenia patients: a naturalistic follow-up study. J Clin Psychiatry. 2003; 64:1370–8.
67. Selten JP, Wiersma D, van den Bosch RJ. Distress attributed to negative symptoms in schizophrenia. Schizophr Bull 2000; 26:737–744.
68. Fakhoury WK, Wright D, Wallace M. Prevalence and extent of distress of adverse effects of antipsychotics among callers to a United Kingdom National Mental Health Helpline. Int Clin Psychopharmacol 2001; 16:153–1623.
69. Caron J, Lecomte Y, Stip E, Renaud S. Predictors of quality of life in schizophrenia. Community Ment Health J. 2005; 41:399–417.
70. Awad AG, Lapierre YD, Angus C, Rylander A. Quality of life and response of negative symptoms in schizophrenia to haloperidol and the atypical antipsychotic remoxipride. Journal of Psychiatry and Neuroscience 1997; 22:244–248.
71. Lasalvia A, Ruggeri M, Santolini N. Subjective quality of life: its relationship with clinician-rated and patient-rated psychopathology. The South-Verona Outcome Project 6.Psychotherapy and Psychosomatics 2002; 71:275–284.
72. Lobana A, Mattoo SK, Basu D, Gupta N. Quality of life in schizophrenia in India: comparison of three approaches. Acta Psychiatrica Scandinavica 2001; 104:51–55.
73. Hooley JM, Campbell C. Control and controllability: 462 beliefs and behavior in high and low expressed emotion relatives. Psychol Med 2002; 32:1091–1099.
74. Mubarak AR, Barber JG. Emotional expressiveness and the quality of life of patients with schizophrenia. Soc Psychiatry Psychiatr Epidemiol. 2003; 38:380–384.
75. Guillem F, Bicu M, Semkovska,M, Debruille . The dimensional symptom structure of schizophrenia and its association with temperament and character. Schizophrenia Research 2002; 56:137–147.
76. Szoke, A., Schurhoff, F., Ferhadian, N., et al.Temperament in schizophrenia: a study of the tridimensional personality questionnaire (TPQ). European Psychiatry 2002; 17:379–383.
77. Diener E, Diener M. Cross-cultural correlates of life satisfaction and self-esteem. J Pers Soc Psychol. 1995; 68:653–663.
78. Roe D. A prospective study on the relationship between self-esteem and functioning during the first year after being hospitalized for psychosis. J Nerv Ment Dis. 2003; 191:45–9.
79. Gureje O, Harvey C, Herrman H. Self-esteem in patients who have recovered from psychosis: profile and relationship to quality of life. Aust N Z J Psychiatry. 2004; 38:334–338.
80. Zissi A, Barry MM, Cochrane R. A mediational model of quality of life for individuals with severe mental health problems. Psychological Medicine 1998; 28:1221–1230.
81. Ruggeri M, Bisoffi G, Fontecedro L, Warner R. Subjective and objective dimensions of quality of life in psychiatric patients: a factor analytical approach: The South Verona Outcome Project 4. British Journal of Psychiatry 2001; 178:268–275.
82. Jerusalem M, Schwarzer R. Self-efficacy as a resource factor in stress appraisal processes. In R. Schwarzer (Ed.), Self-efficacy: Thought control of action. Hemisphere, Washington, DC pp. 195–213, 1992.
83. Strous, RD, Ratner, Y, Gibel, A, Ponizovsky, A, Ritsner, M. Longitudinal assessment of coping abilities at exacerbation and stabilization in schizophrenia. Comprehensive Psychiatry 2005; 46:167–175.
84. Endler NS, Parker JDA, Summerfeldt LJ. Coping with health problems: Conceptual and methodological issues. Can J Behav Sci 1993; 25:384–399.
85. Lazarus RS. Coping theory and research: past, present, and future. Psychosom Med 55:234–247. 1993.
86. Endler NS, Parker JD. Multidimensional assessment of coping: a critical evaluation. Journal of Personality and Social Psychology 1990; 58:844–854.
87. Wilder-Willis, K.E., Shear, P.K., Steffen, J.J., Borkin, J. The relationship between cognitive dysfunction and coping abilities in schizophrenia. Schizophrenia Research 2002; 55:259–267.
88. Higgins JE, Endler NS. Coping, life stress and psychological and somatic distress. European Journal of Personality 1995; 9:253–270.

89. Lysaker PH, Bryson GJ, Marks K, Greig TC, Bell MD. Coping style in schizophrenia: associations with neurocognitive deficits and personality. Schizophrenia Bulletin 2004; 30:113–121.
90. Bengtsson-Tops A, Hansson L. Quantitative and qualitative aspects of the social network in schizophrenic patients living in the community. Relationship to sociodemographic characteristics and clinical factors and subjective quality of life. International Journal of Social Psychiatry 2001; 47:67–77.
91. Young , K.M. Factors predicting overall life satisfaction for people with long-term mental illness factors. International Journal of Psychosocial Rehabilitation. 2004 ; 9:23–35.
92. Marshall M, Lewis S, Lockwood A, et al. Association between duration of untreated psychosis and outcome in cohorts of first-episode patients: a systematic review. Arch Gen Psychiatry 2005; 62:975–83.
93. Ho BC, Andreasen NC, Flaum M, et al. Untreated initial psychosis: its relation to quality of life and symptom remission in first-episode schizophrenia. Am J Psychiatry 2000; 157:808–815.
94. Harris MG, Henry LP, Harrigan SM, et al. The relationship between duration of untreated psychosis and outcome: an eight-year prospective study. Schizophr Res. 2005; 79:85–93.
95. Malla A, Payne J. First-episode psychosis: psychopathology, quality of life, and functional outcome. Schizophr Bull. 2005; 31:650–671.
96. Sands JR, Harrow M. Depression during the longitudinal course of schizophrenia. Schizophr Bull. 1999; 25:157–71.
97. Packer S, Husted J, Cohen S, Tomlinson G. Psychopathology and quality of life in schizophrenia. Journal of Psychiatry and Neuroscience 1997; 22:231–234.
98. Fitzgerald PB, Williams CL, Corteling N, et al. Subject and observer-rated quality of life in schizophrenia. Acta Psychiatrica Scandinavica 2001;103:387–392.
99. Smith KW, Avis NE, Assmann SF. Distinguishing between quality of life and health status in quality of life research: A meta-analysis. Quality of Life Research: International Journal of Quality of Life Aspects Treatment, Care and Rehabilitation 1999; 8:447–459.
100. Huppert JD, Smith TE. Longitudinal analysis of subjective quality of life in schizophrenia: anxiety as the best symptom predictor. Journal of Nervous and Mental Disease 2001; 189:669–675.
101. Huppert JD, Weiss KA, Lim R, et al. Quality of life in schizophrenia: contributions of anxiety and depression. Schizophrenia Research 2001; 51:171–180.
102. Hofer A, Kemmler G, Eder U, et al. Quality of life in schizophrenia: The impact of psychopathology, attitude toward medication, and side effects. J Clin Psychiatry 2004; 65: 932–939.
103. Braga RJ, Mendlowicz MV, Marrocos RP, Figueira IL.Anxiety disorders in outpatients with schizophrenia: prevalence and impact on the subjective quality of life. J Psychiatr Res. 2005; 39:409–14.
104. Karow A, Moritz S, Lambert M, et al. PANSS syndromes and quality of life in schizophrenia. Psychopathology. 2005; 38:320–326.
105. Huppert JD, Smith TE. Anxiety and schizophrenia: the interaction of subtypes of anxiety and psychotic symptoms. CNS Spectr. 2005;10:721–31.
106. Wetherell JL, Palmer BW, Thorp SR, et al. Anxiety symptoms and quality of life in middle-aged and older outpatients with schizophrenia and schizoaffective disorder. Journal of Clinical Psychiatry 2003; 64:1476–1482.
107. Galletly CA, Clark CR, McFarlane AC, Weber DC. Relationships between changes in symptom ratings, neurophysiological test performance and quality of life in schizophrenic patients treated with clozapine. Psychiatry Res 1997; 72:161–166.
108. Kaiser W, Priebe S, Barr W, et al. Profiles of subjective quality of life in schizophrenic in- and out-patient samples. Psychiatry Research 1997; 66:153–166.
109. Palmer BW, Heaton RK, Paulsen JS, Kuck J, Braff D, Harris MJ, Zisook S,Jeste DV. Is it possible to be schizophrenic yet neuropsychologically normal? Neuropsychology 1997;11:437–446.
110. Heslegrave RJ, Awad AG, Voruganti LN. The influence of neurocognitive deficits and symptoms on quality of life in schizophrenia. *J Psychiatry Neurosci* 1997; 22:235–243.

111. Dickerson FB, Ringel NB, Parente. Subjective quality of life in out-patients with schizophrenia: clinical and utilization correlates. Acta Psychiatr Scand 1998; 98:124–127.
112. Kaiser W. Cognitive effects of antipsychotics in schizophrenia and relationship to quality of life. British Journal of Psychiatry 2000; 176:92–93.
113. McDermid Vaz SA, Heinrichs RW. Schizophrenia and memory impairment: evidence for a neurocognitive subtype. Psychiatry Research, 2002, 113, 93–105.
114. Sota TL, Heinrichs RW. Demographic, clinical, and neurocognitive predictors of quality of life in schizophrenia patients receiving conventional neuroleptics. Compr Psychiatry. 2004; 45:415–21.
115. Alptekin K, Akvardar Y, Kivircik Akdede BB, et al. Is quality of life associated with cognitive impairment in schizophrenia? Progress in Neuropsychopharmacology and Biological Psychiatry 2005; 29:239–44.
116. Wegener S, Redoblado-Hodge MA, Lucas S, et al. Relative contributions of psychiatric symptoms and neuropsychological functioning to quality of life in first-episode psychosis. Aust N Z J Psychiatry. 2005; 39:487–492.
117. Ritsner, MS. Predicting quality of life impairment in chronic schizophrenia from cognitive variables. Quality of Life Research (2007; in press).
118. Amador, X., Strauss, D., Yale, S., Gorman, J. 1991. Awareness of illness in schizophrenia. Schizophrenia Bulletin 1991; 17:113–132.
119. Browne S, Garavan J, Gervin M, et al. Quality of life in schizophrenia: insight and subjective response to neuroleptics. Journal of Nervous and Mental Disease 1998; 186:74–78.
120. Lysaker PH, Bell MD, Bryson GJ, Kaplan E. Insight and interpersonal function in schizophrenia. Journal of Nervous and Mental Disease 1998; 186:432–436.
121. Zeitlhofer J, Schmeiser-Rieder A, Tribl G, et al. Sleep and quality of life in the Austrian population. Acta Neurologica Scandinavica 2000; 102:249–257.
122. Leger D, Scheuermaier K, Philip P, et al. SF-36: evaluation of quality of life in severe and mild insomniacs compared with good sleepers. Psychosomatic Medicine 2001; 63:49–55.
123. Doi Y, Minowa M, Okawa M, Uchiyama M. Prevalence of sleep disturbance and hypnotic medication use in relation to sociodemographic factors in the general Japanese adult population. Journal of Epidemiology 2000; 10:79–86.
124. Benca, RM. Consequences of insomnia and its therapies. Journal of Clinical Psychiatry 2001; 62 (Suppl 10): S33-S38.
125. Rosch PJ. Stress and sleep: some startling and sobering statistics. Stress Medicine 1996; 12:207–210.
126. Singareddy RK, Balon R. Sleep and suicide in psychiatric patients. Annals of Clinical Psychiatry 2001; 13:93–101.
127. Harris EC, Barraclough B. Suicide as an outcome for mental disorders: a meta-analysis. Br J Psychiatry 1997; 170:205–228.
128. Harkavy-Friedman JM, Resifo K, et al. Suicidal behavior in schizophrenia: characteristics of individuals who had and had not attempted suicide. Am J Psychiatry 1999; 156:1276–1278.
129. Koivumaa-Honkanen HT, Honkanen R, Viinamaki H, et al. Life satisfaction and suicide: a 20-year follow-up study. Am J Psychiatry 2001; 158:433–439.
130. Wiersma D, Nienhuis FJ, Giel R, Slooff CJ. Stability and change in needs of patients with schizophrenic disorders: a 15- and 17-year follow-up from first onset of psychosis, and a comparison between 'objective' and 'subjective' assessments of needs for care. Soc Psychiatry Psychiatr Epidemiol. 1998; 33:49–56.
131. Slade M, Leese M, Ruggeri M, et al. Does meeting needs improve quality of life? Psychother Psychosom. 2004; 73:183–189.
132. Thornicroft G, Tansella M, Becker T, et al. EPSILON Study Group. The personal impact of schizophrenia in Europe. Schizophr Res. 2004; 69:125–132.
133. Korkeila J, Heikkila J, Hansson L, et al. Structure of needs among persons with schizophrenia. Soc Psychiatry Psychiatr Epidemiol. 2005; 40:233–9.
134. Barry MM, Crosby C. Quality of life as an evaluative measure in assessing the impact of community care on people with long-term psychiatric disorders. Br J Psychiatry. 1996; 168:210–6.

135. Brekke JS, Kohrt B, Green MF. Neuropsychological functioning as a moderator of the relationship between psychosocial functioning and the subjective experience of self and life in schizophrenia. *Schizophr Bull.* 2001; 27:697–708.
136. Becker T, Leese M, Krumm S, et al. EPSILON Study Group.Needs and quality of life among patients with schizophrenia in five European centres: what is the impact of global functioning scores? Soc Psychiatry Psychiatr Epidemiol. 2005; 40:628–634.
137. Priebe S, Warner R, Hubschmid T, Eckle I. Employment, attitudes toward work, and quality of life among people with schizophrenia in three countries. *Schizophr Bull* 1998; 24:469–77.
138. Eklund M, Hansson L, Bejerholm U. Relationships between satisfaction with occupational factors and health-related variables in schizophrenia outpatients. Soc Psychiatry Psychiatr Epidemiol. 2001; 36:79–83.
139. Bond GR, Resnick SG, Drake RE, et al. Does competitive employment improve nonvocational outcomes for people with severe mental illness? J Consult Clin Psychol. 2001; 69:489–501.
140. Marwaha S, Johnson S. Schizophrenia and employment - a review. Soc Psychiatry Psychiatr Epidemiol. 2004; 39:337-349.
141. Kaiser W, Priebe S, Barr W, et al. Profiles of subjective quality of life in schizophrenic in- and out-patient samples. Psychiatry Research 1997; 66:153–66.
142. Franz M, Meyer T, Reber T, Gallhofer B. The importance of social comparisons for high levels of subjective quality of life in chronic schizophrenic patients. Qual Life Res. 2000; 9:481–9.
143. Hobbs C, Tennant C, Rosen A, et al. Deinstitutionalisation for long-term mental illness: a 2-year clinical evaluation. Aust N Z J Psychiatry. 2000; 34:476–83.
144. Lehman AF, Possidente S, Hawker F. The quality of life of chronic patients in a state hospital and in community residences. Hosp Community Psychiatry 1986; 37:901–7.
145. Kemmler G, Meise U, Tasser A. [Subjective quality of life of schizophrenic patients. Effect of treatment setting, psychopathology and extrapyramidal motor drug effects] Psychiatrische Praxis 1999; 26:9–15.
146. Rossler W, Salize HJ, Cucchiaro G, et al. Does the place of treatment influence the quality of life of schizophrenics? Acta Psychiatr Scand. 1999; 100:142–148.
147. Malla A, Williams R, Kopala L, et al. Outcome on quality of life in a Canadian national sample of patients with schizophrenia and related psychotic disorders. Acta Psychiatr Scand. 2006 ; 113:22–28.
148. Revicki DA, Genduso LA, Hamilton SH, et al. Olanzapine versus haloperidol in the treatment of schizophrenia and other psychotic disorders: quality of life and clinical outcomes of a randomized clinical trial. Qual Life Res 1999; 8:417–426.
149. Rovner J, Assuncao S, Gargoloff P, et al.[Functional status and quality of life in Latin American outpatients with schizophrenia treated with atypical or typical antipsychotics: outcomes of the 12 months schizophrenia outpatient health outcomes (IC-SOHO) Study]. Vertex. 2005; 16:332-41 [Article in Spanish].
150. Beasley CM Jr, Sutton VK, Taylor CC, et al. Is quality of life among minimally symptomatic patients with schizophrenia better following withdrawal or continuation of antipsychotic treatment? J Clin Psychopharmacol. 2006; 26:40–44.
151. Mayoral F, Montejo A, Bousono M, et al. Quality of life and social functioning in schizophrenic patients treated with olanzapine: 1 year follow-up naturalistic study. Actas Esp Psiquiatr. 2006; 34:7–15.
152. Rosenheck R, Perlick D, Bingham S, et al. Department of Veterans Affairs Cooperative Study Group on the Cost-Effectiveness of Olanzapine. Effectiveness and cost of olanzapine and haloperidol in the treatment of schizophrenia: a randomized controlled trial. JAMA. 2003; 290:2693–702.
153. Strakowski SM, Johnson JL, Delbello MP, et al. Quality of life during treatment with haloperidol or olanzapine in the year following a first psychotic episode. Schizophr Res. 2005; 78:161–169.
154. Gureje O, Miles W, Keks N, et al. Olanzapine vs risperidone in the management of schizophrenia: a randomized double-blind trial in Australia and New Zealand. Schizophr Res 2003; 61:303–314.

155. Ho BC, Nopoulos P, Flaum M, et al. Two-year outcome in first-episode schizophrenia: predictive value of symptoms for quality of life. American Journal of Psychiatry 1998; 155:1196–1201.
156. Werneke U, Turner T, Priebe S. Complementary medicines in psychiatry: review of effectiveness and safety. Br J Psychiatry. 2006; 188:109–21.
157. Aasen I, Kumari V, Sharma T. Effects of rivastigmine on sustained attention in schizophrenia: an FMRI study. J Clin Psychopharmacol. 2005; 25:311–7.
158. Lenzi A, Maltinti E, Poggi E, et al. Effects of rivastigmine on cognitive function and quality of life in patients with schizophrenia. Clin Neuropharmacol. 2003;; 26:317–21.
159. Peet M, Stokes C. Omega-3 fatty acids in the treatment of psychiatric disorders. Drugs. 2005; 65:1051–1059.
160. Arvindakshan M, Ghate M, Ranjekar PK, et al. Supplementation with a combination of omega-3 fatty acids and antioxidants (vitamins E and C) improves the outcome of schizophrenia. Schizophr Res. 2003; 62:195–204.
161. Rosenthal MH, Bryant SL. Benefits of adjunct modafinil in an open-label, pilot study in patients with schizophrenia. Clin Neuropharmacol. 2004; 27:38–43.
162. Becker PM, Schwartz JR, Feldman NT, Hughes RJ. Effect of modafinil on fatigue, mood, and health-related quality of life in patients with narcolepsy. Psychopharmacology (Berl). 2004; 171:133–9.
163. Ritsner M, Maayan R, Gibel A, et al. Elevation of the cortisol/dehydroepiandrosterone ratio in schizophrenia patients. European Neuropsychopharmacology 2004; 14:267–273.
164. Ritsner M, Gibel A, Maayan R, et al. Cortisol/dehydroepiandrosterone ratio and responses to antipsychotic treatment in schizophrenia. Neuropsychopharmacology. 2005; 30:1913–22.
165. Ritsner M, Gibel A, Ram E, et al. Alterations in DHEA metabolism in schizophrenia: two-month case-control study. Eur Neuropsychopharmacol. 2006; 16:137–146.
166. Ritsner, MS., Gibel, A., Ratner, Y., et al. Improvement of Sustained Attention, Visual and Movement Skills, but not Clinical Symptomatology, following Dehydroepiandrosterone Augmentation in Schizophrenia: a randomized double-blind, placebo-controlled, crossover trial. Journal of Clinical Psychopharmacology 2006; 26:495–499.
167. Voruganti LN, Heslegrave RJ, Awad AG. Quality of life measurement during antipsychotic drug therapy of schizophrenia. Journal of Psychiatry and Neuroscience 1997; 22:267–274.
168. Dernovsek MZ, Prevolnik Rupel V, Rebolj M, Tavcar R. Quality of life and treatment costs in schizophrenic outpatients, treated with depot neuroleptics. European Psychiatry 2001; 16: 474–482.
169. Browne S, Roe M, Lane A, et al. Quality of life in schizophrenia: relationship to sociodemographic factors, symptomatology and tardive dyskinesia. Acta Psychiatr Scand. 1996; 94:118–124.
170. Gerlach J, Larsen EB. Subjective experience and mental side-effects of antipsychotic treatment. Acta Psychiatrica Scandinavica Supplementum 1999; 395:113–117.
171. Allison DB, Mackell JA, McDonnell DD. The impact of weight gain on quality of life among persons with schizophrenia. Psychiatr Serv. 2003; 54:565–567.
172. Olfson M, Uttaro T, Carson WH, Tafesse E. Male sexual dysfunction and quality of life in schizophrenia. J Clin Psychiatry. 2005; 66:331–338.
173. Wirshing DA, Pierre JM, Marder SR, et al. Sexual side effects of novel antipsychotic medications. Schizophr Res. 2002; 56:25–30.
174. Skantze K. Malm U, Dencker S, et al. Comparison of quality of life with standard of living in schizophrenic outpatients. Br J Psychiatry 1992; 161:797–801.
175. Vandiver VL. Quality of life, gender and schizophrenia: a cross-national survey in Canada, Cuba, and U.S.A. Community Ment Health J 1998; 34:501–11.
176. Jarema, M., Konieczynska, Z. Quality of life in schizophrenia: Impact of psychopathology, patients' gender and antipsychotic treatment. Int J Psych Clin Pract 2001; 5:19–26.
177. Patterson TL, Shaw W, Semple SJ, et al. Health-related quality of life in older patients with schizophrenia and other psychoses: relationships among psychosocial and psychiatric factors. Int J Geriatr Psychiatry. 1997; 12:452–461.

178. [No authors listed] Predictors of quality of life in people with severe mental illness. Study methodology with baseline analysis in the UK700 trial. British Journal of Psychiatry, 1999; 175: 426–432.
179. Bechdolf A, Klosterkotter J, Hambrecht M, et al. Determinants of subjective quality of life in post acute patients with schizophrenia. Eur Arch Psychiatry Clin Neurosci. 2003; 253:235.
180. Chan PS, Krupa T, Lawson JS, Eastabrook S. An outcome in need of clarity: building a predictive model of subjective quality of life for persons with severe mental illness living in the community. Am J Occup Ther. 2005; 59:181–190.
181. Ruggeri M, Nose M, Bonetto C, et al. Changes and predictors of change in objective and subjective quality of life: multiwave follow-up study in community psychiatric practice. Br J Psychiatry. 2005; 187:121–130.
182. Gaite L, Vazquez-Barquero JL, Borra C, et al. The EPSILON Study Group. Quality of life in patients with schizophrenia in five European countries: the EPSILON study. Acta Psychiatrica Scandinavica 2002; 105:283–292.
183. Endicott J, Nee J, Harrison W, Blumenthal R. Quality of Life Enjoyment and Satisfaction Questionnaire: a new measure. Psychopharmacology Bulletin 1993; 29:321–326.
184. American Psychiatric Association. Diagnostic and statistical manual of mental disorders. Washington, DC, 1994.
185. Kay, S.R., Fiszbein, A., Opler, L.A., The Positive and Negative Syndrome Scale (PANSS) for schizophrenia. Schizophrenia Bulletin 1987; 13:261–276.
186. White L, Harvey PD, Opler L, Lindenmayer JP. Empirical assessment of the factorial structure of clinical symptoms in schizophrenia. A multisite, multimodel evaluation of the factorial structure of the Positive and Negative Syndrome Scale. The PANSS Study Group. Psychopathology 1997; 30:263–274.
187. Lancon C, Aghababian V, Llorca PM, Auquier P. Factorial structure of the Positive and Negative Syndrome Scale (PANSS): a forced five-dimensional factor analysis. Acta Psychiatrica Scandinavica 1998; 98:369–376.
188. Lindenmayer JP, Bernstein-Hyman R, Grochowski S. Five-factor model of schizophrenia. Initial validation. Journal of Nervous and Mental Disease 1994; 182:631–638.
189. Mohr PE, Cheng CM, Claxton K, et al. The heterogeneity of schizophrenia in disease states. Schizophrenia Research 2004; 71:83–95.
190. McEvoy JP, Freter S, Everett G, et al. Insight and the clinical outcome of schizophrenic patients. Journal of Nervous and Mental Disease 1989; 177:48–51.
191. Birchwood M, Smith J, Drury V, et al. A Self-Report Insight Scale for Psychosis: Reliability, Validity and Sensitivity to Change. Acta Psychiatrica Scandinavica 1994; 89:62–67.
192. Ritsner M, Rabinowitz J, Slyuzberg M. The Talbieh Brief Distress Inventory: a brief instrument to measure psychological distress among immigrants. Comprehensive Psychiatry 1995; 36:448–453.
193. Ritsner M, Ponizovsky A. Psychological symptoms among immigrant population: a prevalence study. Compr Psychiatry 1998; 39:21–27.
194. Derogatis LR, Spencer PM. The Brief Symptom Inventory (BSI): Administration, Scoring and Procedures Manual. Johns Hopkins University, School of Medicine; 1982.
195. Rosenberg M. Society and the Adolescent Self-image. Princeton, NJ: Princeton University Press; 1965.
196. Zimet GD, Dahlem NW, Zimet SG, Farley GK. The Multidimensional Scale of Perceived Social Support. Journal of Personality Assessment 1988; 52:30–41.
197. Cole JD, Kazarian SS. The level of expressed emotion scale: a new measure of expressed emotion. Journal of Clinical Psychology 1988; 44:392–397.
198. Cole JD, Kazarian SS. Predictive validity of the Level of Expressed Emotion (LEE) Scale: readmission follow-up data for 1, 2, and 5-year periods. J Clin Psychol. 1993; 49:216–8.
199. Cloninger CR A systematic method for clinical description and classification of personality variants. Arch Gen Psychiatry 1987; 44:573–588.
200. Hintze, JL. NCSS 2000 Statistical System for Windows. User's Guide. Number Cruncher Statistical Systems: Kaysville, Utah, 2000.

CHAPTER 11

INSIGHT AND QUALITY OF LIFE IN SCHIZOPHRENIA SPECTRUM DISORDERS

An examination of their paradoxical relationship

PAUL H. LYSAKER[1] AND SHIRA LOURIA[2]

[1]*Roudebush VA Medical Center and The Indiana University School of Medicine, Indianapolis, USA*

[2]*University of Indianapolis, School of Psychological Science, Indianapolis, IN*

Abstract: While interest has grown steadily in understanding how persons with schizophrenia appraise their disorder and subsequent needs, the nature of the impact of awareness or admission of disorder on various domains of quality of life has remained a matter of considerable debate. At the level of both theory and empirical study it has been alternately held that acknowledgement of one's mental illness is a detriment and a key to successful adaptation. From one perspective, acceptance of illness has been advanced as a key to making informed decisions about one's future, to free oneself from blame for difficulties linked with illness and to forming bonds with others who are aware of one's difficulties. From another view, however, "awareness of illness" has been suggested to represent the acceptance of a system of social power in which one's individuality and dignity is at risk of being diminished. Indeed empirical studies suggest both awareness and lack of awareness have significant risks associated with them. In this chapter we review this evidence and present data which suggest that the impact of insight on quality of life may be mitigated by the degree to which persons have internalized stigmatizing beliefs about their illness. Clinical and theoretical implications are discussed

Keywords: Awareness, Quality of life, Schizophrenia

Relative to persons with other psychiatric disorders, persons with schizophrenia spectrum disorders are often unaware, or willfully contest that they have what others think to be a mental illness [1,2]. They may dispute the possibility that experiences they have such as hearing or seeing things others do not hear or see, or interpreting events in ways that others find entirely implausible, are symptoms of a mental illness. This may be particularly baffling to those around them given that these same persons may be fully aware of much of what is occurring in their environment [3]. They may, for instance, be fully cognizant of the irrationality of delusions of persecution experienced by someone they know and yet deny that their own belief that they are being persecuted could similarly be linked with a mental illness [4]. They may appear perplexed as to why others think something in their lives has gone wrong.

M.S. Ritsner and A.G. Awad (eds.), Quality of Life Impairment in Schizophrenia, Mood and Anxiety Disorders, 227–240.

Despite acknowledged distress and markers of personal difficulties such as social isolation and unemployment many with schizophrenia may consequently reject well intentioned professional assistance. They may reject such help despite pleading from family and previous personal experiences suggesting psychiatric treatment is beneficial.

Clearly this phenomenon, referred to alternatively as poor insight, denial of illness or unawareness of illness poses many intriguing theoretical issues. How for instance, is it possible that persons can be reasonable observers of others and yet interpret the same issues so differently when applied to themselves? Beyond this, the even larger issue looms regarding non-adherence with treatment and poorer outcome. Most models of help-seeking behavior[5,6] are based on the core assumption that persons who reject the possibility they are ill may not only fail to seek assistance from others but possibly reject assistance when offered. If illness is denied, offers of help are likely to be seen as absurd and possibly unwarranted intrusions. Thus if we assume current treatments such as medication are helpful then it appears a matter of intuition that persons who are unaware of their illness may be at even greater risk for prolonged and possibly catastrophic deficits and difficulties.

As an illustration, consider Ravel (this name and accompanying information is disguised to protect confidentiality), a man with schizophrenia in his 40s. When first seen in the outpatient psychiatry clinic of a comprehensive medical center, he had just been hospitalized with delusions of reference, thought blocking and considerable rage and anxiety. He was divorced and homeless and had been unemployed for several years due to worsening symptoms. He experienced multiple hospitalizations as well as lengthy incarcerations as a result of instances in which he had attacked a variety of people with a potentially lethal weapon. In an initial interview when he enrolled in the clinic he noted that he felt frightened, and wished to work again and to find a stable living situation. He firmly denied though that these difficulties were linked to a mental illness. He knew and acknowledged that during his last period when he had worked and had stable housing, that he had been taking antipsychotic medication. He denied experiencing any significant side effects of that medication and he recalled that he felt better emotionally during that time. Interestingly, Ravel nevertheless insisted that this was all a coincidence and he wanted no part of taking medication again.

Ravel finally adhered to an injectable antipsychotic medication following another hospitalization and began to experience a profound reduction in symptoms. With difficulty, he was able to describe how he no longer felt in such terrible distress and was led through the Socratic method to the possibility that medication helped him because he had a mental illness. Concurrent with these improvements Ravel married his long standing romantic partner, reestablished contact with his two adult sons from whom he was previously estranged, obtained full time competitive employment, and purchased a condominium and a car all over the course of 8 months.

From here Ravel began to cancel or fail to attend his counseling appointments and missed multiple doses of his injectable antipsychotic medication. He did not return phone calls and finally after an absence of many months from treatment he was hospitalized again, acutely paranoid, believing that others with demonic powers were reading his mind. He was again denying that he had a mental illness and was separated from his wife and was in danger of losing his job. He had become again a man who radiated a sense of fear, discomfort and a consumption with an overwhelming sensation that he was in mortal danger. While back in the hospital, his medication was restarted and he quickly experienced a reduction in symptoms though he again resisted acknowledging that he had a mental illness and refused to accept any link between improvements and medication. A sense of humor returned but he had no story plausible in the eyes of others of what had happened.

While a wealth of anecdotal evidence points to a widespread acceptance in the psychiatric community that examples such the vignette portrayed above are common place, what evidence is there to support this view? Is unawareness as clearly linked to poorer quality of life as one might expect? As will be reviewed below, research on the link between insight and function has been strikingly inconsistent. While some studies have linked awareness of illness to a range of undesirable outcomes or experiences that no one would want to have in their lives such as hospitalization, heightened symptoms and greater difficulties in vocational settings, other studies have suggested the opposite. Specifically awareness of illness also appears to serve as a risk factor for a range of equally undesirable outcomes such as a reduction in well-being, dysphoria and poorer self-esteem. Paradoxically, awareness of illness, which is often an implicit goal of most psychiatric treatment programs thus seems confusingly to both help persons avoid certain risks for poorer quality of life while exposing them to a new set of risk factors for poor quality of life. A corollary of informed consent for medication is awareness of the condition for which it is prescribed. Yet, just as lack of awareness leads to undesirable outcomes, awareness of illness also seems to lead to undesirable outcomes.

In this chapter we seek to evolve and develop an understanding of these seeming contradictions and to discuss the implications of these contradictions for treatment. In particular it appears essential that if psychiatric treatment continues to promote services that assist persons to better understand their conditions that the field also develop a more thorough understanding of the complex effects of awareness of illness. Accordingly this chapter is divided into five sections. In the first we first review literature on the risks that accompany lack of awareness of illness. In contrast to this, in the second section the risks to quality of life that accompany awareness of illness are examined. In the third section we suggest that the effects of awareness of illness on quality of life are mediated by the meaning which severe mental illness holds for persons with schizophrenia. Here we will draw on evidence from the study of the impact that the meanings of non-psychiatric medical conditions have on quality of life of persons who suffer from them. In the fourth section we will review one of our recent studies which examined the interaction of insight with one type of meaning often associated with schizophrenia namely societal stereotypes of mental illness or stigma. In addition to this review we will also present analyses of data from this study not previously published and suggest that unawareness and awareness coupled with the acceptance of stigma are both viable paths to dysfunction. We will lastly offer some thoughts regarding the clinical implications of this material.

UNAWARENESS OF ILLNESS AS A PREDICTOR OF POORER QUALITY OF LIFE

To begin, as might be expected there is some evidence that appears to confirm that persons who do not believe they are ill resist taking antipsychotic medication and, therefore, fare more poorly over time. Bartko et al.[7], for instance, reported that patients diagnosed with schizophrenia who were non-adherent with medications (n = 32) tended to have lesser insight, to lack a 'feeling of illness,' to have greater levels of grandiose ideation, and difficulties with social adjustment compared with others rated as adherent (n = 26). Without awareness of illness perhaps persons with schizophrenia reject treatment and become more disabled. Consistent with this, Cuffel et al.[8] found that awareness of illness was significantly linked with perceived need for treatment among persons hospitalized for schizophrenia. Interestingly, while insight predicted compliance when assessed concurrently, insight while hospitalized did not predict compliance nor did it predict persons' appraisals of whether or not they were ill six months later. This may possibly suggest that personal awareness of illness is malleable and changes appreciably over time. A more recent study[9] explored the relationship of awareness of illness to non-adherence to antipsychotic medications. In this study the authors sent surveys to over 700 psychiatrists, of whom 534 responded. The psychiatrists were asked to report on the presentation, management, and course of illness for an adult patient with schizophrenia under their care for at least one year and who had been non-adherent to oral anti-psychotics. From these responses 300 patients under the psychiatrists' care were classified according to degree of insight. Ninety-seven patients were classified according to their psychiatrists report as not aware that they had a mental illness. This group, relative to the group of participants classified as more aware of their condition, had significantly longer episodes of antipsychotic non-adherence, more severe levels of positive symptoms, and were more likely to be hospitalized after periods of non-adherence.

Consistent with this, denial of illness has been linked to poorer clinical outcome and prognosis[10]. For instance in an early study Heinrichs et al.[11] retrospectively examined the outcomes linked with the presence of awareness of illness. They studied a sample of 38 patients with schizophrenia in an acute state and found that the 63% whom they labeled as having "early insight" into the acuity of their condition had significantly lower rates of re-hospitalization than the 37% labeled as not having insight. McEvoy et al.[12] followed 46 patients with schizophrenia for several years after they were hospitalized also in an acute phase of illness. They found that patients with better insight at the time of hospitalization were significantly less likely to be readmitted to the hospital at a later point within their window of follow-up. One cross sectional study of over 500 persons in the early phases of illness found poorer insight correlated with higher levels of positive symptoms, negative symptoms, neurocognitive deficits and general psychopathology[13]. Another longitudinal study assessed the symptoms of 180 persons with early psychosis at admission and at the following 3, 6 and 12 months. Analyses revealed enduring associations between lower levels of insight and higher levels of both positive and negative symptoms[14].

Exploring the relationship of insight and function, several studies have similarly found that unawareness of illness predicts greater difficulties in social and community interactions. In a study by Francis and Penn[15], persons with schizophrenia engaged in social interactions with a research confederate and were observed by a trained rater. Coding of those interactions suggested that better insight was associated with better social skills, self-disclosure of suffering from a mental illness and less observed social strangeness. In a study by Lysaker et al.[16] the authors explored the link of insight with level of interpersonal function as rated through an interview. Participants rated as having impaired insight (n = 57) had fewer social relationships and lesser capacity for interpersonal function than those with unimpaired insight (n = 44). Of note in addition to these studies at least two other investigators have reported that poor insight may also be related to poorer overall social function prior to illness[13,17], suggesting any deficits linked to insight may precede the onset of symptoms or deficits that emerge with the full blown presentation of the disorder itself.

To explore the association of insight with vocational function Lysaker et al.[18] divided a sample of 121 persons with schizophrenia spectrum disorders who had completed vocational rehabilitation program into two groups; those with good insight (n = 65) and those impaired insight (n = 56). Analyses comparing averaged blind ratings of work performance across the first 7 weeks of a work placement indicated that the group with impaired insight had significantly greater difficulties in the area of cooperativeness, work habits, work quality, and personal presentation. When neurocognitive ability was controlled for, further analyses revealed that the unaware group still demonstrated poorer performance on cooperativeness and personal presentation. This was taken to suggest that unawareness of illness may specifically limit persons' abilities to appraise their areas of need and thereby to form relationships with others who may see those areas of need more clearly.

AWARENESS OF ILLNESS AS A PREDICTOR OF POORER QUALITY OF LIFE

While lack of insight has been found to predict poorer medication adherence and poorer outcomes in terms of clinical social and vocational outcomes, other research has suggested that greater awareness of schizophrenia is also linked with a range of negative outcomes as well. Mintz et al.[19] conducted a meta-analysis of 40 studies conducted from 1992 to 2001 which assessed both psychopathology and insight in schizophrenia. Inclusion criteria for this meta analysis were study publication in an English language journal and the use of objective, reliable and valid measures of insight. The authors abstracted measures of five dimensions of insight: "individual's awareness of the mental illness, awareness of the social consequences of the disorder, awareness of the need for treatment, awareness of symptom and attribution of symptoms to the disorder" from studies which met inclusion criteria and utilized those to create an overall insight index. A weighted effect size was then calculated which estimated the relationship of insight with positive, negative and depressive symptoms. This analysis revealed that consistent with some of the work cited above, that as insight increased positive and negative symptoms were observed to decrease. At the same time though as insight increased, depressive symptoms were also noted to increase across this significant body of studies.

Examples of studies in this meta analysis included a study by Pyne et al.[20] in which the authors sought to distinguish the characteristics of those who did and did not believe that they were mentally ill from among 177 inpatients and outpatients with schizophrenia. The authors found that 37% of their sample did not believe they were mentally ill and this group tended to not only perceive their medication as having lesser efficacy but also reported fewer depressive symptoms. Moore et al.[21], also included in the Mintz et al. meta analysis, found that among 46 patients with schizophrenia, those with lesser awareness of their mental illness similarly reported fewer depressive symptoms but possibly also used greater levels of self deception. Depression, they speculated was connected to awareness of illness and was perceived as the social 'price' of having a mental disorder in the absence of self-deception as a "denial defense." As a third example of work included in the Mintz et al. analyses is a study conducted by Warner et al.[22]. Here 42 patients with psychosis were asked whether they believed they had a mental illness and then their self-esteem, locus of control and perceptions of self and others with and without mental illness was assessed. Results indicated that persons who accepted the mentally ill label ($n = 28$) had significantly lower self-esteem and tended to perceive life events as related to forces outside of their control when compared with those who rejected the mental illness label ($n = 14$).

From a somewhat different angle, but consistent with the studies included in the Mintz et al. meta analysis, Thompson[23] explored the changes in self-concept of young adult patients with psychosis. In this study participants were asked to indicate whether they viewed themselves more as a typical community resident or more as a typical mental patient. Two thirds of the participants described themselves as community residents while the remaining third described themselves as patients. Comparisons of these groups on a wide range of variables revealed that those who described themselves as patients were more likely to report psychological anguish, to be older, not married, and to use the support of limited and homogeneous social networks. The group who thought of themselves as patients first also reported greater adherence with medication but held little hope that they could ever achieve a fulfilling life.

In a study examining acceptance of the label of mental illness and the impact on quality of life, Kravetz et al.[24] found that psychiatric self-labeling was negatively correlated with quality of life, while perceived control of the mental illness was found to be positively correlated with quality of life. The authors interpreted these findings as suggesting that while the acceptance of a disability may improve adherence to treatment, it can also represent a barrier to recovery as it may involve an individual's surrender to the 'inferiority' status embedded within socially prevalent stereotypes of mental illness. In a more recent study by the same group[25], a new sample of 128 participants with schizophrenia or schizoaffective disorder were administered assessments of insight and quality of life. Results indicated that greater levels of awareness of illness were related to lesser emotional well being, poorer vocational function and lesser economic satisfaction.

AWARENESS OF ILLNESS: THE ROLE OF THE MEANING WITHIN THE REALM OF NON-PSYCHIATRIC DISORDERS

Taken at face value, research on insight and quality of life in schizophrenia presents a striking paradox. Lack of awareness of illness seems to place persons at risk for rejecting potentially helpful treatments, as well as symptom exacerbation and poorer social and vocation function. At the same time awareness of illness puts persons at risk for depression, low self- esteem and helplessness. One possible explanation for these seemingly contradictory findings is that the impact of the acceptance of schizophrenia depends on the meanings persons attach to schizophrenia[26]. Awareness of illness is, after all, not merely an isolated belief but represents a complex personal construction that is established within a life story and which may be imbued with any number of particular meanings given the context in which it is placed[27]. For example, returning to the vignette presented at the outset of this chapter, if the meaning of illness in the eyes of Ravel is that his future will hold no opportunities for the achievement of valued social roles, might not he resist telling a story of his life in which he as the protagonist has a mental illness? Clearly his unawareness might cost him work and housing yet allow a future and some possible meaning in daily life events. If he thinks he is not ill he may, for example, see himself facing adversity (e.g. homelessness) but his actions in a given day could be seen as steps towards a better future.

While the impact of the meaning of mental illness upon those with mental illness has been relatively neglected by the literature in psychiatry, some work examining medical illness may provide a context for thinking about this issue. For instance, as summarized by Lambert et al.[28], the impact of physical illnesses on quality of life may be a function not only of the severity of that illness but of the extent to which that illness is experienced and interpreted as disrupting each individual persons sense of personal biography. In other words, as illness is experienced as a force which interrupts the flow of one's life story, as a phenomenon that dissolves the coherence of links between one's previous story and the future, perhaps more profound disability ensues and along with it previous levels of quality of life are eroded.

Consistent with this, some researchers have designed interventions for non-psychiatric illness which seek to impact the disruptive meanings illnesses can have on one's life. Carter et al.[29] for instance, have examined the effects of pain workshops for children with chronic pain. In these workshops, an emphasis is placed on evolving a sense of the meaning of the pain experienced, taking into account children's different developmental and verbal abilities. In this intervention children are encouraged to verbalize their pain, talk about the effect that it has on them, and how they cope with pain. This process is intended to aid the children in organizing and re-organizing the meanings they attach to pain. As a result of the workshop, the authors found that children developed a sense that they were not their pain, that pain did not subsume their identity and that pain was a "separate entity" from them. Pain was also understood as an aspect of human experience and did not mean one was lesser for experiencing it. Though unable to eliminate the pain, and acknowledging the social isolation experienced as a result, children learned to guide themselves, stay calm and relaxed when the pain returned, allowing for a better quality of life.

Also consistent with the contention that the meaning made of one's illness qualifies the impact of awareness of it is a qualitative study which examined perceptions, experiences and quality of life among 17 persons with AIDS[30]. In this study quality of life was linked not only with the concrete impact of AIDS but also with its effect of derailing and "shattering" persons' dreams of a sense of self as projected into the future. The impact of this catastrophic illness on quality of life was understood largely by the authors as involving the meanings that individuals were evolving and developing in the midst of their illness regarding their past and their attitudes towards death. Specifically as persons were able to accept their disabling condition as an entity which did not erase the previous meanings of their lives, their quality of life improved. Similarly, a study of cervical cancer[31] found that quality of life was deeply impacted by whether or not participants understood and viewed their cancer as affecting their sense of self and sense of being a woman. Persons who did not perceive this illness as nullifying their femininity reported the lowest levels of sexual dysfunction.

MEANING OF AWARENESS OF ILLNESS AND QUALITY OF LIFE IN SCHIZOPHRENIA: AN EXPLORATORY STUDY

As a possible means to understand how awareness and unawareness of illness in schizophrenia could both have negative consequences we have suggested that the meaning persons assign to that illness could explain why acceptance could have a devastating effect on quality of life even though acceptance may lead to useful treatment outcomes and social exchange. Following the medical literature noted above it seems likely that if persons understand schizophrenia as biographical disruption then acceptance of illness is likely to result in despair and hopelessness.

And one does not have to look far to find reasons why schizophrenia may have particularly destructive meanings attached to it, particularly ones that might disrupt sense of personal narrative. Population surveys indicate that despite increased awareness of the nature of schizophrenia many among the general public hold stereotyped beliefs about persons with this condition[32]. Categorically referred to as stigma, these beliefs include the expectation of violent and unsettling behavior as well as an inability to work or make informed decisions[33,34]. Beyond being a matter of misinformation, stigma may incline persons in the general public to avoid or seek social distance from those with schizophrenia[35].

Accordingly we have recently suggested[36] that internalized stigma or the acceptance of stereotypic beliefs about mental illness may be the factor which mitigates the impact of whether awareness of illness can lead to improved quality of life. To investigate this we have sought to determine whether we could: a) distinguish groups among persons with schizophrenia enrolled in outpatient treatment that varied according to level of insight and internalized stigma and b) determine whether those different groups differed in hopefulness, self-esteem and social function. To examine this hypothesis, following informed written consent, we gathered measures of insight, internalized stigma, hope, self-esteem and social function among 75 persons with schizophrenia spectrum disorders. We then performed a cluster analysis using insight and internalized stigma scores. A cluster analysis is similar to a factor analysis but instead of separating test scores into groups it separates participants into groups based on patterns of scores. Insight was measured using the Scale to Assess Awareness of Mental Disorders[37]. Stigma was assessed using the The Internalized Stigma of Mental Illness Scale[38]. Hope was assessed using the Beck Hopelessness Scale[39]. Self esteem was assessed using the Multidimensional Self-Esteem Inventory[40] and social function was assessed using the Quality of Life Scale[41]. Ratings of insight were made blind to responses on all questionnaires

We hypothesized that this cluster analyses would result in three groups: (a) those who held stigmatizing beliefs about mental illness and believed they had a mental illness; (b) those who did not endorse having internalized stigmatizing beliefs about mental illness and believed they had a mental illness; and (c) those who endorsed stigmatizing beliefs about mental illness but did not have believe they had a mental illness. We secondly predicted that the group who rejected stigma but acknowledged their illness and that the group who accepted stigma but rejected their illness would have more hope and self-esteem than the group who accepted stigma and acknowledged their illness. Third, we predicted that the group who rejected stigma and acknowledged their illness would have the best social function of all three groups.

Consistent with our hypotheses the cluster analysis produced three groups. Based on insight and stigma levels we have labeled these groups as: Low insight/ Mild stigma ($n = 23$), High insight/ Minimal stigma ($n = 25$) and High insight/ Moderate Stigma ($n = 27$). These groups did not differ significantly in age, education, or hospitalization history. Turning to the link between insight, stigma and self experience, Analyses of variance (ANOVA) revealed that the High insight/Moderate stigma group reported statistically significantly poorer self-esteem and lesser hope than either the group with insight but not stigma or the group with stigma but no insight. The High insight/ Minimal stigma group had statistically significantly better interpersonal function on the Quality of Life Scale than either of the other two groups.

To explore these findings more deeply in this chapter we have analyzed differences between groups on the subscales of our self-esteem measure. The MSEI is a 116-item self-report measure that assesses individuals' self-perception of their overall social value. Respondents rate items on a 5-point scale according to the degree or frequency with which each item applies to them. While we previously reported group differences on the total score we have, as noted above returned to the data and compared groups on select subscales: Competence, Lovability, Self control, Personal power, Moral self approval and Body image. The results of ANOVA comparing these groups are reported in Table 1. As revealed in this table, the group who accepted their illness yet endorsed stigmatizing beliefs about mental illness, had lower ratings than either the high Insight/ Minimal stigma or Low insight/Mild stigma on assessments of their own lovability, moral self approval and body image. Notably there were no differences on participants report of their sense of Competence or Personal Power.

Taken together we suggest that these results may offer some way of making sense of the contradictory findings regarding the negative effect of both awareness and unawareness of illness. In particular it may be that awareness of illness among a subgroup of persons who endorse self stigma leads directly to poorer self esteem and hope and thus poorer adaptation to illness. Concurrent with this it may also be that persons who deny they are ill, while they avoid demoralization also become socially isolated and thus also have equivalent difficulties adapting to their illness. Thus both awareness and unawareness are risk factors or possible pathways for persons who have assigned profoundly negative meanings to schizophrenia.

Table 1. Self esteem component among groups clustered by level of insight and stigma endorsement

	Group 1 low insight/mild stigma (n = 21)	Group 2 high insight/ minimal stigma (n = 24)	Group 3 high insight/ moderate stigma (n = 26)	ANOVA F =	Group comparisons p < .05
MSEI Competence	33.19 (7.38)	32.88 (6.72)	29.81 (6.17)	1.89	na
MSEI Lovability	29.81 (5.42)	31.92 (6.40)	24.38 (6.48)	9.90***	3< 1, 2
MSEI Self control	33.81 (7.05)	34.29 (4.84)	30.04 (7.42)	3.51*	3< 2
MSEI Personal power	32.67 (6.47)	31.92 (4.75)	30.46 (6.61)	0.83	na
MSEI Moral self approval	36.14 (6.54)	38.88 (4.80)	31.88 (7.60)	7.45***	3< 1, 2
MSEI Body appearance	31.76 (6.67)	29.50 (6.22)	24.85 (7.66)	6.22**	3< 1, 2

* p < .05; ** p < .01, *** p < .001

CONCLUSIONS AND FINAL DIRECTIONS

In this chapter we have explored the confusing and contradictory research finding that suggest that unawareness that one has schizophrenia complicates quality of life just as accepting that one has schizophrenia complicates quality of life. To resolve this we have proposed that it may be that the impact of accepting illness is mitigated by the meanings associated with that illness. Drawing on literature on non-psychiatric illness, we have asked whether the effects of insight are mitigated by the meanings assigned to illness, particularly meanings that might disrupt person's personal sense of biography. As evidence of this possibility we have presented results of a recent study along with further analyses of the data of that study which suggest that insight is linked with lower self-esteem only when coupled with the endorsement of stigma and that lack of insight and endorsement of stigma may be equivalents paths to social dysfunction.

While much more research is necessary, with replication, these findings may have important clinical implications. First, along with education about mental illness it may be useful to consider interventions that decrease internalized stigma. Warner[22], for instance, has suggested that it is just as important for interventions to assist in developing a sense of mastery as it is to help enhance insight. This is consistent with a recent intensive case study, which suggested that as a person with schizophrenia recovered in psychotherapy he first evolved a greater sense of personal agency before developing a more complex grasp of his illness[42]. It is also consistent with a study of a different sample that suggested hope was more closely tied to a sense of agency than illness awareness[43]. Perhaps if dysfunctional beliefs stemming from stigma impact quality of life in such an enduring manner, tailored interventions could be devised to help persons combat these self-stigmatizing beliefs. Future interventions and research could be directed to help persons with schizophrenia overcome their negative beliefs and find newer and more adaptive ways to think of themselves and their futures, thus allowing for the acceptance of mental illness to have fewer devastating effects. One possibility offered is for further research to explore ways to facilitate the efforts of persons with schizophrenia to replace self-stigmatizing beliefs and transform their narratives and experience themselves in such a manner that they see themselves as active protagonists in their own lives with realistic appraisals of their strengths and deficits as well as capacities to survive devastating grief[44,45].

In short there are many unanswered questions. For one, how should we go about helping anyone accept any disability without seeing it as a disruption? Elliot et al.[46] have investigated the relationship between the stability of personal goals and the acceptance and adjustment of persons with spinal chord injury. Interestingly they have suggested that acceptance of a disability for individuals led to poorer quality of life among persons who tended to have fewer stable life goals and who relied on external sources for self-worth. Perhaps the same is true among persons with schizophrenia and interventions are needed in order to help persons develop both stable goals over time as well as internal sense of worth.

REFERENCES

1. Amador, XF, Strauss DH, Yale SA, & Gorman JM. Awareness of illness in schizophrenia. Schizophr Bull, 1991; 17: 113–132.
2. David AS. Insight and psychosis. Br J Psychiatry, 1990; 156: 798–805.
3. Lysaker PH, Wickett AM, Wilke N & Lysaker JT. Narrative incoherence in schizophrenia: The absent agent-protagonist and the collapse of internal dialogue. Am J Psychotherapy, 2003; 57, 153–166.
4. Startup M. Awareness of own and others' schizophrenic illness. Schizophr Res 1997; 29, 203–211.
5. Pescosolido BA. Beyond rational choice: The social dynamics of how people seek help. American Journal of Sociology, 1997; 97: 1096–1138.
6. Pescosolido BA. Illness careers and network ties: A conceptual model of utilization and compliance. Adv Med Sociol, 1991; 2: 161–184.
7. Bartko G, HerczegI, Zador G. Clinical symptomatology and drug compliance in schizophrenia. Acta Psychiatr Scand, 1988; 77: 74–76.
8. Cuffel BJ, Alford J, Fischer EP, Owen RR. Awareness of illness in schizophrenia and outpatient treatment compliance. J Nerv Ment Dis 1996; 184: 653–659.
9. Olfson M, Marcus SC, Wilk J, West JC. Awareness of illness and nonadherence to antipsychotic medications among persons with schizophrenia. Psychiatr Serv, 2006; 57: 205–211.
10. Schwartz RC. The relationship between insight, illness and treatment outcome in schizophrenia. Psychiatr Q 1998; 69: 1–22.
11. Heinrichs DW, Cohen BP, & Carpenter WT (1985) Early insight and the management of schizophrenic decompensation. J Nerv Ment Dis, 1985; 173: 133–138.
12. McEvvoy JP, Freter S, Everett G, Geller JL, Appelbaum P, Apperson LJ, & Roth L Insight and clinical outcome of schizophrenic patients. J Nerv Ment Dis, 1989;177(1): 48–51.
13. Keshavan MS, Rabinowitz J, DeSmedt G, Harvey PD & Schooler N. Correlates of insight in first episode psychosis. Schizophr Res, 2004; 70: 187–194.
14. Mintz AR, Addington J & Addington D. Insight in early psychosis: A one year follow up. Schizophr Res, 2004; 67: 213–218.
15. Francis JL, Penn, DL. The relationship between insight and social skill in persons with severe mental illness. J Nerv Ment Dis, 2001; 189: 822–829.
16. Lysaker PH, Bell MD, Bryson GJ, Kaplan EZ. Insight and interpersonal function in schizophrenia. J Nerv Ment Dis 1998; 186: 432–436.
17. Debowska G, Grzywa A & Kucharska-Pieture K . Insight in paranoid schizophrenia: Its relationship to psychopathology and premorbid adjustment. Compr Psychiatry, 1998; 39: 255–260.
18. Lysaker PH, Bryson GJ, Bell MD. Insight and work performance in schizophrenia. J Nerv Ment Dis, 2002; 190: 142–146.
19. Mintz AR, Dobson KS, Romney, DM. Insight in schizophrenia: A meta-analysis. Schizophr Res, 2003; 61: 75–88.
20. Pyne JM, Bean D, Sullivan G. Characteristics of patients with schizophrenia who do not believe they are mentally ill. J Nerv Ment Dis, 2001; 189: 146–153.
21. Moore O, Cassidy E, Carr A, O'Callaghan E. Unawareness of illness and its relationship with depression and self-deception in schizophrenia. Eur Psychiatry, 1999; 14: 264–269.
22. Warner R, Taylor D, Powers M, Hyman, R. Acceptance of the mental illness label by psychotic patients: effects on functioning. Am J Orthopsychiatry, 1989; 59: 389–409.
23. Thompson EH. Variations in the selfconcept of young adult chronic patients: Chronicity reconsidered. Hosp Comm Psychiatry, 1988; 39: 771–775.

24. Kravetz, S., Faust, M., & David, M. Accepting the mental illness label, perceived control over the illness, and quality of life. Psychiatr Rehabil J, 2000; 23: 323–332.
25. Hasson-Ohayon I, Kravetz S, Roe D, Weiser M. Insight Into Severe Mental illness, Perceived Control Over the Illness, And Quality of Life. Compr Psychiatry, 2006; 47: 265–269.
26. Roe D, Kravetz S. Different Ways of Being Aware of and Acknowledging a Psychiatric Disability. A Multifunctional NarrativeApproach to Insight into Mental Disorder. J Nerv Ment Dis, 2003; 191:417–24.
27. Lysaker, P. H., Clements, C. A., Plascak-Hallberg, C. D., Knipscheer, S. J., & Wright, D.E. Insight and personal narratives of illness in schizophrenia. Psychiatry, 2002; 65(3): 197–206.
28. Lambert, BL, Levy, N A, & Winer, J. Keeping the Balance and Monitoring the Self-System: Towards a More Comprehensive Model of Medication Management in Psychiatry. In D. Brashers and D. Goldsmith (Eds.), Communication and the Management of Health and Illness. New Jersey, Erlbaum, (in press).
29. Carter, B, Lambrenos, K, Thursfielf, J. A pain workshop: an approach to eliciting the views of young people with chronic pain. J Clin Nurs, 2002; 11: 753–762.
30. Witt Sherman, D. The perceptions and experiences of patients with AIDS: implications egarding quality of life and palliative care. Journal of Hospice and Palliative Nursing, 2001; 3(1):7–16.
31. Juraskova, I, Butow, P, Robertson, R, Sharpe, L, McLeod, C, & Hacker, N. Post- treatment sexual adjustment following cervical and endometrial cancer: a qualitative insight. Psycho-Oncology, 2003; 2: 267–279.
32. Swindle, R., Heller, K., Pescosolido, BA. & Kikuzawa, S. Responses to nervous breakdowns in America over a 40-year period: Mental health policy implications. Am Psychol, 2000; 55: 740–749.
33. Link, BG., Phelan, M., Bresnahan, M., Stueve, A. & Pescosolido, BA. Public conceptions of mental illness: Labels, causes, dangerousness and social distance. Am J of Public Hea, 1999; 89: 1328–1333.
34. Pescosolido, BA., Monahan, J., Link, B.G., Stueve, A. & Kikuzawa, S. The public's view of the competence, dangerousness, and need for legal coercion of person with mental health problems. Am J of Public Hea, 1999; 89: 1339–1345.
35. Martin, JK, Pescosolido, BA & Tuch, SA. Of fear and loathing: The role of "disturbing behavior," labels and causal attributions in shaping public attitudes toward persons with mental illness. J Health Soc Behav, 2000; 41(2): 208–233.
36. Lysaker PH, Roe D & Yanos PT. Toward Understanding the Insight Paradox: Internalized Stigma Moderates the Association Between Insight and Social Functioning, Hope and Self-Esteem Among People with Schizophrenia Spectrum Disorders. Schizophr Bull, (in press).
37. Amador XF, Strauss DH, Yale SA, Flaum MM, Andreasen NC, Yale SA, Clark SC, Gorman JM. Awareness of illness in schizophrenia. Arch Gen Psychiatry 1994; 51: 826–836.
38. Ritsher JB, Otilingam PG, Grajales M. Internalized stigma of mental illness: Psychometric properties of a new measure. Psychiatric Res 2003; 121: 31–49.
39. Beck AT, Weissman A, Lester D, Trexler L. The measurement of pessimism: The Hopelessness Scale. J Consult Clin Psychol 1974; 42: 861–865.
40. O'Brien EJ, Epstein S. MSEI: The multidimensional self-esteem inventory Professional Manual. Psychological Assessment Resources, Inc., Lutz FL, 1998.
41. Heinrichs, DW, Hanlon TE, Carpenter WT. The Quality of Life Scale: An instrument for assessing the schizophrenic deficit syndrome. Schizophr Bull 1984; 10: 388–396.
42. Lysaker PH, Davis LD, Eckert GJ, Strasburger A, Hunter N, Buck, KD. Changes in narrative structure and content in schizophrenia in long term individual psychotherapy: A single case study. Clin Psychol Psychotherapy 2005; 12: 406–416.
43. Lysaker PH, Buck KD, Hammoud, K, Taylor, AC, Roe D. Associations of symptom remission, psychosocial function and hope with qualities of self experience in schizophrenia: Comparisons of objective and subjective indicators of recovery. Schizophr Res 2006; 82: 241–249.

44. Roe D, Davidson L. Self and narrative in schizophrenia: time to author a new story, J. Med. Humanit 2005; 31: 89–94.
45. Lysaker PH, Lysaker, JT. A typology of narrative impoverishment in schizophrenia: Implications for understanding the processes of establishing and sustaining dialogue in individual psychotherapy. Couns Psychol Q, 2006; 19: 57–68.
46. Elliot, TR, Uswatte, G, Lewis, L, Palmatier, A. (2000). Goal instability and adjustment to physical disability. J Couns Psychol. 2000; 47(2): 251–265.

CHAPTER 12

QUALITY OF LIFE AND MAJOR DEPRESSION

Current findings and future perspectives

MARCELO T. BERLIM[1] AND MARCELO P.A. FLECK[2]

[1]*Depressive Disorders Program, Douglas Hospital Research Centre, FBC Pavilion, Rm. F-3116-B, 6875 LaSalle Blvd., Montréal, Québec, H4H 1R3, Canada*

[2]*Department of Psychiatry and Forensic Medicine; Head, Mood Disorders Program Hospital de Clínicas de Porto Alegre, Brazil*

Abstract: Major depression (MD) is a public health problem that is associated with grave consequences in terms of excessive mortality, disability, and secondary morbidity. Indeed, it ranked fourth in 1990 and could rise to second by 2020 in terms of the overall burden of all diseases worldwide. Therefore, it is now clear that current research on the health impact of depression should go beyond estimating its prevalence, symptoms severity, and complications to include studies that seek to establish how it influences the quality of life (QOL) of the affected individuals. In the present chapter we will outline how measures of QOL may reveal differences between patients with depression and control groups, be sensitive to change in status during treatment, have predictive value for outcome measures and provide additional information about timelines for improvement in psychosocial functioning, which may occur at a different rate than changes in other depressive symptoms. More specifically, we will summarize recent investigations that have generally shown that: (1) depressed patients have QOL deficits that are directly attributable to the mood disturbance, (2) the degree of the decrement in QOL is proportional to the severity of depressive symptoms, (3) the negative relation between depression and QOL is as great as (or worse than) that observed in chronic medical disorders such as rheumatoid arthritis and diabetes, and (4) the adequate treatment of depression is usually associated with a significant improvement in the QOL of patients. Finally, we will discuss future perspectives involved in the evaluation of QOL in populations of depressed subjects

DEPRESSIVE DISORDERS: EPIDEMIOLOGY, SYMPTOMATOLOGY AND PSYCHOSOCIAL IMPACT

Unipolar major depression has a lifetime prevalence of around 21-24% in women and 12-15% in men[1]. It is characterized by the presence of depressed mood and/or lack of interest in activities that would usually be pleasurable, and generally associated with appetite and sleep disorders, with feelings of guilt and/or depreciation, among others, all present for at least two weeks[2,3].

M.S. Ritsner and A.G. Awad (eds.), Quality of Life Impairment in Schizophrenia, Mood and Anxiety Disorders, 241–252.

Depressive disorder represents, worldwide, a significant public health problem not only due to its high prevalence[4], but also to its high annual direct and indirect costs which, in the United States alone, reached around 43 billion dollars in 1990[5,6]. In fact, projections for the year 2020 indicate that major depression will occupy the second place in terms of impact on human health, being second only to ischemic heart disease. If incapacity alone is taken into consideration, major depression occupied the first position in 1990[7].

Depressive disorders are also associated with serious consequences in terms of mortality and secondary morbidity leading, for example, to a lack of work productivity and interpersonal problems[8]. Furthermore, they can adversely influence longevity and well being during the episode and, potentially, for the rest of the afflicted person's life and its functional effects in the long term are as devastating as those found in chronic medical diseases, such as diabetes mellitus and cardiovascular disease[6,9]. High mortality rates are also a major problem, since almost 15% of depressed patients commit suicide during their lives[10].

For these reasons, current research on the impact of depressive disorders should go beyond estimating its prevalence, the seriousness of its symptoms and complications, and begin to include studies that attempt to establish how they affect the quality of life (QOL) of the affected individuals[11–13].

DEPRESSIVE DISORDERS AND QUALITY OF LIFE

In the specialized literature, there is strong evidence that depressed patients present a significant reduction in their QOL[11,13,14]. This is especially due to the fact that depressive disorders affect various domains that are part of the global assessment of the QOL[15].

Below, we summarize the main conceptual aspects – methodological and empirical, involved in the assessment of the QOL in depression.

How to Assess the QOL in Depressed Patients

Present controversies

Present controversies regarding the measurement of QOL in depression include the following dilemmas[11,13,15]:

- *Subjective scales* (i.e., self-rated) versus *objective scales* (i.e., applied by the clinician). Many studies have demonstrated that there are disagreements between physicians and patients over the severity of the symptoms and the success of medical treatments. Clinicians generally base their assessment of treatment results on the improvement of the symptoms or, preferably on the "non-progression" of the disease process. This reveals that the focus of the clinician is, in general, on the status of the patient's health. On the contrary, patients most commonly evaluate the results of the treatment in relation to feeling more comfortable or being able to do daily activities satisfactorily again. In any event, the sensation of the patient's well-being is the main point of reference here. Since the psychosocial elements of the patient's experience cannot be easily assessed by the physician, one can argue that the patient is the best judge of his/her health status. As well, numerous studies indicate that the search for medical attention is much more related to the subjective impact of the disease on patient's life than with the presence, by itself, of symptoms. Additionally, the effectiveness of any kind of treatment depends mainly on whether the patient considers or not that his/her health status showed improvement after beginning a therapeutic approach. Finally, it is a current consensus that, to the extent that interventions in the health area have as an objective making life more comfortable (instead of "curing" diseases), the most valid source of information is the patient him/herself. In short, it is inappropriate to value only the clinician's evaluation, particularly when the degree of disagreement between the physician and the patient is taken into consideration, and the point of view of the latter should be emphasized whenever possible. However, "objective" assessments can be useful and should be complementary to subjective ones.

- *Generic scales* (i.e., those developed to be used with varying diseases and virtually applicable to all people) versus *disease-specific scales* (i.e., those developed for people with specific diseases). Amongst the arguments that defend the disseminated use of disease-specific scales, the most important one is that probably special aspects of each disease uniquely contribute to the perception of the QOL in the affected individuals and cannot be captured by generic instruments. As well, some authors believe that a more specific measurement would be more sensitive to symptomatic changes in a certain disease. However, if the objective is to assess the influence of a disease (or of their symptoms) on the QOL understood in a broader way (i.e., involving a series of domains not specifically linked only to health problems) there are strong arguments for the use of generic scales (even if disease-specific measurements can be used as complementary strategies). Additionally, another limitation of specific instruments is that they are not effective in comparing the QOL of different clinical conditions.
- *Medical* versus *mediational models*. At present, there is still a belief that the QOL is primarily a product of the symptoms of a disease and of the side effects of medications (*medical model*). However, there is increasing evidence that two patients can have different levels of QOL even if they present the same seriousness of the disease and/or the same degree of treatment side effects. In this way, individual characteristics of each patient can, in fact, mediate the relation between QOL and symptoms/para-effects (*mediational model*).
- Scales that reflect a *functionalistic model* (i.e., assessing the individual's capacity of fulfilling certain functions that are considered "normal" for the average person in western society [e.g. physical mobility, conducting a job, socialization], with a divergence from the norm indicating a reduced QOL) versus scales that reflect a model based on *basic needs* (i.e., assessing if the individual is capable of accomplishing his/her basic needs, such as shelter, feeding and safety and his/her psychological needs, such as autonomy, friendship and pleasure). In short, the proponents of the functionalistic model believe that the disease only becomes a

problem when it affects the fulfillment of the individual's functions, while the proponents of the basic needs model suggest that life gains quality through the ability and the capacity of the individual in satisfying his/her own needs.

Theoretical implications of the relationship between quality of life and depression

Even though there is no definitive consensus on the concept of QOL, emphasis on the subjective dimension (i.e., the perception of the individual regarding the different domains of his/her life) is undoubtedly a central element [16]. To the extent that in mental disorders the sick "organ" is the mind, it is expected that perception and/or the processing of such perception (cognition) will be altered [11]. Because of its prevalence and nature, depression presents an extra challenge [17].

In depression the possibility of superposition of the measurements of depression and QOL can occur in at least three levels: (a) *conceptual* (i.e., depression and QOL could be representations of the same phenomenon); (b) *mediational* (the affective state could lead to a distorted perception of reality), and (c) *metric* (even though they are supposedly different,constructs, there are some items that are common) [17].

From the *conceptual* point of view, "well being" and "degree of satisfaction" can be understood as antagonistic constructs and considered as "antonyms" of depression. When someone says he/she is "depressed," he/she is implicitly communicating that they "do not feel well" and that they are "not satisfied."

From the *mediational* point of view, the cognitive model illustrates this possible relationship. Beck [18] proposed that depression could be defined through a cognitive triad according to which the individual (1) sees him/herself in a negative way, (2) interprets the majority of the events in his/her life in an unfavorable way, and (3) believes that his/her future has no hope. Thus, it is expected that a depressed person will assess the different domains of his/her life, based on the evaluation of the quality of life, negatively and that once his/her state has improved, that same objective reality will be assessed differently (i.e., more positively).

From the *metric* point of view, many items present in the assessment of QOL are also present in the scales of depression. For example, in the case of the World Health Organization's Quality of Life Instrument - Brief Version (WHOQOL BREF) [19], using item response theory, 11 of the 26 items present DIF for depression and 11 of the 26 items are also conceptually assessed with the Hamilton Rating Scale for Depression (HAM-D) [20] (positive feelings, spirituality, thoughts, energy, leisure, sleep, daily life activities, work, self-esteem, sexual life and negative feelings) [21]. Various studies have shown high correlation coefficients, especially in the psychological and physical domains, with the Beck Depression Inventory (BDI) [22], and lower ones for the other domains [23–25].

Even though the concepts of depression and QOL are closely related and some authors consider them as "tautologic" measurements [26], there is empirical evidence that those two concepts do not measure the same phenomenon: (1) QOL and depression are synchronic measurements, but correlation coefficients are only moderate in some domains [24,27,28]; (2) there seems to be a gap between the improvement of depression and the improvement in some domains of the QOL, especially in relation to antidepressant use, which shows a more precocious improvement of the symptoms of depression followed by an improvement in the QOL[29]; (3) the quality of life measured by the Quality of Life in Depression Scale (QLDS)[30] and not intensity of depression was found as a predictive factor for complete remission within 9 month follow-up[31]; (4) for the same intensity of depression it is possible to have different QOL scores and this finding can have important clinical implications[32,33].

Assessment instruments

Traditional scales used to assess depressive symptoms, such as the HAM-D[20], the BDI[22], do not encompass important aspects of the QOL in that they are restricted to descriptive investigations of symptomatology and not necessarily to its repercussions in the psychosocial life of the depressed patient[15]. Thus, should the clinician or researcher use them exclusively to assess the results of the treatment, they will not capture important aspects of the QOL perception of the patient (e.g., the presence or quality of interpersonal relations are not verified)[11,14]. Because of that, many authors have argued for the use of additional scales that aim at investigating the QOL in a broader way, i.e., not limited only to secondary aspects of the depressive symptomatology[13].

Below are examples of psychometric instruments that have been largely used in the assessment of QOL in patients presenting with depressive disorders.

WHOQOL BREF[19,27,28] According to what has been seen before, the QOL assessment can be based both on the generic and the specific models. As an example of the generic model we can cite the WHOQOL-BREF, a shorter version of the WHOQOL-100, which is composed of 26 items that encompass four domains of QOL (i.e., physical, psychological, social and environmental). The WHOQOL BREF also contains two items that are examined separately: question 1 asks about an individual's "overall perception of QOL," and question 2 asks about an individual's "overall perception of their general health." Items are rated on a 5-point Likert scale where 1 indicates low and negative perceptions, and 5 indicates high and positive perceptions. Respondents judge their QOL over the previous 2 weeks, and high scores demonstrate good QOL. The mean score of items within each domain is used to calculate the domain score.

Quality of Life in Depression Scale (QLDS)[30,34] The QLDS is a disease-specific instrument used to assess QOL in depressed patients. It is based on the model that measures the QOL as arising from the ability and capacity of the patients to meet their basic needs. This model arose from qualitative non-structured interviews that the authors of the QLDS applied with individuals exhibiting depressive disorders. In those interviews, a consistent finding was that the depressive subjects described the impact of the disease in their lives in terms of their personal needs (for example work, love, conversation, pleasure, self-care and nutrition) being frustrated by depression.

In its final format, the QLDS is a self-applicable questionnaire comprised of 34 items (with dichotomous answers, i.e., yes/no or true/false) that describe each one of the basic physical and psychological needs of the depressed patients. They, in turn, answer the questions of the QLDS based on a defined period of time (i.e., the previous week), with the result that the higher the scores on the QLDS, the worst is the QOL.

The QLDS is very responsive to changes in the QOL. In its original version it was found to be very trustworthy, consistent internally and valuable.

Quality of Life Enjoyment and Satisfaction Questionnaire (Q-LES-Q)[35] The Quality of Life Enjoyment and Satisfaction Questionnaire (Q-LES-Q) is a self-administered questionnaire developed to measure the degree of contentment and satisfaction in relation to various aspects of daily life. Even though the Q-LES-Q has not been developed specifically for depressed patients, it was originally tested in a sample for depressed patients and has been used with this aim. It is comprised of 93 items, 91 of them grouped into 8 sub-scales. Five of those scales are to be filled out by all individuals: physical health (13 items), subjective feelings (14 items), time of leisure activities (6 items), social relations (11 items) and general activities (14 items). The other three sub-scales are to be filled out by the individuals depending on their particular activities: work (13 items), domestic services (10 items), and activities in school/courses (10 items).

The Q-LES-Q was developed for use not only in depression, but also in other psychiatric conditions or even in other medical areas. The items are proposed as questions and the respondent assesses the degree of satisfaction on Likert-type scale of 5 options. The original version of the instrument presents good internal consistency with all coefficients α of Cronbach above 0.90. The test-retest trustworthiness, however, obtained low values (between 0.63 and 0.89). The different sub-scales of the Q-LES-Q presented correlation coefficients that vary between −0.34 to −0.68 with Clinical Global Impression (CGI)[36], and correlations in the same order with two scales of depression, namely the HAM-D[20] and the BDI[22].

SmithKline Beecham Quality of Life Scale (SBQOL)[37] Even though the SmithKline Beecham Quality of Life Scale (SBQOL) has not been developed as a specific measure for depression, its validation study was based on patients presenting with major depression and general anxiety. It is a self-report questionnaire composed of 28 items. The items are scored on a decimal scale with an extreme positive anchor point on one side and a negative anchor point on the opposite side. The questionnaire includes psychological and physical well-being, social relations, activities/interests/hobbies, humor, control, sexual function, work/job, religion and finances. The answers are given from three perspectives: "me now", "the ideal me", and "the sick me", and the instrument generates three total scores ("me now," "the ideal me," "the sick me") and two comparisons ("me now" vs. "the sick me" and "me now" vs. "the ideal me"). The SBQOL demonstrated adequate internal consistency. Additionally, it was shown that its total scores had good correlations with clinical improvement (as measured by the HAM-D[20]), and with comparison scores of generic QOL instruments (i.e., the Sickness Impact Profile[38], and the General Health Questionnaire[39]).

Depression and QOL: Summary of Current Evidence

QOL measurement in depression

Recent studies on the QOL in depression have demonstrated that depressed patients present with deficits in interpersonal, psychological and even physical functioning that is only partially explained by the variation in the intensity of the depressive symptoms[11,13–15].

Recent evidence indicates that mood disorders are associated with important deficits in the QOL and the global functioning of their bearers. In fact, patients with major depression presented QOL scores inferior not only to that of individuals with subsyndromal depressive disorders, but also to that of non-depressive subjects in the general population[40]. Moreover, studies have consistently shown that depressed patients present with significant deficits in many areas of social functioning (e.g., leisure, work, interpersonal relations, health status and academic performance) when compared with healthy controls[13].

A recent study[41] evaluated the impact of major depression, double depression and dysthymia (and other anxiety disorders) on the QOL of the affected subjects (as assessed by the Q-LES-Q) and compared them to a control group. Demographic variables (e.g., age, sex), co-morbidities, duration and seriousness of the specific symptoms of each disease were assessed as predictive QOL factors. The authors demonstrated that all groups with psychiatric disorders presented reduced scores on the Q-LES-Q when compared to control individuals, but that patients with major depression and double depression (as well as posttraumatic stress disorder) were the ones with the lower QOL scores. In addition, they observed that the depression scores were responsible for less than 10% of the variance on the Q-LES-Q[41].

In sum, current findings reinforce the impression that QOL is a measurement that is semi-independent from the perception of the patient regarding his/her disease, and that it is important for researchers and clinicians to use not only scales that take into account the severity of the depressive symptoms, but also non-medical aspects of the patient's life (e.g., his/her subjective QOL).[13,15]

QOL deficits in depressed patients versus patients with general medical conditions

The long term effects of depressive disorders are as serious as those observed in many general medical conditions[13]. However, when one refers to the immediate impact, those effects are even more noteworthy. This was highlighted in the important Medical Outcomes Study (MOS)[42], which caught the attention of researchers and clinicians at the end of the 1980s and led the way for many of the current studies on the relationship between QOL and depression. In the MOS, patients with depressive disorders were compared to individuals with diabetes mellitus, hypertension, coronary arterial disease, arthritis, lumbar problems, pulmonary and gastrointestinal diseases. The results of this study clearly demonstrated that depression is associated with higher physical and social deficits, a poorer quality of life, more absenteeism, less pain-free days, higher treatment costs and poorer perception of health status when compared to other chronic physical diseases[42,43]. During the last few years, dozens of studies have corroborated the pioneer findings of the MOS.

Impact of antidepressants on the QOL of depressed subjects

During a depressive episode, patients experience difficulties in accomplishing physical activities and find that their level of energy is reduced. The perception of their health status is also altered, as is their ability to relate socially, work effectively and conduct day-to-day activities adequately. Therefore, to globally assess the impact of any particular antidepressant treatment it is important to estimate the physical, social and psychological status of the patient[14]. Since this status is generally assessed by QOL instruments, we can infer that the impact of antidepressant treatment can be measured in part by the QOL scales – with the exception that the QOL encompasses other areas than only those of health.[13] As well, QOL instruments have been capable of differentiating depressed patients from control groups[44]. Various studies comparing respondent and non-respondent individuals to many antidepressants demonstrated a significant improvement in social functioning of those presenting a clinical response after the acute phase of treatment[13]. This finding confirms the impression that QOL questionnaires can detect the response to the treatment in depressive subjects[14].

Even though it is largely accepted in the literature that the improvement of depressive symptoms during treatment leads to an improvement in the QOL[11], few clinical studies of antidepressants have specifically assessed the impact of the treatment on the QOL[45]. In fact, knowing the impact of antidepressants on the QOL is as important as knowing if those treatments are effective in the reduction of depressive symptoms. Moreover, the search for treatment is generally motivated by the subjective impact of a disease on the QOL of the patient, and his/her adherence to any type of treatment is strongly influenced by the subjective perception of improvement. Thus, in depression it is important to emphasize the patient's point of view, and this is better assessed by QOL scales than by any objective assessment of the symptoms[15,16,46].

Even though the majority of studies indicate that an improvement in the depressive symptoms from treatment leads to a significant improvement in the QOL[13,14], the correlation between the two measurements is only moderate, indicating that there is no total superposition between those two domains[47]. As well, improvement in the QOL can occur even after the recuperation of the depressive symptoms[29].

The QOL questionnaires have already shown sensitivity in detecting residual symptoms in patients that have responded partially to the treatment with antidepressants. In a study patients who had recently become depressed, De Lisio and collaborators[48] found important changes in the subjects' work performance, as well as in the social and leisure activities. An assessment performed one year later demonstrated that deficits in leisure activities remained even in patients that presented remission of the central depression symptoms[48]. Studies like this support the clinical impression that the psychosocial functioning of the patients can take more time to return to normal than other depressive symptoms. Furthermore, a maintenance study of patients treated with sertraline showed that QOL continues to improve after the remission of symptoms, and that this did not occur with the patients that received a placebo after remitting from the acute clinical condition[49].

In another study[50] with depressed patients, it was demonstrated that even though fluoxetine and amitriptyline equally improved the core symptoms of depression, the former was superior in terms of benefits in social functioning and in the general perception of the individual in relation to health itself. Fluoxetine was also superior to clomipramine in those aspects. According to the authors of this study, the superiority of fluoxetine was probably related with the different profiles of drugs' side effects and the consequent level of patients' compliance[50].

Generally speaking, the studies that compared tricyclic antidepressants (TCAs) with selective serotonin reuptake inhibitors (SSRsI) tended to show better QOL for the patients treated with SSRIs, even when significant differences were not detected between the groups in relation to the improvement of depressive symptoms[11]. Comparisons among different SSRIs (e.g., sertraline, paroxetine and fluoxetine) did not show significant differences between them in relation to their effects on QOL[13].

Additionally, there is evidence that some instruments for assessing QOL could be capable of predicting response to the antidepressant treatment, at least to a certain extent. For example, Pyne and collaborators[51] showed that patients with more deficits in social activities presented a worse response to treatment, while patients with significant deficits in physical activities presented a better response. Another study has identified that the level of positive thoughts at the beginning of treatment was associated with higher levels of response after 8 weeks[47].

In short, instruments for the assessment of QOL seem to be sensitive to changes in the patient's health status during treatment, have predictive value, and also offer information on the time for improvement of psychosocial functioning[14]. Thus, they represent an advance in the assessment of the efficacy of antidepressants, as they can lead to more focused interventions and provide more specific and sensitive measurements of the outcome of treatment.

CONCLUSION

The assessment of QOL in depression is a promising endeavour that will probably lead to a better understanding of depressed patients and to the development of more rational and individualized treatments.

To the extent that the individual perceives him/herself, his/her environment and his/her future in a negative and distorted way during the depressive episode, it is expected that his/her QOL will be negatively affected. However, there is evidence that depression and QOL are constructs with areas of intersection, but are not necessarily redundant.

Unfortunately, there is a lack of definitive theoretical models in the literature satisfactorily establishing the relationship between those two constructs. Thus, the development and validation of current and new models is fundamental for a better understanding of the existing relationship between depression and QOL.

REFERENCES

1. Bland, R.C. Epidemiology of affective disorders: a review. *Can J Psychiatry* **42**, 367–77 (1997).
2. Parikh, S.V. & Lam, R.W. Clinical guidelines for the treatment of depressive disorders, I. Definitions, prevalence, and health burden. *Can J Psychiatry* **46 Suppl 1**, 13S–20S (2001).
3. Doris, A., Ebmeier, K. & Shajahan, P. Depressive illness. *Lancet* **354**, 1369–75 (1999).
4. Andrews, G., Sanderson, K., Slade, T. & Issakidis, C. Why does the burden of disease persist? Relating the burden of anxiety and depression to effectiveness of treatment. *Bull World Health Organ* **78**, 446–54 (2000).
5. Hall, R.C. & Wise, M.G. The clinical and financial burden of mood disorders. Cost and outcome. *Psychosomatics* **36**, S11–8 (1995).
6. Sartorius, N. The economic and social burden of depression. *J Clin Psychiatry* **62 Suppl 15**, 8–11 (2001).
7. Murray, C.J. & Lopez, A.D. Global mortality, disability, and the contribution of risk factors: Global Burden of Disease Study. *Lancet* **349**, 1436–42 (1997).
8. Ballenger, J.C. et al. Consensus statement on the primary care management of depression from the International Consensus Group on Depression and Anxiety. *J Clin Psychiatry* **60 Suppl 7**, 54–61 (1999).
9. Greden, J.F. The burden of recurrent depression: causes, consequences, and future prospects. *J Clin Psychiatry* **62 Suppl 22**, 5–9 (2001).
10. Angst, J., Angst, F. & Stassen, H.H. Suicide risk in patients with major depressive disorder. *J Clin Psychiatry* **60 Suppl 2**, 57–62; discussion 75–6, 113–6 (1999).
11. Demyttenaere, K., De Fruyt, J. & Huygens, R. Measuring quality of life in depression. *Current Opinion in Psychiatry* **15**, 89–92 (2002).
12. Berlim, M.T., Mattevi, B.S. & Fleck, M.P. Depression and quality of life among depressed Brazilian outpatients. *Psychiatr Serv* **54**, 254 (2003).
13. Papakostas, G.I. et al. Quality of life assessments in major depressive disorder: a review of the literature. *Gen Hosp Psychiatry* **26**, 13–7 (2004).
14. Kennedy, S.H., Eisfeld, B.S. & Cooke, R.G. Quality of life: an important dimension in assessing the treatment of depression? *J Psychiatry Neurosci* **26 Suppl**, S23–8 (2001).
15. Berlim, M.T. & Fleck, M.P. "Quality of life": a brand new concept for research and practice in psychiatry. *Rev Bras Psiquiatr* **25**, 249–52 (2003).
16. Orley, J., Saxena, S. & Herrman, H. Quality of life and mental illness. Reflections from the perspective of the WHOQOL. *Br J Psychiatry* **172**, 291–3 (1998).
17. Fleck, M.P.A. Avaliação de qualidade de vida. in *Depressões em Medicina Interna e em Outras Condições Médicas* (eds. Fráguas, R. & Figueiró, J.A.B.) 235–257 (Editora Atheneu, São Paulo, 2001).
18. Beck, A.T., Rush, A.J., Shaw, B.F. & Emery, G. *Cognitive therapy of depression: A treatment manual*, (Guilford Press, New York, 1979).
19. Development of the World Health Organization WHOQOL-BREF quality of life assessment. The WHOQOL Group. *Psychol Med* **28**, 551–8 (1998).

20. Hamilton, M. A rating scale for depression. *J Neurol Neurosurg Psychiatry* **23**, 56–62 (1960).
21. Rocha, N., Power, M., Fleck, M.P. & Bushnell, D. What are we measuring using WHOQOL-Bref in depressed patients? The LIDO experience. *Qual of Life Res* **14** 2016 (2003).
22. Beck, A.T., Ward, C.H., Mendelson, M., Mock, J. & Erbaugh, J. An inventory for measuring depression. *Arch Gen Psychiatry* **4**, 561–71 (1961).
23. Berlim, M.T. et al. Psychache and suicidality in adult mood disordered outpatients in Brazil. *Suicide Life Threat Behav* **33**, 242–8 (2003).
24. Fleck, M.P. et al. [Application of the Portuguese version of the instrument for the assessment of quality of life of the World Health Organization (WHOQOL-100)]. *Rev Saude Publica* **33**, 198–205 (1999).
25. Aigner, M. et al. What does the WHOQOL-Bref measure? Measurement overlap between quality of life and depressive symptomatology in chronic somatoform pain disorder. *Soc Psychiatry Psychiatr Epidemiol* **41**, 81–6 (2006).
26. Katschnig, H. & Angermeyer, M. Quality of Life in Depression. in *Quality of Life in Mental Disorders* (eds. Katschnig, H., Freeman, H. & Sartorius, N.) 137–147 (John Wiley & Sons, New York, 1997).
27. Fleck, M.P. et al. [Application of the Portuguese version of the abbreviated instrument of quality life WHOQOL-bref]. *Rev Saude Publica* **34**, 178–83 (2000).
28. Berlim, M.T., Pavanello, D.P., Caldieraro, M.A. & Fleck, M.P. Reliability and validity of the WHOQOL BREF in a sample of Brazilian outpatients with major depression. *Qual Life Res* **14**, 561–4 (2005).
29. McCall, W.V., Reboussin, B.A. & Rapp, S.R. Social support increases in the year after inpatient treatment of depression. *J Psychiatr Res* **35**, 105–10 (2001).
30. McKenna, S.P. et al. International development of the Quality of Life in Depression Scale (QLDS). *J Affect Disord* **63**, 189–99 (2001).
31. Fleck, M.P. et al. Longitudinal Investigation of Depression Outcomes Group: Major depression and its correlates in primary care settings in six countries. 9-month follow-up study. *Br J Psychiatry* **186:**(2005).
32. Berlim, M.T., Mattevi, B.S., Pavanello, D.P., Caldieraro, M.A. & Fleck, M.P. Suicidal ideation and quality of life among adult Brazilian outpatients with depressive disorders. *J Nerv Ment Dis* **191**, 193–7 (2003).
33. Berlim, M.T. et al. Quality of life in unipolar and bipolar depression: are there significant differences? *J Nerv Ment Dis* **192**, 792–5 (2004).
34. Hunt, S.M. & McKenna, S.P. The QLDS: a scale for the measurement of quality of life in depression. *Health Policy* **22**, 307–19 (1992).
35. Endicott, J., Nee, J., Harrison, W. & Blumenthal, R. Quality of Life Enjoyment and Satisfaction Questionnaire: a new measure. *Psychopharmacol Bull* **29**, 321–6 (1993).
36. Guy, W. *ECDEU Assessment Manual for Psychopharmacology*, (U.S. Department of Health, Education, and Welfare, Washington, D.C.,, 1976).
37. Stoker, M.J., Dunbar, G.C. & Beaumont, G. The SmithKline Beecham 'quality of life' scale: a validation and reliability study in patients with affective disorder. *Qual Life Res* **1**, 385–95 (1992).
38. Bergner, M., Bobbitt, R.A., Carter, W.B. & Gilson, B.S. The Sickness Impact Profile: development and final revision of a health status measure. *Med Care* **19**, 787–805 (1981).
39. Goldberg, D.P. & Hillier, V.F. A scaled version of the General Health Questionnaire. *Psychol Med* **9**, 139–45 (1979).
40. Goldney, R.D., Fisher, L.J., Wilson, D.H. & Cheok, F. Major depression and its associated morbidity and quality of life in a random, representative Australian community sample. *Aust N Z J Psychiatry* **34**, 1022–9 (2000).
41. Rapaport, M.H., Clary, C., Fayyad, R. & Endicott, J. Quality-of-life impairment in depressive and anxiety disorders. *Am J Psychiatry* **162**, 1171–8 (2005).
42. Wells, K.B. et al. The functioning and well-being of depressed patients. Results from the Medical Outcomes Study. *Jama* **262**, 914–9 (1989).

43. Hays, R.D., Wells, K.B., Sherbourne, C.D., Rogers, W. & Spritzer, K. Functioning and well-being outcomes of patients with depression compared with chronic general medical illnesses. *Arch Gen Psychiatry* **52**, 11–9 (1995).
44. Miller, I.W. et al. The treatment of chronic depression, part 3: psychosocial functioning before and after treatment with sertraline or imipramine. *J Clin Psychiatry* **59**, 608–19 (1998).
45. Barge-Schaapveld, D.Q. & Nicolson, N.A. Effects of antidepressant treatment on the quality of daily life: an experience sampling study. *J Clin Psychiatry* **63**, 477–85 (2002).
46. Skevington, S.M. Advancing cross-cultural research on quality of life: observations drawn from the WHOQOL development. World Health Organisation Quality of Life Assessment. *Qual Life Res* **11**, 135–44 (2002).
47. Skevington, S.M. & Wright, A. Changes in the quality of life of patients receiving antidepressant medication in primary care: validation of the WHOQOL-100. *Br J Psychiatry* **178**, 261–7 (2001).
48. De Lisio, G. et al. Impairment of work and leisure in depressed outpatients. A preliminary communication. *J Affect Disord* **10**, 79–84 (1986).
49. Kocsis, J.H. et al. Psychosocial outcomes following long-term, double-blind treatment of chronic depression with sertraline vs placebo. *Arch Gen Psychiatry* **59**, 723–8 (2002).
50. Souetre, E., Martin, P., Lozet, H. & Monteban, H. Quality of life in depressed patients: comparison of fluoxetine and major tricyclic antidepressants. *Int Clin Psychopharmacol* **11**, 45–52 (1996).
51. Pyne, J.M. et al. Health-related quality-of-life measure enhances acute treatment response prediction in depressed inpatients. *J Clin Psychiatry* **62**, 261–8 (2001).

CHAPTER 13

QUALITY OF LIFE IMPAIRMENT IN BIPOLAR DISORDER

ERIN E. MICHALAK[1], GREG MURRAY[2], ALLAN H. YOUNG[1] AND RAYMOND W. LAM[1]

[1] *Department of Psychiatry, University of British Columbia, Vancouver, Canada*
[2] *Faculty of Life and Social Sciences, Swinburne University of Technology, Hawthorn, Australia*

Abstract: This chapter will present an overview of what is currently known about quality of life (QoL) in bipolar disorder (BD). While there is growing consensus that the QoL concept provides an important counterpoint to the objective symptom measures that dominate psychiatry, the implications of this position are yet to be systematically addressed in the BD field. The literature reviewed in this chapter is not informed by any model or theory of QoL as it applies to BD. The data presented are largely empirical in the negative sense of this term, drawn from studies making different assumptions, addressing different questions in BD. In many cases, the measurement of QoL has been incidental to the major thrust of the research design. Indeed, even at the empirical level there is a lack of agreement about the best measure of QoL in this population, and no disorder-specific measure currently exists.

In this context, the present chapter aims to capture the range of applications of the QoL concept in BD, and highlight any reliable findings. To structure the presentation, it is useful to categorize data into three general areas: studies that have looked at QoL impairment in BD compared with non-clinical and other diagnostic groups, studies that have looked at QoL differences across the states of BD and studies that have looked at QoL changes as an outcome of treatment. We will briefly set the scene for these three investigations by presenting the findings of four manuscripts that have reviewed research into QoL and BD.

The final section of this chapter will make some recommendations about future QoL research in BD, highlighting the need for work at both the theoretical and measurement levels. At the theoretical level, the implications of the vulnerability model of health-related QoL (HRQOL) (e.g., Ritsner, Chapter 1 of this collection) will be briefly reviewed. At the measurement level, we will present some data on the ongoing development of a patient-centred disorder-specific QoL measure for BD

Keywords: Bipolar disorder, Quality of life, Functioning

M.S. Ritsner and A.G. Awad (eds.), Quality of Life Impairment in Schizophrenia, Mood and Anxiety Disorders, 253–274.

INTRODUCTION

Bipolar disorder is a highly complex and heterogeneous psychiatric condition that is characterised by both a variety of symptoms and marked variability in course. For example, a patient with BD can experience episodes of depression, hypomania, mania, or psychosis, and indeed, can experience a mixture of emotional states, or cycle rapidly between them. Marked variability exists between patients in terms of the length, number and type of episodes, severity and type of symptoms, and the degree of interepisode recovery experienced. Overall, however, patients with BD appear to spend more time in episodes of depression than mania or hypomania[1,2], and unsurprisingly, the condition is associated with marked disability. In the year 2000, the World Health Organization estimated that BD was the 6th leading cause of disability worldwide amongst young adults (i.e., 15–44 years of age)[3]. A woman who develops BD at the age of 25 may lose 9 years in life expectancy (due to cardiovascular and other medical problems), 14 years of productivity, and 12 years of normal health[4]. While the condition obviously has considerable ramifications at an individual level, it also has a significant impact at a societal level. Bipolar disorder is estimated to affect at least 1% of individuals worldwide[5], making it a serious public health concern. The broader category of bipolar spectrum disorders (encompassing, for example, BD type II, which is characterised by episodes of depression and hypomania) has been less extensively studied, but is likely to be more prevalent, affecting up to 8% of the population[6]. In one widely cited study, the direct and indirect costs associated with BD were estimated to be $45 billion in the United States during 1991, of which only $7 billion was due to actual treatment costs[7]. Lost productivity within salaried employees and homemakers, however, accounted for costs of approximately $20 billion.

Patient outcome in BD has traditionally been determined by the assessment of objectively measured clinical information, such as rates of relapse, the number of times a patient is hospitalised or clinician-rated symptom reduction. In recent years, however, we have seen a shift towards the concomitant assessment of more subjective, patient-centered measures of well-being such as functioning and QoL. As leading researchers Colom and Vieta (2004)[8] have stated: "*A very important change of paradigm in the treatment of bipolar disorders started a few years ago, when crucial findings on the impact of bipolar disorders on quality of life and social, cognitive and occupational functioning suggested that therapy targets should be changed from symptomatic recovery to functional recovery*". Whilst this statement is laudable, we would go one step further to suggest that restoration of QoL (which encompasses more than restoration of functioning per se) should be a primary goal for treatment, over and above minimizing the symptomatic burden experienced by the individual.

An Overview of QoL Research in BD

Compared with the study of QoL in unipolar depression, research into QoL in BD has been sparse. To our knowledge, four systematic reviews of previous research in this area have been conducted[9–12]. In the first of these, Namjoshi and colleagues (1999)[9] assessed all relevant English-language articles published prior to 1999, identifying 10 studies for inclusion. The studies proved to be quite heterogeneous, and used a wide variety of instruments to assess HRQOL. They also tended to be relatively small, most often conducted in depressed or euthymic (rather than hypo/manic) patients, and rarely included descriptions of the psychometric properties of the instruments they utilized. The authors of the review made a number of suggestions for future research, including the need for the development of a disorder-specific measure of QoL for BD, more assessments in manic patients, and more longitudinal research.

The second review, conducted by Dean and colleagues (2004)[10] examined studies published prior to November 2002 that had assessed HRQOL, work-impairment or healthcare costs and utilization in patients with BD. This review applied a very broad definition of HRQOL, including studies that had assessed social or physical functioning in isolation (for example, the Global Assessment of Functioning (GAF) scale was considered a measure of HRQOL). Using this broad definition, the review identified 65 HRQOL articles. The authors concluded, (i) that the HRQOL of patients with BD is similar to that of patients with unipolar depression and equal or lower than that observed in patients with chronic medical conditions and, (ii) that treatment interventions for BD have been shown to have a beneficial impact on HRQOL. The third review, conducted by Revicki and colleagues (2005)[11] focused on pre-2003 studies measuring HRQOL outcomes of BD treatment and the burden of BD on HRQOL. The authors identified just three clinical trials that had included HRQOL assessments. On the basis of this initial data, they concluded that treatment interventions for BD hold the potential to improve HRQOL, but that there is limited evidence for differences between the mood stabilizers in terms of HRQOL outcomes, and "that the time may be ripe for developing disorder-specific HR-QOL instruments for use in clinical trials" (pg. 592).

In the most recent of this series of reviews[12], we searched for studies published prior to November 2004 which had examined QoL in BD. This search, which applied slightly more stringent inclusion criteria than some of the earlier reviews, identified 28 studies in total, 7 (25%) of which were published before 1999. The remaining 21 (75%) were published between 2000 and 2004, underscoring the developing interest in this field of research. The studies we identified formed a heterogeneous set. Several undertook to assess QoL during different phases of the disorder, for example, cross-sectional research that compared perceived QoL in euthymic, manic or depressed patients with BD. Other studies compared the QoL of patients with BD to that of other patient populations, both with other psychiatric disorders and with chronic physical conditions. Another area of research examined the psychometric properties of a few HRQOL instruments in BD populations. Finally, we identified a small number of studies that had used a QoL instrument to assess outcome in trials of treatment inventions (mostly pharmacological) for the condition. The studies were also of variable scientific quality. Methodological shortcomings included small sample sizes, cross-sectional designs, idiosyncratic diagnostic methods or poorly differentiated diagnostic groups, use of poorly validated QoL instruments and lack of control for chronicity of illness or the number of previous episodes experienced by the individual. However, the overall scientific quality of the research in this field does appear to be improving. Of the 10 studies identified in the review by Namjoshi and colleagues, only one had a sample size of greater than 100. In comparison, we identified eleven studies that had enrolled more than 100 patients, and it was particularly encouraging to see that some of the large pharmacological trials of treatment interventions for BD are now using QoL measures as secondary outcome measures.

The preceding section provides an outline of existing literature on QoL in BD via synopsis of four review studies. As a group, these reviews serve to highlight the significant gaps and deficits in this body of literature. In particular, there appear to be widely held concerns about the valid measurement of QoL in patients with BD, with several reviewers calling for the development of a disorder-specific QoL instrument for this patient population. In the following sections, we will describe this body of literature in more detail, clustering the studies into three subgroups: i) degree of QoL impairment in patients with BD; ii) QoL across mood states within BD; and, iii) QoL as an outcome measure in BD treatment studies.

Degree of QoL Impairment in BD

Several studies have attempted to ascertain the degree of impairment in QoL experienced by patients with BD. Unsurprisingly, QoL in bipolar populations appears to fall far below that observed in general population samples, at least in the realms of emotional or psychosocial well-being. For example, one study utilizing the MOS SF-36 [13], to date the most widely used HRQOL measure in this patient population, compared scores between patients with BD (N = 44) with previously reported norms for a general population sample (N = 2,474) [14]. The SF-36 contains eight sub-scales assessing physical functioning, social functioning, role limitations (physical), role limitations (emotional), pain, mental health, general health and vitality. These yield an overall domain score on a 0-100 scale, where 0 represents worst possible health and 100 best possible health. The results of the study indicated that, compared with the general population, HRQOL was significantly compromised in patients with BD in all SF-36 domains except physical functioning. While the study provided a useful initial comparison, its findings should be interpreted with some caution owing to the disparate sample sizes involved and the fact that it utilised previously published norms for the HRQOL instrument.

More recent data was provided by Yatham and colleagues, who reported on SF-36 scores in patients with BD type I (N = 920) who were either currently depressed, or had experienced a recent episode of depression [15]. Scores were significantly lower across all scale domains in the bipolar group compared with norms reported for the US general population, with markedly lower scores in the mental health, vitality, social functioning and role emotional domains. The authors went on to compare these scores with SF-36 scores derived from seven large studies of HRQOL in patients with unipolar depression. Scores on four domains (general health, social functioning, role-physical and role-emotional) were consistently lower than those observed in unipolar depression, however, the unipolar samples tended to exhibit higher scores in the bodily pain domain. Whilst this study was robust in terms of its large sample size and well-described clinical population, it did not control for depression severity or demographic variables in between-group comparisons. Also, diagnosis of BD was made by clinical interview, whereas unipolar depression was diagnosed via a range of subjective and objective methods. A smaller Brazilian study that compared patients with unipolar MDD (N = 89) and those with bipolar depression (N = 25) on the WHOQOL-BREF (a carefully developed 26-item scale assessing four domains of wellbeing: physical, psychological, social relationships and environment[16] found poorer psychological functioning in the bipolar group[17].

Several other studies have compared HRQOL in patients with BD with that of patients with other psychiatric conditions. For example, the NEMESIS study conducted in the Netherlands compared SF-36 scores in 136 adults with DSM-III-R lifetime BD (93 with BD type I and 43 with BD NOS, as diagnosed by the Composite International Diagnostic Interview or CIDI interview) with that observed in a variety of other psychiatric disorders[18]. Participants with BD showed significantly more impairment in most SF-36 domains compared with other NEMESIS subjects. For example, in the domain of mental health, participants with BD type I experienced significantly lower scores (62.3) than people with other mood (75.2), anxiety (74.0), substance use (80.2) or no psychiatric disorders (85.8). BD type I subjects also reported significantly lower SF-36 scores than patients with BD NOS in the domains of mental health, role limitations (emotional), social functioning and pain. However, there remains some controversy about the accuracy with which the CIDI detects BD NOS, limiting somewhat the inferences that can be made on the basis of these sub-group results. Furthermore, more recent results from the NEMESIS study suggest that the CIDI interview might over-diagnose BD compared to the Structured Clinical Interview for DSM (SCID) interview schedule[19].

Other research has compared QoL in patients with schizophrenia with that observed in patients with BD. For example, Chand[20] and colleagues in India measured QoL via the Quality of Life Enjoyment and Satisfaction Questionnaire (Q-LES-Q)[21] and the WHOQOL-BREF[16] in patients with BD who were in remission and stabilized on lithium prophylaxis (N = 50), patients with schizophrenia (N = 21) and healthy controls (N = 20). The Q-LES-Q is a 93-item self-report measure of the degree of enjoyment and satisfaction in various areas of daily living. The questionnaire was developed and validated for use in depressed outpatients and has eight summary scales that reflect major areas of functioning: physical health, mood, leisure time activities, social relationships, general activities, work, household duties and school/coursework. Mean Q-LES-Q scores can be derived from the eight summary scales and range from 0-100, where higher scores indicate better QoL. The bipolar group reported significantly better QoL than the schizophrenia group in all domains of the scale, and in general well-being, physical health and psychological health on the WHOQOL-BREF[22].

Research findings in this nascent area of research, however, have been inconsistent. For example, Atkinson and colleagues[23] used the Quality of Life Index[24], to assess QoL in patients with BD (N = 37), MDD (N = 35) and schizophrenia (N = 69). Controversially, the authors found that subjectively reported QoL was lower in patients with BD and MDD than in those with schizophrenia, a trend that was reversed for objectively assessed QoL, which included measures such as medical history, health risk behaviors, educational and financial levels and social functioning. These findings led the authors to speculate about the validity of subjective measures of QoL, particularly in people with affective disorders, and the study became an oft-quoted example of the potential difficulties of using patient-centered outcome assessments in psychiatric conditions that may be characterized by factors such as loss of insight (for example, see[25,26]).

More recently, Goldberg and Harrow (2005)[27] compared subjective life satisfaction and objectively measured functioning in patients following hospitalization for an episode of bipolar mania (N = 35), unipolar psychotic depression (N = 27) and unipolar nonpsychotic depression (N = 95). Both life satisfaction (five major domains, work satisfaction, economic security, satisfaction with social activities and relationships, satisfaction with living situation and perceived mental health were assessed via non-standardised semi-structured interview questions) and functioning (objectively measured via a global outcome scale based on factors such as rehospitalization, psychiatric illness, self-support, role performance and social relationships) were assessed longitudinally in the sample at approximately 2, 4.5 and 7–8 years. Their data indicated, firstly, that subjectively assessed life satisfaction was not significantly different between the bipolar/unipolar groups (although objectively assessed work functioning scores were poorer in the bipolar group compared to the unipolar groups). Secondly, it was found that concordance between subjective and objective measures of functioning was higher in the nonpsychotic depression sample than in the bipolar or unipolar psychotic depression samples, consistent with Atkinson et al.'s earlier finding that more severely ill affective disorder patients' subjectively assessed life satisfaction may not be equivalent to objectively assessed functioning. The authors posited some tentative explanations for this finding, including the possibility that more severely ill patients may experience diminished insight, insensitivity to changing life circumstances as a consequence of demoralization or desensitization to stress, or altered life expectations. An alternative explanation for these findings is that discrepancies between objective and subjective measures of HRQOL signify a genuine difference rather than an anomaly related to the patient's psychiatric condition (see Ritsner, Chapter 1 of this collection). Clearly, better designed research is required to help unpack the complex and probably multifaceted relationship between subjective interpretations of life quality and objectively assessed functioning in individuals who are experiencing severe and chronic forms of BD.

Finally, other research has applied a 'health utility' or 'health preference' model to assess health perceptions in patients with BD. The concept of health utility refers to an individual's preferences for different health states under conditions of uncertainty[28]. Health preferences are values that reflect an individual's level of subjective satisfaction, distress or desirability associated with various health conditions, and are frequently assessed by the 'time tradeoff' (TTO) and 'standard gamble' (SG) approaches[29]. TTO refers to the years of life a person would be hypothetically willing to exchange for perfect health. For example, patients might be asked to imagine that a treatment exists that would allow them to live in perfect physical and mental health, but reduces their life expectancy. They might then be asked to indicate how much time they would give up for a treatment that would permit them to live in perfect health, if they had ten years to live. SG refers to the required chance for successful outcome to accept a treatment that could result in either immediate death or perfect health. For example, patients might be asked to imagine that they had ten years to live in their current state of health, and that a treatment existed that could either give them perfect health, or kill them immediately. Patients might then be asked to indicate what chance of success the treatment would have to have before they would accept it.

In a recent study, Revicki et al. (2005)[30] assessed health utility values in N = 96 clinically stable outpatients with BD type I. Standard gamble data indicated that the least preferred bipolar state was inpatient mania-related states, on a par with severe depression states ($M = 0.26$, $SD = 0.29$ and $M = 0.29$, $SD = 0.28$ respectively). Importantly, 96% of their sample was able to complete the SG tasks, indicating that (clinically stable and well-educated) patients with BD are capable of completing these relatively challenging interpretations of health status. In earlier research, Wells and colleagues (1999)[31] assessed functioning and health utility in patients with depression or chronic medical conditions within seven managed care organizations in the United States. HRQOL was assessed via the global mental and physical scales of the SF-12 and health utility was measured via TTO and SG. Patients with depression were categorized as those with BD ($N = 331$), MDD ($N = 3,479$), double depression ($N = 944$), dysthymia ($N = 151$) or brief subthreshold depressive symptoms ($N = 987$). In terms of HRQOL, the bipolar group showed levels of impairment second only to patients with double depression. Health utility followed the same pattern. Specifically, patients with BD were willing to give up on average 17% of their life expectancy in return for perfect health, and would accept on average an 11% risk of death in exchange for perfect health. In comparison, patients with MDD were willing to give up 11% of their life expectancy, and accept a 6% risk of death.

Tsevat and colleagues[32] examined functional status and health utility in 53 outpatients with BD recruited from one site of the multicenter Stanley Foundation Bipolar Network study. The authors aimed to assess how patients with BD rated their current overall health versus their current mental health, and to determine the extent to which health utility correlated with disease state. TTO scores for current overall health were 0.71, but were significantly higher than scores for current mental health, which averaged 0.61. In other words, patients with BD were willing to give up, on average, 39% of their life expectancy in return for perfect mental health. SF-36 data was also collected in the study, and the authors reported that certain SF-36 domains (general health, vitality and role-emotional) were significantly correlated with mental health TTO and SG scores, but levels of mania were not correlated with utilities for either overall health or mental health. The authors concluded that health utilities may be related to certain health status attributes and to level of depression, but may not be related to level of mania. One advantage of the health utility/preference approach to QoL assessment is that it allows the calculation of quality-adjusted life years (QALYs) (for example, [33]). QALYs are a commonly used outcome measure in cost-effectiveness studies (for example, see [34]), but we are unaware of any studies that have taken advantage of this metric in assessing BD populations at the time of writing.

In summary, studies using a range of questionnaire and more complex measures have generally confirmed that QoL is, in a range of domains and to a marked extent, lower amongst patients with BD than in the general population. Beyond this commonsense finding, there is some evidence that QoL is poorer in BD than in other mood and anxiety disorders (perhaps with the exception of double depression). Comparison of BD and psychotic disorders has highlighted a fundamental challenge in comparing QoL across patient groups, namely, that QoL when assessed subjectively in BD might be systematically influenced by the disorder's varying mood and cognitive symptoms. In the next section, we look more closely at state effects on QoL in BD.

QoL Across Mood States in BD

Individuals with BD can experience a number of different mood states or episodes (for example, severe depressions through to highly elated mood) during the course of their disorder, or indeed a combination of these mood states (for example, mixed episodes of depression and mania). Bipolar disorder provides a unique window for the QoL researcher to gaze through; it is of significant scientific and clinical relevance to know what impact these mood shifts have upon individuals' perceived life quality.

It is reasonable to expect that QoL would be negatively affected by the depressive episodes of BD. Indeed, as noted above, the large-N study of Yatham et al.[15] found remarkably low self-reported QoL amongst patients with BD who were either depressed or had experienced a recent episode of depression. The potential relationship between QoL and hypo/mania in BD, however, is not so straightforward. The 'father of modern psychiatry' Emil Kraepelin recognized a wide variety of presentations of hypomania and mania, including episodes that were characterised by significant dysphoria or depressive symptomatology. The notion that hypo/mania can be an uncomfortable and unpleasant experience for some individuals, however, lost favour in the mid 20th century, and was almost entirely replaced with an understanding of hypo/mania as a primarily euphoric, positive and driving state. However, more recent research has provided further evidence for Kraepelin's more flexible conceptualization of mania and both dysphoric mania (which is characterised by significant depressive symptomatology) and mixed episodes (where the patient meets criteria for both manic and depressive episodes simultaneously) are recognized.

The observation that manic states can be accompanied by symptoms of depression led Vojta and colleagues to hypothesize that patients with manic symptoms would report significantly lower QoL than would patients who were euthymic[35]. To test this theory, the authors administered the SF-12 and the EuroQoL visual analog scale [36] in bipolar patients in a manic/hypomanic episode (N = 16), depressive episode (N = 26), mixed episode (N = 14) or who were euthymic (N = 30). In keeping with their prediction, patients with mania/hypomania showed significantly lower SF-12 mental health scores than euthymic patients, with depressed or mixed patients showing significantly poorer HRQOL again. Mean EuroQoL scores ran in the same direction, although the difference between euthymic and manic/hypomanic patients was not significant. This research was supported in part by the standard gamble data provided by Revicki and colleagues[30] which indicated that the condition least preferred by patients with BD is an inpatient mania-related state, although this was rated in their study as being as unpleasant as being in a severe depressive state.

Although the research by Vojta and colleagues produced some interesting preliminary data, the study did not control for confounding variables such as psychiatric comorbidity, sociodemographic factors, personality type or other psychosocial variables. The complex issue of QoL across clinical states in BD has recently been addressed more comprehensively in a cross-sectional study of the first 2000 participants enrolled in the Systematic Treatment Enhancement Program for Bipolar Disorder, or STEP-BD[37]. STEP-BD, a large multicentre prospective, naturalistic study that features several embedded randomized-controlled trials is measuring QoL via the long version of the Q-LES-Q and the SF-36[38]. In addition, the study is assessing a comprehensive battery of possible confounders, including demographic and socioeconomic factors, family history, psychiatric comorbidity, clinical characteristics (including age of onset), personality (as assessed by the 60-item NEO-Five Factor Inventory), social support, negative life events and attributional style. The primary finding from the analysis was that depressive symptoms were strongly associated with poorer emotional QoL, even after relevant confounding variables were controlled for. Conversely, apparent "supranormal" QoL reported in patients with hypo/mania (compared to that of euthymic patients) disappeared after statistical control. These results represent early data from a study that has the potential to address several important questions regarding QoL in BD.

The above sections have summarized our current understanding of how episodes of depression and hypo/mania impact upon HRQOL. However, it is also of interest for us to know how patients function *between* these episodes, or when they are considered to be in remission. The notion that patients with BD usually achieve complete functional recovery between episodes has been brought into question by a growing body of evidence indicating that inter-episode functioning in the condition often remains compromised, with marked residual functional deficits often remaining after symptomatic recovery[39]. Available research indicates that 25-35% of patients continue to experience partial impairment in their occupational and social functioning; a similar proportion will exhibit extreme functional problems[40–43]. Furthermore, degree of functional impairment appears to be strongly correlated with residual depressive symptoms (for example,[44,45]) during periods of syndromal remission.

Several studies have now assessed QoL in inter-episode patients with BD. For example, a Canadian research group generated a series of interrelated reports on QoL in euthymic patients with BD. Three publications[46–48] describe various aspects of QoL in a single sample of outpatients (N =∼ 68) with BD type I or II who had been clinically euthymic for at least one month. First, Cooke and colleagues[46] examined levels of HRQOL using the MOS SF-20, a short version of the SF-36. Mean scores on the scale were comparable to those reported for patients with MDD by Wells and colleagues in the large RAND Corporation MOS Study[49]. Analysis of SF-20 scores by type of BD showed that patients with BD type II reported significantly poorer HRQOL than BD type I in the areas of social functioning and mental health. Robb and colleagues[47] then reported on QoL in the context of the 'Illness Intrusiveness Model'[50,51], which addresses the impact a disorder and/or its treatment has upon an individual's activities across 13 life domains: health, diet, active/passive recreation, work/financial status, self expression/improvement, family relations, relations with spouse, sex life, other relationships, religious expression and community involvement. Illness intrusiveness occurred in several life areas, with more intrusion being associated with higher Hamilton Depression Rating Scale (Ham-D)[52] scores, patients having experienced a recent episode of depression and having BD type II.

In a separate sample of euthymic BD type I patients (N = 62), MacQueen and colleagues[53] reported that SF-20 scores did not appear to be impacted by whether or not the patient had experienced psychosis during their index episode of mania, although the psychotic sample may have been too small (N = 16) to detect statistically significant differences between sub-groups. Finally, MacQueen and colleagues[54] focused upon the effect of number of manic and depressive episodes on SF-20 and GAF scores in euthymic patients (N = 64), finding that number of past episodes of depression was a stronger determinant of HRQOL than number of previous manic episodes. In this study, good correlation was reported between subjective (SF-20) and objective (GAF) ratings.

A study by Ozer and colleagues[55] assessed 100 interepisode patients with BD in Turkey with the aim of examining the impact of 'history of illness' and 'present symptomatology' factors upon a variety of outcome measures including the Schedule for Affective Disorder and Schizophrenia (SADS) and Q-LES-Q. Using multivariate analysis, Ozer and colleagues found that none of the historical variables (including age at first episode, number of previous depressive/manic episodes, duration of illness, number of hospitalizations, age at first hospitalization, or number of symptoms during first episode) were predictive of mean Q-LES-Q scores. Of the current symptoms assessed, only the depression subscale of the SADS interview significantly predicted lower Q-LES-Q scores, accounting for only 13% of the observed variance. When the patient population was subdivided into three groups (low, moderate and high) according to severity of SADS depression scores, mean Q-LES-Q scores were 39%, 38% and 35%, respectively. In comparison, mean Q-LES-Q scores have been reported to be 42% in hospitalized psychiatric inpatients[56], 42% in outpatients with MDD[57], 44% in patients with seasonal affective disorder (SAD)[57], 53% in patients with chronic MDD[58], and 83% in the general population[22].

One potential problem with existing research addressing HRQOL in euthymic patients with BD concerns the differing research methodologies used to determine euthymia. Whereas clinical trials of maintenance treatment interventions for BD tend to apply rigorous determinants of euthymia on the basis of standardized psychiatric rating scales such as the Young Mania Rating Scale (YMRS)[59] and Ham-D, many of the studies in the HRQOL arena have applied less stringent criteria, such as clinician impression. Given that even mild residual symptoms (particularly of depression) appear to be strongly associated with impaired QoL, it is important for future research to use more sophisticated and thorough assessments of residual symptomatology and more stringent criteria for euthymic state.

In summary, a number of studies have attended to the QoL implications of the different syndromes that fall under the rubric of BD. In terms of the degree of QoL impact, there is evidence to suggest that both the depressive and hypo/manic syndromes are associated with reduced QoL relative to euthymic states. However, further investigation of QoL in euthymic states within BD highlights the importance of considering the course by which the patient arrived at euthymia (past episodes of depression, hypomania versus mania) and the current features of the euthymic state (especially the quasi-ubiquitous presence of depressive symptoms). Bipolar disorder is a heterogeneous condition both between and within individuals: not surprisingly, then, the QoL consequences of BD are measurably impacted by both current state and individual course of the disorder.

QoL as an Outcome Measure in BD Treatment Research

In a review of clinical trials published prior to 2002, Revicki and colleagues[11] identified three studies that had included HRQOL assessments. In our more recent literature review[12] that included studies published prior to 2005, we identified eight studies that had incorporated a QoL outcome measure: five clinical trials that examined pharmacological interventions for BD and three studies that assessed non-pharmacological interventions. At the time of writing,[1] a Medline search reveals that results from two further large clinical trials have been published that included a HRQOL assessment[60,61]. The following section will provide an overview of this nascent body of research.

Namjoshi and colleagues from a Lilly research group have conducted a series of studies examining the efficacy of olanzapine as a treatment intervention for

[1] June 2006

BD[62–66]. In the first of these, Namjoshi et al., (2002)[62] evaluated the impact of acute (3-week) treatment with olanzapine or placebo and long-term (49-week open label) treatment of BD type I (manic/mixed). During the acute-phase treatment period, treatment-related improvements in HRQOL were only apparent for the physical functioning domain of the SF-36. Improvement in other aspects of HRQOL (specifically, bodily pain, vitality, general health and social functioning) occurred during the open label treatment period, indicating that olanzapine may have a relatively rapid effect in terms of improving physical functioning in patients with acute mania, but other HRQOL domains may be slower to respond to treatment.

Shi and colleagues compared the treatment effects of olanzapine and haloperidol in patients with acute mania (N = 453)[63,64]. During the acute treatment phase of the study, significantly greater improvement in five of the SF-36 domains (general health, physical functioning, role limitations – physical, social functioning and vitality) was apparent in the olanzapine group. Superiority of olanzapine over haloperidol persisted over the study's 6-week continuation phase, during which time improvements in work and household functioning also became apparent. Other research has examined the effects of adding olanzapine to lithium or valproate in patients with BD (N = 224)[65]. Combination therapy was associated with better outcome in several Quality of Life Interview (QOLI[66]) domains compared to lithium or valproate monotherapy. The SF-36 and Quality of Life in Depression Scale (QLDS[67]) have been used in a study comparing the benefits of olanzapine alone versus an olanzapine-fluoxetine combination or placebo[68]. Compared with placebo, patients who received olanzapine showed greater improvement at 8 weeks in SF-36 mental health summary scores, and in mental health, role-emotional and social functioning domain scores. The combination group fared significantly better in terms of HRQOL improvement than the olanzapine-alone group, showing improvement in five of the SF-36 domain scores and in QLDS total score.

The Q-LES-Q has been administered at baseline (hospital discharge), 6 and 12 weeks in a comparison of divalproex sodium and olanzapine in the treatment of acute mania[60]. No significant treatment effects were detected in Q-LES-Q scores in the study, although only 52 (43%) of the 120 patients randomized to either divalproex or olanzapine completed the QoL instrument. The authors reported an association between weight gain being reported as an adverse event and poorer change scores in the physical, leisure, and general activities domains of the Q-LES-Q at 6 weeks (but not at 12 weeks). Negative correlations were reported between increased weight (at 6 weeks) and overall life satisfaction, physical health, mood, general activities and satisfaction with medication on the Q-LES-Q. In a more recent study, Revicki and colleagues (2005)[69] reported on data from a pragmatic, randomized clinical trial of divalproex versus lithium in 201 patients hospitalized for a BD type I manic or mixed episode. Life quality was assessed via the SF-36 at 1, 3, 6, 9 and 12 months, with no significant differences emerging between intervention groups in HRQOL outcome. An analysis of HRQOL in patients who continued mood stabilizing therapy beyond 3 months versus those who did not found no significant between group differences in SF-36 physical functioning scores, although there was a non-significant trend towards improved mental functioning scores in the continuation group at the 6 and 12 month assessment points.

Finally, some initial HRQOL data has recently been published from the BOLDER study[61]. BOLDER is a large, 8-week, multicenter, double-blind, randomized, fixed-dose, placebo-controlled monotherapy study of quetiapine (600 or 300 mg/day) versus placebo in outpatients with DSM-IV bipolar I or bipolar II disorder, with or without rapid cycling, in a major depressive episode. The study administered the 16-item short form of the Q-LES-Q at baseline, weeks 4 and 8 to assess QoL, finding 12 and 11 point increase in Q-LES-Q scores at last assessment in the high and low dose groups respectively, compared to a 7-point change in the placebo group (i.e. significantly greater improvement in QoL in both groups after 8 weeks of treatment compared to placebo). These data are encouraging; elsewhere, the Q-LES-Q has been used to assess outcome in the treatment of dysthymia with sertraline, imipramine or placebo, where after 12 weeks of treatment an 8-point change was reported in the two active intervention arms, with a 4-point change in the placebo arm[70]. In another study of the treatment of chronic depression with either sertraline or imipramine, a 9-point change was seen after 4 weeks of treatment, and a 14-point change after 12 weeks[58]. The 12-point change in Q-LES-Q score observed in the BOLDER study is in keeping, then, with outcome data from other studies in unipolar depression, indicating that QoL in patients with BD is also amenable to relatively rapid change, even when study inclusion criteria are broadened to include greater diagnostic heterogeneity, such as rapid cycling and BD type II patients. It will be interesting to see the impact of treatments upon long-term QoL outcomes in the BOLDER trial, and other well-designed pharmacological intervention studies.

Although pharmacology forms the bedrock of treatment for BD[71], there is a clear need for other treatment modalities that augment the effects of medication in this complex psychiatric condition. Over the last decade, we have seen an upsurge of interest in examining the role of psychological interventions as an adjunct to the pharmacological treatment of BD[72,73]. Most of this research has concentrated upon examining the efficacy of psychotherapy (in a variety of guises) as a treatment intervention. Several multi-modal psychotherapeutic interventions for BD have been developed, such as Family Focused Treatment (FFT)[74,75], Interpersonal and Social Rhythm Therapy[76] and Cognitive-Behavioural Therapy (CBT)[73]. Surprisingly few studies of psychosocial treatment interventions for BD, however, have used QoL measures to assess outcome. In one study, Patelis-Siotis and colleagues[77] used the SF-36 in a feasibility study of group cognitive behavior therapy (CBT) in patients with BD. Although baseline SF-36 data was available for 42 patients, pre and post intervention data was only available for a proportion of participants (N = 22) as completion of the QoL questionnaires was optional. Nevertheless, SF-36 vitality and role emotional scores were significantly improved following CBT, with an accompanying trend towards improved social functioning. A more recent study examined the effects of providing three sessions of psychoeducation (PE) about lithium treatment to patients (N = 26) with BD[78]. In addition to assessing the effects of PE upon medication adherence, the authors examined the impact of the intervention upon QoL, as measured by the WHOQOL-BREF. Following PE, patients in the intervention arm of the study showed significant improvement in two of the WHOQOL-BREF's four domains (physical health and social functioning) and in overall perceived health. Patients in the control arm of the study, in comparison, showed no significant changes in their perceived QoL. The results of the study indicate that even relatively brief psychosocial treatment interventions can have a positive impact upon HRQOL – a finding supported by our own non-randomised study of the effects of PE upon Q-LES-Q scores in euthymic or mildly symptomatic outpatients (N = 57) with BD type I or II[79].

In summary, research using QoL as an outcome measure attests to the fact that QoL measures provide additional important information over that provided by symptom measures. Emerging research into adjunctive psychosocial treatments suggests that even relatively brief interventions may have effects on QoL.

Future Research Directions

Over the past decade, we have seen an upsurge of interest in examining QoL in BD. This emerging body of research has provided some initial evidence that it is both feasible and important to assess life quality in this complex psychiatric condition. Our own qualitative data garnered from in-depth interviews with patients diagnosed with BD suggests that QoL is a meaningful indicator of well-being in this patient population[80]. Importantly, when measuring outcome in response to treatment, QoL assessment scales hold the potential to provide information over and above that provided by traditional outcome measures. However, this chapter has also served to illustrate that the existing literature on QoL in BD is relatively immature, lagging well behind the study of QoL in relation to other psychiatric conditions such as schizophrenia. In particular, there is no current consensus about which QoL instruments are most appropriate for use in this patient population, very little replication of core findings across research studies, and as yet there is no disorder-specific measure to assess QoL in patients with BD.

How, then, should the field progress? At a theoretical level, it is useful to draw direction from the substantial research into QoL in other psychiatric disorders. Ritsner's Distress/Protection Vulnerability model (Chapter 1, this collection), for example, is based on a number of studies showing that immediate illness factors are only one aspect of the QoL picture. His model is framed around two groups of factors – distress (e.g., negative symptoms of schizophrenia) and protective (e.g., social support). Most importantly, perhaps, his model parallels the prominent diathesis-stress model of psychopathology, in positing a primary role for stable temperamental traits that act as vulnerabilities (e.g., neuroticism, emotion-focused coping), or protective factors (e.g., extraversion, high self-esteem) for QoL. He suggests, therefore, that QoL is not usefully seen as an outcome of illness, but as a stable feature of the person that is impacted by illness (as well as other features of the environment). The evidence he presents to support this framing of QoL includes the generally low correlations between illness-related variables and QoL, the absence of progressive alteration of QoL over time in patients with serious mental disorders, and the presence of QoL impairments in patients who ultimately will, but are not currently, exhibiting signs of serious mental disorder.

In Chapter 1 of this collection, Ritsner also discusses the potential interface between neurobiological and quality of life research. Numerous neurobiological factors may conceivably be related to QoL in BD. These include genetic polymorphisms, structural brain abnormalities (e.g., white matter lesions), hypothalamic-pituitary adrenal (HPA) axis function and neurocognition. Although little work has been done relating these factors to QoL in BD there is particular interest in the latter two factors at present. HPA axis function is now understood to be abnormal in patients with BD and it has also been shown that this abnormality persists into euthymia[81]. In cancer patients pilot studies have shown enhanced QoL and decreased stress symptoms with possibly beneficial changes in HPA axis functioning after stress reduction programs[82]. We await the application of such an approach to BD with interest. Neurocognitive impairment has also been described in euthymic BD, including widespread impairments in executive function[83]. Such deficits may clearly contribute to impaired QoL particularly in the euthymic phase and indeed are commonly a source of complaint in patients with BD. However the temporal relationship between HPA axis function, cognition and QoL in BD remains to be fully established and is an area of active research at present.

Ritsner further highlights that, in contrast to QoL research in schizophrenia, the BD QoL field is still driven by a strongly medical model conception, with the illness itself dominating explanations of QoL. No studies in QoL of BD have yet measured the psychological variables (specifically, personality, coping style, cognitive style) that are components of Ritsner's more holistic model of HRQL. In a fundamental way, QoL research in the field of BD still starts with the disorder, while QoL research as championed by Ritsner starts with the person. Following the lead of schizophrenia research, future research into QoL in BD should include important "third variables", such as measures of normal personality, which are almost certain to moderate and/or mediate the effect of life events (including mental disorder) on QoL.

At an empirical level, the widespread adoption of a disease-specific QoL instrument for BD is critical for the advancement of the field. Although some existing QoL instruments are likely to capture key aspects of QoL in this condition, they may be less sensitive to some of the more unique aspects of this complex psychiatric disorder. QoL scales developed for patients with unipolar depression (such as the QLDS) or for unspecified psychiatric populations (such as the Q-LES-Q) will overlap in content somewhat with a scale developed specifically for patients with BD. However, there are likely to be some bipolar symptoms (for example, financial indiscretion, hypersexuality, interpersonal behaviour when hypo/manic) that have a unique impact upon QoL that existing scales do not tap. Indeed, part of the rationale for developing disorder-specific measures lies in the understanding that such scales can be more sensitive to change in response to treatment than their more generic counterparts[84]. However, the validity of disorder-specific scales should be maximized through thorough consultation with individuals with the condition themselves, a core part of the development of our own disorder specific scale for BD, the Quality of Life in Bipolar Disorder (QoL.BD).

In order to generate an initial set of items for the QoL.BD, we performed a series of in-depth qualitative interviews, aiming a priori to garner the viewpoints of a fairly representative sample of people diagnosed with the disorder (for example, BD type I and II, low functioning versus high functioning individuals). We also aimed to interview caregivers of people severely affected by BD (in the belief that the individuals themselves might be unable to undergo a lengthy qualitative interview), healthcare workers with expertise in BD (in the belief that they would be able to describe the impact of BD upon QoL at a broader group level) and international experts in the BD research. In total, we conducted 52 interviews with people with BD ($N = 35$), their caregivers ($N = 5$) and healthcare professionals or experts ($N = 12$) identified by both convenience and purposive sampling. Clinical characteristics of the affected sample ranged widely between individuals who had been clinically stable for several years through to inpatients who were recovering from a severe episode of depression or mania.

The results of this initial study provided some interesting initial data concerning the impact of BD upon QoL[80,85]. The majority of the affected individuals we interviewed described how the condition had had a profoundly negative effect upon their QoL, often having serious and enduring effects on their ability to have good education, meaningful vocation, financial independence and healthy social and intimate relationships. Having said this, we also interviewed a number of people who were functioning exceptionally well despite their diagnosis; a minority of people even espoused the view that their condition had opened up new doors of opportunity for them, for example, in terms of positively changing their career paths or social networks. On the whole, however, even these individuals described having undergone several years of hardship and adjustment before getting 'back on track'. Respondents described a wide variety of factors that influenced their QoL, including but not limited to: side effects of medications, occupation, education, physical functioning, environment, healthcare factors, leisure activities, routine and sexuality. Some of the factors discussed (for example, independence, stigma and disclosure, identity, and spirituality) are not frequently examined in relation to QoL; yet, they appear to have a significant impact upon peoples' ability to live their lives to the full in the context of BD. We are continuing to develop the QoL.BD in close consultation with both individuals with BD and healthcare professionals in the hope of maximizing the validity of the resulting scale.

CONCLUSIONS

In terms of empirical findings to date, and recognizing the limitations described above, it can be provisionally concluded that:

a) QoL is seriously affected in BD. The QoL impacts of BD might be more serious than seen in other mood and anxiety disorders (except perhaps double depression), and may even be comparable to the consequences of the deteriorating clinical course seen in schizophrenia.

b) Consistent with recent thinking and epidemiological data, there is little support for the notion that there are states of BD that have purely positive QoL implications. Both interepisode functioning and mania are commonly flavoured with depression, and the depressed states and depressive symptoms of BD are clearly negative impacts on QOL

c) Consistent with an emphasis on broader outcome measures in psychiatry generally, and a more patient-centred perspective, the small number of studies with relevant data have found that symptom variables and QoL variables respond differently to both pharmacotherapy and psychotherapy. Consistent with the definition of QoL, there is some evidence that effective treatment is measured later in QoL than symptom variables.

d) Future QoL research in BD is likely to benefit from adopting a broader focus, and including measures of the range of variables that undoubtedly interact with life events (including disorder) to produce individual differences in QoL.

e) Pace the prescription above to be less disorder-centric, the field also requires a disorder-specific measure that is sensitive to the range of QoL impacts that result from the interaction between predispositions (both vulnerabilities and strengths), positive and negative features of the present environment and the specific symptoms of BD. Given the immature state of the literature, such an instrument is best developed "bottom-up", with the raw data being descriptions of QoL deriving from patients, carers and relevant others.

The challenge of investigating QoL in BD is sizeable but worthwhile. It is sizeable because BD is heterogeneous both within and between individuals, and the variety of states implied by the label will have different QoL implications. It is a worthwhile challenge not only because of the importance of QoL understandings for the comprehensive management of the disorder, but also because the complexity of the disorder provides a useful stimulus for thinking about significant issues in QoL research. The next few years will see no doubt see exciting developments in this burgeoning field.

ACKNOWLEDGEMENTS

Erin Michalak is supported by a Michael Smith Scholar Award from the Michael Smith Foundation for Health Research and a Canadian Institutes of Health Research New Investigator Award.

REFERENCES

1. Judd, L. L., Akiskal, H. S., Schettler, P. J., Endicott, J., Maser, J., Solomon, D. A., Leon, A. C., Rice, J. A., and Keller, M. B. The Long-Term Natural History of the Weekly Symptomatic Status of Bipolar I Disorder. Archives of General Psychiatry 2002;59(6):530–7.

2. Judd, L. L., Schettler, P. J., Akiskal, H. S., Maser, J., Coryell, W., Solomon, D., Endicott, J., and Keller, M. Long-Term Symptomatic Status of Bipolar I Vs. Bipolar II Disorders. Int.J.Neuropsychopharmacol. 2003;6(2):127–37.
3. Murray, C. J. and Lopez, A. D. Global Mortality, Disability, and the Contribution of Risk Factors: Global Burden of Disease Study. Lancet 5-17-1997;349(9063):1436–42.
4. US DHEW Medical Practice Project. A State of the Service Report for the Office of the Assistant Secretary for the US Dept of Health, Education and Welfare. Policy Research. 1979.
5. Weissman, M. M., Bland, R. C., Canino, G. J., Faravelli, C., Greenwald, S., Hwu, H. G., Joyce, P. R., Karam, E. G., Lee, C. K., Lellouch, J., Lepine, J. P., Newman, S. C., Rubio-Stipec, M., Wells, J. E., Wickramaratne, P. J., Wittchen, H., and Yeh, E. K. Cross-National Epidemiology of Major Depression and Bipolar Disorder. JAMA 7-24-1996;276(4):293–9.
6. Angst, J. The Emerging Epidemiology of Hypomania and Bipolar II Disorder. J.Affect.Disord. 1998;50(2–3):143–51.
7. Wyatt, R. J. and Henter, I. An Economic Evaluation of Manic-Depressive Illness–1991. Soc.Psychiatry Psychiatr.Epidemiol. 1995;30(5):213–9.
8. Colom, F. and Vieta, E. A Perspective on the Use of Psychoeducation, Cognitive-Behavioral Therapy and Interpersonal Therapy for Bipolar Patients. Bipolar Disord. 2004;6(6):480–6.
9. Namjoshi, M. A. and Buesching, D. P. A Review of the Health-Related Quality of Life Literature in Bipolar Disorder. Qual.Life Res. 2001;10(2):105–15.
10. Dean, B. B., Gerner, D., and Gerner, R. H. A Systematic Review Evaluating Health-Related Quality of Life, Work Impairment, and Healthcare Costs and Utilization in Bipolar Disorder. Curr.Med.Res.Opin. 2004;20(2):139–54.
11. Revicki, D. A., Matza, L. S., Flood, E., and Lloyd, A. Bipolar Disorder and Health-Related Quality of Life: Review of Burden of Disease and Clinical Trials. Pharmacoeconomics. 2005;23(6):583–94.
12. Michalak, E. E., Yatham, L. N., and Lam, R. W. Quality of Life in Bipolar Disorder: a Review of the Literature. Health Qual.Life Outcomes. 2005;3:72.
13. Stewart, A. L., Hays, R. D., and Ware, J. E., Jr. The MOS Short-Form General Health Survey. Reliability and Validity in a Patient Population. Med.Care 1988;26(7):724–35.
14. Arnold, L. M., Witzeman, K. A., Swank, M. L., McElroy, S. L., and Keck, P. E., Jr. Health-Related Quality of Life Using the SF-36 in Patients With Bipolar Disorder Compared With Patients With Chronic Back Pain and the General Population. J.Affect.Disord. 2000;57(1–3):235–9.
15. Yatham, L. N., Lecrubier, Y., Fieve, R. R., Davis, K. H., Harris, S. D., and Krishnan, A. A. Quality of Life in Patients With Bipolar I Depression: Data From 920 Patients. Bipolar Disord. 2004;6(5):379–85.
16. Development of the World Health Organization WHOQOL-BREF Quality of Life Assessment. The WHOQOL Group. Psychological Medicine 1998;28(3):551–8.
17. Berlim, M. T., Pargendler, J., Caldieraro, M. A., Almeida, E. A., Fleck, M. P., and Joiner, T. E. Quality of Life in Unipolar and Bipolar Depression: Are There Significant Differences? J.Nerv.Ment.Dis. 2004;192(11):792–5.
18. ten Have, M., Vollebergh, W., Bijl, R., and Nolen, W. A. Bipolar Disorder in the General Population in The Netherlands (Prevalence, Consequences and Care Utilisation): Results From The Netherlands Mental Health Survey and Incidence Study (NEMESIS). J.Affect.Disord. 2002;68(2–3):203–13.
19. Williams, J. B. W., Gibbon, M., and First, M. B. The Structured Clinical Interview for DSM-III-R (SCID), I: History, Rationale, and Description. Archives of General Psychiatry 1992;49:624–9.
20. Chand, P. K., Mattoo, S. K., and Sharan, P. Quality of Life and Its Correlates in Patients With Bipolar Disorder Stabilized on Lithium Prophylaxis. Psychiatry Clin.Neurosci. 2004;58(3):311–8.
21. Endicott, J., Nee, J., Harrison, W., and Blumenthal, R. Quality of Life Enjoyment and Satisfaction Questionnaire: a New Measure. Psychopharmacol.Bull. 1993;29(2):321–6.
22. Schechter, D, Endicott, J., and Nee, J. Quality of Life of 'Normal' Controls: Association With Lifetime History of Mental Illness. In press, Psychiatry Research.
23. Atkinson, M., Zibin, S., and Chuang, H. Characterizing Quality of Life Among Patients With Chronic Mental Illness: a Critical Examination of the Self-Report Methodology. Am.J.Psychiatry 1997;154(1):99–105.

24. Ferrans, C. E. and Powers, M. J. Psychometric Assessment of the Quality of Life Index. Res.Nurs.Health 1992;15(1):29–38.
25. Dell'Osso, L., Pini, S., Cassano, G. B., Mastrocinque, C., Seckinger, R. A., Saettoni, M., Papasogli, A., Yale, S. A., and Amador, X. F. Insight into Illness in Patients With Mania, Mixed Mania, Bipolar Depression and Major Depression With Psychotic Features. Bipolar.Disord. 2002;4(5):315–22.
26. Ghaemi, S. N., Stoll, A. L., and Pope, H. G., Jr. Lack of Insight in Bipolar Disorder. The Acute Manic Episode. J.Nerv.Ment.Dis. 1995;183(7):464–7.
27. Goldberg, J. F. and Harrow, M. Subjective Life Satisfaction and Objective Functional Outcome in Bipolar and Unipolar Mood Disorders: a Longitudinal Analysis. J.Affect.Disord. 2005;89 (1–3):79–89.
28. Torrance, G. W. Measurement of Health State Utilities for Economic Appraisal. J.Health Econ. 1986;5(1):1–30.
29. Torrance, G. W. Preferences for Health Outcomes and Cost-Utility Analysis. Am. J. Manag. Care. 1997;3 Suppl:S8–20.
30. Revicki, D. A., Hanlon, J., Martin, S., Gyulai, L., Nassir, Ghaemi S., Lynch, F., Mannix, S., and Kleinman, L. Patient-Based Utilities for Bipolar Disorder-Related Health States. J.Affect.Disord. 2005;87(2–3):203–10.
31. Wells, K. B. and Sherbourne, C. D. Functioning and Utility for Current Health of Patients With Depression or Chronic Medical Conditions in Managed, Primary Care Practices. Arch.Gen.Psychiatry 1999;56(10):897–904.
32. Tsevat, J., Keck, P. E., Hornung, R. W., and McElroy, S. L. Health Values of Patients With Bipolar Disorder. Qual.Life Res. 2000;9(5):579–86.
33. Rasanen, P., Roine, E., Sintonen, H., Semberg-Konttinen, V., Ryynanen, O. P., and Roine, R. Use of Quality-Adjusted Life Years for the Estimation of Effectiveness of Health Care: A Systematic Literature Review. Int.J.Technol.Assess.Health Care 2006;22(2):235–41.
34. Peveler, R., Kendrick, T., Buxton, M., Longworth, L., Baldwin, D., Moore, M., Chatwin, J., Goddard, J., Thornett, A., Smith, H., Campbell, M., and Thompson, C. A Randomised Controlled Trial to Compare the Cost-Effectiveness of Tricyclic Antidepressants, Selective Serotonin Reuptake Inhibitors and Lofepramine. Health Technol.Assess. 2005;9(16):1–134, iii.
35. Vojta, C., Kinosian, B., Glick, H., Altshuler, L., and Bauer, M. S. Self-Reported Quality of Life Across Mood States in Bipolar Disorder. Compr.Psychiatry 2001;42(3):190–5.
36. EuroQol–a New Facility for the Measurement of Health-Related Quality of Life. The EuroQol Group. Health Policy 1990;16(3):199–208.
37. Zhang, H., Wisniewski, S. R., Bauer, M. S., Sachs, G. S., and Thase, M. E. Comparisons of Perceived Quality of Life Across Clinical States in Bipolar Disorder: Data From the First 2000 Systematic Treatment Enhancement Program for Bipolar Disorder (STEP-BD) Participants. Compr.Psychiatry 2006;47(3):161–8.
38. Sachs, G. S., Thase, M. E., Otto, M. W., Bauer, M., Miklowitz, D., Wisniewski, S. R., Lavori, P., Lebowitz, B., Rudorfer, M., Frank, E., Nierenberg, A. A., Fava, M., Bowden, C., Ketter, T., Marangell, L., Calabrese, J., Kupfer, D., and Rosenbaum, J. F. Rationale, Design, and Methods of the Systematic Treatment Enhancement Program for Bipolar Disorder (STEP-BD). Biol.Psychiatry 2003;53(11):1028–42.
39. MacQueen, G. M., Young, L. T., and Joffe, R. T. A Review of Psychosocial Outcome in Patients With Bipolar Disorder. Acta Psychiatr.Scand. 2001;103(3):163–70.
40. Dion, G. L., Tohen, M., Anthony, W. A., and Waternaux, C. S. Symptoms and Functioning of Patients With Bipolar Disorder Six Months After Hospitalization. Hosp.Community Psychiatry 1988;39(6):652–7.
41. Harrow, M., Goldberg, J. F., Grossman, L. S., and Meltzer, H. Y. Outcome in Manic Disorders. A Naturalistic Follow-Up Study. Arch.Gen.Psychiatry 1990;47(7):665–71.
42. Tohen, M., Waternaux, C. M., and Tsuang, M. T. Outcome in Mania. A 4-Year Prospective Follow-Up of 75 Patients Utilizing Survival Analysis. Arch.Gen.Psychiatry 1990;47(12):1106–11.
43. Goldberg, J. F., Harrow, M., and Grossman, L. S. Course and Outcome in Bipolar Affective Disorder: a Longitudinal Follow-Up Study. Am.J.Psychiatry 1995;152(3):379–84.

44. Fagiolini, A., Kupfer, D. J., Masalehdan, A., Scott, J. A., Houck, P. R., and Frank, E. Functional Impairment in the Remission Phase of Bipolar Disorder. Bipolar.Disord. 2005;7(3):281–5.
45. Bauer, M. S., Kirk, G. F., Gavin, C., and Williford, W. O. Determinants of Functional Outcome and Healthcare Costs in Bipolar Disorder: a High-Intensity Follow-Up Study. J.Affect.Disord. 2001;65(3):231–41.
46. Cooke, R. G., Robb, J. C., Young, L. T., and Joffe, R. T. Well-Being and Functioning in Patients With Bipolar Disorder Assessed Using the MOS 20-ITEM Short Form (SF-20). J.Affect.Disord. 1996;39(2):93–7.
47. Robb, J. C., Cooke, R. G., Devins, G. M., Young, L. T., and Joffe, R. T. Quality of Life and Lifestyle Disruption in Euthymic Bipolar Disorder. J.Psychiatr.Res. 1997;31(5):509–17.
48. Robb, J. C., Young, L. T., Cooke, R. G., and Joffe, R. T. Gender Differences in Patients With Bipolar Disorder Influence Outcome in the Medical Outcomes Survey (SF-20) Subscale Scores. J.Affect.Disord. 1998;49(3):189–93.
49. Wells, K. B., Stewart, A., Hays, R. D., Burnam, M. A., Rogers, W., Daniels, M., Berry, S., Greenfield, S., and Ware, J. The Functioning and Well-Being of Depressed Patients. Results From the Medical Outcomes Study. JAMA 1989;262(7):914–9.
50. Flanagan, J. C. A Research Approach to Improving Our Quality of Life. American Psychology 1978;33:138–147.
51. Devins, G. M., Edworthy, S. M., Seland, T. P., Klein, G. M., Paul, L. C., and Mandin, H. Differences in Illness Intrusiveness Across Rheumatoid Arthritis, End-Stage Renal Disease, and Multiple Sclerosis. J.Nerv.Ment.Dis. 1993;181(6):377–81.
52. Hamilton, M. Development of a Rating Scale for Primary Depressive Illness. British Journal of the Society of Clinical Psychology 1967;6:278–96.
53. MacQueen, G. M., Young, L. T., Robb, J. C., Cooke, R. G., and Joffe, R. T. Levels of Functioning and Well-Being in Recovered Psychotic Versus Nonpsychotic Mania. J.Affect.Disord. 1997;46(1):69–72.
54. MacQueen, G. M., Young, L. T., Robb, J. C., Marriott, M., Cooke, R. G., and Joffe, R. T. Effect of Number of Episodes on Wellbeing and Functioning of Patients With Bipolar Disorder. Acta Psychiatr.Scand. 2000;101(5):374–81.
55. Ozer, S., Ulusahin, A., Batur, S., Kabakci, E., and Saka, M. C. Outcome Measures of Interepisode Bipolar Patients in a Turkish Sample. Social Psychiatry and Psychiatric Epidemiology 2002;37(1):31–7.
56. Rapaport, M. H.; Clary, C. M.; Judd, L. L. The Impact of Depression and Its Treatment. Presented at the 154th Annual Meeting of the American Psychiatric Association, New Orleans, La. 2001.
57. Michalak, E. E., Tam, E. M., Manjunath, C. V., Solomons, K., Levitt, A. J., Levitan, R., Enns, M., Morehouse, R., Yatham, L. N., and Lam, R. W. Generic and Health-Related Quality of Life in Patients With Seasonal and Nonseasonal Depression. Psychiatry Res. 2004;128(3):245–51.
58. Miller, I. W., Keitner, G. I., Schatzberg, A. F., Klein, D. N., Thase, M. E., Rush, A. J., Markowitz, J. C., Schlager, D. S., Kornstein, S. G., Davis, S. M., Harrison, W. M., and Keller, M. B. The Treatment of Chronic Depression, Part 3: Psychosocial Functioning Before and After Treatment With Sertraline or Imipramine. J.Clin.Psychiatry 1998;59(11):608–19.
59. Young, R. C., Biggs, J. T., Ziegler, V. E., and Meyer, D. A. A Rating Scale for Mania: Reliability, Validity and Sensitivity. Br.J.Psychiatry 1978;133:429–35.
60. Revicki, D. A., Paramore, L. C., Sommerville, K. W., Swann, A. C., and Zajecka, J. M. Divalproex Sodium Versus Olanzapine in the Treatment of Acute Mania in Bipolar Disorder: Health-Related Quality of Life and Medical Cost Outcomes. J.Clin.Psychiatry 2003;64(3):288–94.
61. Calabrese, J. R., Keck, P. E., Jr., Macfadden, W., Minkwitz, M., Ketter, T. A., Weisler, R. H., Cutler, A. J., McCoy, R., Wilson, E., and Mullen, J. A Randomized, Double-Blind, Placebo-Controlled Trial of Quetiapine in the Treatment of Bipolar I or II Depression. Am.J.Psychiatry 2005;162(7):1351–60.
62. Namjoshi, M. A., Rajamannar, G., Jacobs, T., Sanger, T. M., Risser, R., Tohen, M. F., Breier, A., and Keck, P. E. Economic, Clinical, and Quality-of-Life Outcomes Associated

With Olanzapine Treatment in Mania. Results From a Randomized Controlled. J.Affect.Disord. 2002;69(1–3):109–18.

63. Shi, L., Namjoshi, M. A., Zhang, F., Gandhi, G., Edgell, E. T., Tohen, M., Breier, A., and Haro, J. M. Olanzapine Versus Haloperidol in the Treatment of Acute Mania: Clinical Outcomes, Health-Related Quality of Life and Work Status. Int.Clin.Psychopharmacol. 2002;17(5):227–37.
64. Tohen, M., Goldberg, J. F., Gonzalez-Pinto Arrillaga, A. M., Azorin, J. M., Vieta, E., Hardy-Bayle, M. C., Lawson, W. B., Emsley, R. A., Zhang, F., Baker, R. W., Risser, R. C., Namjoshi, M. A., Evans, A. R., and Breier, A. A 12-Week, Double-Blind Comparison of Olanzapine Vs Haloperidol in the Treatment of Acute Mania. Arch.Gen.Psychiatry 2003;60(12):1218–26.
65. Namjoshi, M. A., Risser, R., Shi, L., Tohen, M., and Breier, A. Quality of Life Assessment in Patients With Bipolar Disorder Treated With Olanzapine Added to Lithium or Valproic Acid. J.Affect.Disord. 2004;81(3):223–9.
66. Lehman, A. F. A Quality of Life Interview for the Chronically Mentally Ill. Evaluation and Program Planning 1988;(11):51–62.
67. Hunt, S. M. and McKenna, S. P. The QLDS: a Scale for the Measurement of Quality of Life in Depression. Health Policy 1992;22(3):307–19.
68. Shi, L., Namjoshi, M. A., Swindle, R., Yu, X., Risser, R., Baker, R. W., and Tohen, M. Effects of Olanzapine Alone and Olanzapine/Fluoxetine Combination on Health-Related Quality of Life in Patients With Bipolar Depression: Secondary Analyses of a Double-Blind, Placebo-Controlled, Randomized Clinical Trial. Clin.Ther. 2004;26(1):125–34.
69. Revicki, D. A., Hirschfeld, R. M., Ahearn, E. P., Weisler, R. H., Palmer, C., and Keck, P. E., Jr. Effectiveness and Medical Costs of Divalproex Versus Lithium in the Treatment of Bipolar Disorder: Results of a Naturalistic Clinical Trial. J.Affect.Disord. 2005;86(2–3):183–93.
70. Kocsis, J. H., Zisook, S., Davidson, J., Shelton, R., Yonkers, K., Hellerstein, D. J., Rosenbaum, J., and Halbreich, U. Double-Blind Comparison of Sertraline, Imipramine, and Placebo in the Treatment of Dysthymia: Psychosocial Outcomes. Am.J.Psychiatry 1997;154(3):390–5.
71. Suppes, T., Dennehy, E. B., Swann, A. C., Bowden, C. L., Calabrese, J. R., Hirschfeld, R. M., Keck, P. E., Jr., Sachs, G. S., Crismon, M. L., Toprac, M. G., and Shon, S. P. Report of the Texas Consensus Conference Panel on Medication Treatment of Bipolar Disorder 2000. J.Clin.Psychiatry 2002;63(4):288–99.
72. Scott, J. and Colom, F. Psychosocial Treatments for Bipolar Disorders. Psychiatr.Clin.North Am. 2005;28(2):371–84.
73. Otto, M. W., Reilly-Harrington, N., and Sachs, G. S. Psychoeducational and Cognitive-Behavioral Strategies in the Management of Bipolar Disorder. J.Affect.Disord. 2003;73(1–2):171–81.
74. Miklowitz, D. J., Goldstein, M. J., Nuechterlein, K. H., Snyder, K. S., and Mintz, J. Family Factors and the Course of Bipolar Affective Disorder. Arch.Gen.Psychiatry 1988;45(3):225–31.
75. Miklowitz, D. J. and Goldstein, M. J., Bipolar Disorder: A Family-Focused Treatment Approach. New York: Guilford Press; 1997.
76. Frank, E., Swartz, H. A., and Kupfer, D. J. Interpersonal and Social Rhythm Therapy: Managing the Chaos of Bipolar Disorder [In Process Citation]. Biol.Psychiatry 2000;48(6):593–604.
77. Patelis-Siotis, I., Young, L. T., Robb, J. C., Marriott, M., Bieling, P. J., Cox, L. C., and Joffe, R. T. Group Cognitive Behavioral Therapy for Bipolar Disorder: a Feasibility and Effectiveness Study. J.Affect.Disord. 2001;65(2):145–53.
78. Dogan, S. and Sabanciogullari, S. The Effects of Patient Education in Lithium Therapy on Quality of Life and Compliance. Arch.Psychiatr.Nurs. 2003;17(6):270–5.
79. Michalak, E. E., Yatham, L. N., Wan, D. D., and Lam, R. W. Perceived Quality of Life in Patients With Bipolar Disorder. Does Group Psychoeducation Have an Impact? Can.J.Psychiatry 2005;50(2):95–100.
80. Michalak, E. E., Yatham, L. N., Kolesar, S., and Lam, R. W. Bipolar Disorder and Quality of Life: a Patient-Centered Perspective. Qual.Life Res. 2006;15(1):25–37.
81. Watson, S., Gallagher, P., Ritchie, J. C., Ferrier, I. N., and Young, A. H. Hypothalamic-Pituitary-Adrenal Axis Function in Patients with Bipolar Disorder. Br.J.Psychiatry 2004;184:496–502.

82. Carlson, L. E., Speca, M., Patel, K. D., and Goodey, E. Mindfulness-Based Stress Reduction in Relation to Quality of Life, Mood, Symptoms of Stress and Levels of Cortisol, Dehydroepiandrosterone Sulfate (DHEAS) and Melatonin in Breast and Prostate Cancer Outpatients. Psychoneuroendocrinology 2004;29(4):448–74.
83. Thompson, J. M., Gallagher, P., Hughes, J. H., Watson, S., Gray, J. M., Ferrier, I. N., and Young, A. H. Neurocognitive Impairment in Euthymic Patients With Bipolar Affective Disorder. Br.J.Psychiatry 2005;186:32–40.
84. Patrick, D. L. and Deyo, R. A. Generic and Disease-Specific Measures in Assessing Health Status and Quality of Life. Med.Care 1989;27(3 Suppl):S217–S232.
85. Michalak, E. E., Yatham, L. N., Maxwell, V., Hale, S., and Lam, R. The Impact of Bipolar Disorder Upon Work Functioning: A Qualitative Analysis. In press, Bipolar Disord.

CHAPTER 14

QUALITY OF LIFE IMPAIRMENT IN ANXIETY DISORDERS

MARGARET A. KOURY AND MARK HYMAN RAPAPORT

Department of Psychiatry, Cedars-Sinai Medical Center, 8730 Alden Drive, C-301 Los Angeles, California 90048

Abstract: There is a growing consensus that practitioners must broaden the scope of assessment for anxiety disorders from signs and symptoms to include the measurement of quality of life (QOL) appraisals and social functioning. However, there is not a consensus about the definition of quality of life nor is there agreement about how to operationalize the construct of quality of life. Most measures of QOL assess an individual's perceptions about social relationships, physical health, work and activity functioning, economic status, and an overall sense of well-being[8]. Objective functional impairment can be quantified by measuring work productivity, mental health functioning, and physical health functioning[3]. One of the challenges faced by the field is understanding the relationship between objective measures of functioning and subjective measures of QOL. In this chapter, we will present findings from epidemiologic studies and empirical research investigations on QOL and functional impairment in generalized anxiety disorder, panic disorder, posttraumatic stress disorder, and generalized social phobia. We will discuss the findings, address strengths and weakness of the current research, and suggest future areas of focus to expand our knowledge and competencies including appropriating data from clinical populations in order to better understand and treat this growing population

Keywords: Quality of life, Functional impairment, Anxiety disorders, Generalized anxiety disorder, Panic disorder, Posttraumatic stress disorder, Social phobia

PREVALENCE AND IMPACT OF ANXIETY DISORDERS

Anxiety disorders are the most prevalent class of psychiatric disorders in the United States with lifetime prevalence estimated at 28.8% (NCS-R)[1]. Annual direct and indirect costs in the United States associated with anxiety disorders are estimated at over $42 billion, 10% of which represents indirect workplace costs[2]. Due to high prevalence rates and a deleterious economic impact, there is a growing consensus that practitioners must increase the scope of assessment for anxiety disorders from signs and symptoms to subjective quality of life (QOL) appraisals and objective social functioning measurements.

M.S. Ritsner and A.G. Awad (eds.), Quality of Life Impairment in Schizophrenia, Mood and Anxiety Disorders, 275–291.

Anxiety disorders carry a profound burden of disability for patients in the primary care setting[3]. Reduced life enjoyment and overall life satisfaction are often the motivation for patients to seek treatment. Therefore, strikingly high rates of prevalence coupled with limitations in subjective life enjoyment suggest that mental health care professionals need to consider both functional impairment as well as QOL when assessing and treating individuals with anxiety disorders[4].

DEFINITION OF QUALITY OF LIFE AND DATA PROCUREMENT

Mental health clinicians and researchers have employed many different definitions of QOL when studying people suffering from anxiety disorders. The World Health Organization's definition of health is "a state of complete physical, mental, and social well-being and not merely the absence of disease"[5]. Kaplan and Anderson[6] have suggested that evidence for decreased QOL and morbidity involves clinical outcomes (i.e., clinical judgment, physical findings), subjective evidence of limited functioning (i.e., symptoms, complaints) and behavioral dysfunction or disruption in role performance (i.e., work loss, confinement to hospitals). More recently in a comprehensive review article, Mendlowicz and Stein[7] proposed that the minimal requirements for an operational definition for QOL assessment should include: 1) the patient's subjective perception of the quality of his or her own life, 2) view of QOL as a multidimensional construct covering a certain number of conventionally defined domains and 3) focus on aspects of personal experience that are related to health and health care. Domains suggested to be fundamental to the assessment of QOL include the patients' perceptions of social relationships, physical health, functioning in daily activities and work, economic status and an overall sense of well-being[8].

There are three main approaches employed to study QOL in individuals with anxiety disorders. Epidemiological studies provide a wealth of information about QOL. These large studies use specifically designed instruments to gather prevalence and incidence data about psychiatric and medical syndromes. This approach generates data about the presence of disorders and dysfunction in general populations. A second approach is to study a sample seen as part of routine clinical care. This gives one an opportunity to assess the QOL of individuals who seek treatment either in a primary care or psychiatric setting. The third venue used to provide data about QOL is the research clinic. Here, carefully characterized and often fairly homogeneous samples of patients are subject to intensive study.

There are a variety of validated instruments used in the assessment of QOL in mental health disorders. Some of these are disorder specific and others are more general and can be applied to both psychiatric and non-psychiatric populations. The instruments vary greatly in length and the thoroughness of the assessment. A number of these instruments attempt to measure both QOL and disability. The most commonly used instruments employed in the study of patients with anxiety disorders are the: Medical Outcomes Study 36-item Short-Form Health Survey (SF-36)[9], Illness Intrusiveness Ratings Scale (IIRS)[10], Quality of Life Enjoyment and Satisfaction Questionnaire (Q-LES-Q)[11], Social Adjustment Scale – Self-Report (SAS)[12], Sheehan Disability Scale (SDS)[13], Disability Profile[14], Liebowitz Self-Rated Disability Scale (LSRDS)[14], Quality of Life Inventory (QOLI)[15], Quality of Well-Being Scale[16], and Lancashire Quality of Life Profile (LQL)[17].

FUNCTIONAL IMPAIRMENT AND ANXIETY DISORDERS

Anxiety disorders are associated with substantial impairment in health-related quality of life (HR-QOL). Spitzer and colleagues[18] found subjects with anxiety disorders to have impairment in three major domains of living: social functioning, role functioning, and mental health as well as an increase in disability days. Ormel and colleagues[19] analyzed results from the WHO Collaborative Study on Psychological Problems in General Health Care and found that panic disorder, generalized anxiety disorder, and major depressive disorder were associated with higher levels of disability in occupational role dysfunction, self-reported physical disability, and number of disability days than all other diagnostic categories after controlling for physical disease severity.

Subjective QOL assessment is intended to measure one's level of enjoyment and satisfaction associated with various activities of life while assessment of functioning usually attempts to assess objective information such as the number of days lost from work. This can lead to a fundamental disjuncture between these two related yet quite distinct constructs. In a recent study, Rapaport and colleagues[4] found a significant disconnection between the patients' perceptions of the impact of illness and objective measures of functional impairment. Studies employing objective measures of functional impairment have found that patients with social anxiety disorder and obsessive compulsive disorder are more impaired than they perceive themselves to be[4,14,20]. Thus, it is imperative that the relationship between anxiety disorders and functional outcomes in patients be assessed and better understood.

REVIEW PAPERS SUMMARIZING QUALITY OF LIFE DATA FOR ANXIETY DISORDERS

In 2000, Mendlowicz and Stein[7] systematically reviewed epidemiological and clinical studies assessing QOL of individuals with anxiety disorders. Their analyses found anxiety disorders to be illnesses that markedly compromised QOL and psychosocial functioning[7]. They found that panic disorder and posttraumatic stress disorder (PTSD) seemed to exert a greater toll on QOL than other anxiety disorders. Their review determined that effective pharmacologic and psychotherapeutic treatment improved QOL for patients with panic disorder, social phobia, and PTSD.

In a review of QOL in the anxiety disorders, Mogotsi and colleagues[21] concluded that future research should employ a more consistent definition of QOL and include this in more robust measures of symptom-related disability and functional impairment in a range of life domains as well as ratings of general well-being and life satisfaction[21]. They deduced that research on QOL in the anxiety disorders was most advanced for panic disorder and suggested the inclusion of caregiver QOL assessment as an outcome measure in treatment studies.

Rapaport and colleagues[4] analyzed Q-LES-Q data from 11 pharmacologic treatment trials that included studies of major depressive disorder, chronic depression/double depression, dysthymic disorder, panic disorder, obsessive-compulsive disorder (OCD), social phobia, premenstrual dysphoric disorder, and PTSD. They found the proportion of patients reporting QOL impairment two standard deviations lower than a control sample were 20% for panic disorder, 21% for social phobia, and 59% for PTSD. Subjects with panic disorder, social phobia, and OCD showed significant impairment on the social relationship, family relationship, leisure, ability to function, and vision items. Individuals with major depressive disorder, chronic depression/double depression, and PTSD had the lowest mean Q-LES-Q scores that were associated with profound global impairments in QOL across all items in the Q-LES-Q.

STUDIES OF QUALITY OF LIFE AND FUNCTIONAL IMPAIRMENT IN ANXIETY DISORDERS

Generalized Anxiety Disorder (GAD)

GAD is characterized by excessive and difficult to control anxiety and worry occurring more days than not for at least 6 months that causes significant distress and impairment and is associated with symptoms such as restlessness, fatigue, difficulty concentrating, irritability, muscle tension, and sleep disturbance[22]. GAD is one of the most common anxiety disorders with lifetime prevalence rates in the U.S. of 5.7%[1].

Community samples

Massion and colleagues[23] examined the effects of GAD and panic disorder on QOL using questions derived from the National Comorbidity Survey (NCS). Both groups demonstrated role functioning and social life impairment as well as decreased overall life satisfaction. GAD was associated with a reduction in overall emotional health; however, the majority of the patients had other comorbid anxiety disorders. The authors concluded that even though noncomorbid GAD is relatively rare, it is associated with substantial impairment in functioning and QOL. These findings were replicated and extended by Kessler and colleagues[24] who analyzed data from the NCS and the Midlife Development in the United States and determined that people with GAD had greater social and work impairment and poorer perceived health than those without a mental disorder.

Wittchen and colleagues[25] examined results from a German community survey of 4181 respondents 18-65 years of age and found that after controlling for age, gender, and other psychopathology, pure GAD (n = 33), pure MDD (n = 344), and comorbid GAD and MDD (n = 40) were each associated with poor self-perceived health (at least 3 days dysfunction in the past month) and low QOL scores on the SF-36. In this sample, pure GAD respondents reported the lowest QOL scores of any group.

Stein and Heimberg[26] used data from the Ontario Health Survey (OHS; 8000 residents 15-64 years) and found that both current and lifetime GAD diagnoses increased the likelihood of reporting poor QOL, dissatisfaction with family relationships, and one's main activity. They reported that GAD was associated with an increased likelihood of poor global well-being and life satisfaction and that the long-term benefits of treating GAD were substantial.

Clinical research samples

Bourland and colleagues[27] used the Quality of Life Inventory (QOLI) and Life Satisfaction Index (LSI-Z) in 59 older adults with GAD and compared and contrasted them with 19 age-matched normal individuals. Subjects with GAD reported significantly lower QOL that correlated with the severity of depression and severity of anxiety experienced by subjects.

Jones and colleagues[28] examined the relationship between GAD, QOL, and medical utilization among low-income patients attending primary care clinics using structured psychiatric interviews, HR-QOL self-report measures, and medical chart reviews. They concluded that patients with GAD utilized the emergency department more and reported poorer QOL than patients with other psychiatric disorders and patients without psychopathology.

More recently, Wetherell and colleagues[29] used the SF-36 to investigate the impact of late-life GAD on HR-QOL. Seventy-five older GAD patients were compared and contrasted to 32 healthy controls and the GAD patients reported worse HR-QOL than healthy controls. GAD and symptoms of anxiety or depression were significantly correlated with impairment in every QOL domain. Comparisons suggest that older GAD patients report overall worse QOL than individuals with recent acute myocardial infarction or type II diabetes, and are comparable in impairment in QOL to individuals with major depressive disorder.

Treatment outcome studies

Stanley and colleagues[30] examined the impact of cognitive-behavioral therapy (CBT) versus minimal contact control (MCC) in a sample of 85 adults aged 60 years and older with GAD. Results showed that subjects in the active therapeutic group revealed significant improvement in worry, anxiety, depression, and QOL following CBT relative to MCC and most gains were maintained or enhanced over one year follow-up.

Davidson and colleagues[31] studied the safety and efficacy of escitalopram in the long-term treatment of moderate-to-severe GAD in three 8-week, double-blind, placebo-controlled trials of 526 subjects. Of 299 completers, 92% were responders and treatment led to continuing improvement on all anxiety and QOL scores using the short form of the Q-LES-Q.

Dahl and colleagues[32] performed a 12-week double-blind study with sertraline on outpatients with GAD. They reported improvement of both psychic and somatic anxiety factors of the Hamilton Anxiety Rating scale (HAM-A) as well as improvement in QOL as measured by the Q-LES-Q.

Blank and colleagues[33] performed a 32-week trial of citalopram with 30 subjects aged 60 years and older diagnosed with an anxiety disorder and found that those who completed the trial had significant improvements in sleep and QOL in the SF-36. Improvements were noted in social functioning, vitality, mental health, and role difficulties due to emotional problems.

In summary, both epidemiologic and clinical studies have found that individuals with GAD experience significant impairments in QOL even after controlling for medical and psychiatric comorbidities. Specific impaired areas of focus include social relationship problems and lowered perceived health. Pharmacologic treatments and psychotherapy have both proven effective in decreasing symptoms of anxiety and improving QOL in this population.

Panic Disorder

Individuals with panic disorder experience recurrent unexpected panic attacks where at least one of the attacks is followed by a month or more of persistent concern about having additional attacks, worries about the implications of the attacks and a significant behavior change related to the attacks[22]. Panic disorder has a prevalence rate of 4.7% in the U.S. and is often chronic in nature[1]. Panic disorder patients typically perceive their physical and mental/emotional health to be worse and tend to utilize healthcare services at higher rates in comparison to others.

Community samples

The Epidemiologic Catchment Area (ECA) study is an integral source of information regarding the epidemiology and impact of panic disorder on QOL. Markowitz and colleagues[34] found that both panic disorder and major depression were associated with subjective feelings of poor physical and emotional health, alcohol and other drug abuse, suicide attempts, decreased time devoted to hobbies, poorer marital functioning, increased financial problems, increased use of general medical and psychiatric professionals, increased use of minor tranquilizers and antidepressants, and increased use of the emergency department when compared to subjects with any psychiatric condition.

The European Study of the Epidemiology of Mental Disorders (ESEMeD) project was conducted in Belgium, France, Germany, Italy, the Netherlands, and Spain using in-home computer assisted interviews. It reported that panic disorder, PTSD, and social phobia were among 10 disorders with the greatest independent negative impact on days lost from work and reduced QOL[35].

Clinical research samples

Sherbourne and colleagues[36] compared QOL in patients with panic disorder to those with depression, chronic medical conditions, and general population norms using the SF-36. Results showed panic disorder to be associated with high psychological distress and greater limitations in role functioning with a minor impact on physical functioning.

Katerndahl and Realini[37] conducted a community survey of 97 subjects with panic attacks and matched controls. They found that both infrequent panic attacks and panic disorder negatively impacted QOL and increased disability; however, individuals with panic disorder reported a greater decrement in QOL. Predictors of work disability included panic attack frequency, illness attitudes, family dissatisfaction, and gender.

Hollifield and colleagues[38] studied variables predictive of functional impairment in panic disorder among 62 symptomatic subjects and 61 comparison control subjects. Findings suggested that variables which added to greater impairment in panic disorder subjects were the panic disorder diagnosis, age, education level, gender, and ethnicity. Specifically, they reported that subjects with panic disorder were more likely to endorse higher levels of impairment as age increased.

In a study of 73 patients with a primary diagnosis of panic disorder from an outpatient anxiety disorders program, Candilis and colleagues[39] found that regardless of the presence of additional anxiety and mood disorders, SF-36 mental and physical health subscale scores were worse in patients with panic disorder than in the general population.

Rubin and colleagues[40] analyzed 56 patients with panic disorder using the Sheehan Disability Scale (SDS), Anxiety Sensitivity Index, Spielberger State Trait Anxiety Scale, and Quality of Well-Being Scale. They found that patients with panic disorder lost 39 quality-adjusted days for each year that they lived with the disorder and displayed significant work-related disability. Diminished QOL was correlated with number of panic attacks, state anxiety, and depressive symptoms. This decrease in QOL was similar to what is observed in patients with non-insulin dependent diabetes.

In an investigation of 39 individuals with a primary diagnosis of panic disorder, Gregor and colleagues found that poorer perceived health predicted higher levels of negative affectivity and increased impairment in both family/home responsibilities and social functioning[41]. Evidence also suggested that perceived health was related to life impairment (i.e., poor perceptions of health led to greater impairment and less participation in life activities).

Treatment outcome studies

Telch and colleagues[42] randomly assigned 156 patients with panic disorder and significant impairment in QOL to cognitive-behavioral treatment (CBT) verses a delayed-treatment control. CBT treated patients showed significant reductions in impairment that were maintained at 6-month follow-up. In this study, anxiety and phobic avoidance were significantly associated with QOL impairment but panic attacks were not.

Rollman and colleagues[43] compared telephone-based collaborative care versus usual care provided by primary care physicians in a randomized controlled trial of 191 adults with panic disorder and/or GAD. At 12-month follow-up, intervention subjects reported reduced anxiety, improved mental health-related QOL, larger improvements in baseline hours worked per week, and fewer days absent in the past month. Additionally, if employed at baseline, more intervention subjects remained working at 12-month follow-up.

Jacobs and colleagues[44] found that clonazepam significantly improved mental health-related QOL and work productivity in panic disorder patients. Improvement scores on the SF-36 scale were significantly greater with clonazepam treated individuals than placebo.

Rapaport and colleagues[45] examined data from two multi-center, randomized, double-blind, parallel group, flexible dose studies of panic disorder (n = 302) using seven definitions of response in panic disorder to compare patient-rated improvements in QOL. They found significant differences in QOL between patients who responded to sertraline versus those who responded to placebo across all definitions of clinical response suggesting that although patients may experience symptom relief with placebo, they may not experience improvement in QOL.

In conclusion, epidemiological research suggests that panic disorder adversely affects patients' QOL in domains of poor physical and emotional health, reported work loss days, and emergency service utilization. Similarly, clinical research with this population has reported a negative impact on perceived physical health and functioning as well as impaired emotional well-being. Studies have shown that treatment utilizing either psychotropic or psychotherapeutic interventions have produced favorable results in reducing various detriments in QOL.

Posttraumatic Stress Disorder (PTSD)

More than 60% of men and 51% of women experience at least one traumatic event in their lifetimes but only 8% and 20% respectively develop PTSD[46]. In the United States, the most recent lifetime prevalence rates of PTSD are 6.8%[1]. The definition of PTSD requires that an individual have the subjective experience of intense fear, helplessness, or horror resulting from exposure to real or threatened death, serious injury or a threat to the physical integrity of self or others[7]. This experience leads to the development of a cluster of symptoms exemplified by persistent re-experiencing of the event, persistent avoidance of stimuli and emotional numbing, and persistent symptoms of hyper arousal[22]. The suffering related to PTSD frequently goes beyond signs and symptoms and extends to functional impairment and disability in many areas.

Community samples

Zatzick and colleagues[47] measured the impact of PTSD on QOL and functioning by analyzing archived data from the National Vietnam Veterans Readjustment Study consisting of 1,200 male Vietnam veteran subjects. Six domains were measured: diminished well-being, physical limitations, bed days in the past 2 weeks, compromised physical health status, currently not working, and perpetration violence. Veterans diagnosed with PTSD reported more life impairment on 5 out of 6 of the domains. These findings held even after adjusting for demographic differences and comorbid psychiatric disorders and other medical conditions.

Zatzick and colleagues[48] also analyzed the data from 432 female Vietnam Veterans (mainly nurses). Functional impairment and diminished QOL were assessed in six areas: bed days in the past 3 months, role functioning, subjective well-being, self-reported physical health status, current physical functioning, and perpetration of violent interpersonal acts in the past year. A diagnosis of PTSD was significantly associated with elevated odds of poorer functioning in 3 of the 6 domains: role functioning, self-reported physical health status, and bed days in the past 3 months. These differences were present after adjusting for demographic differences, comorbid psychiatric disorders, and other medical conditions. The researchers concluded that PTSD caused significant functional impairment in both men and women and that the degree of dysfunction was more similar than different.

Clinical research samples

Warshaw and colleagues[49] compared SF-36 scores of patients with anxiety with and without a comborbid PTSD diagnosis. The comorbid PTSD group showed significantly higher psychosocial impairment in the domains of health, role, social, and emotional functioning than the comparison group.

In a study by Zayfert and colleagues[50], subjects with PTSD showed significant impairment on the physical health subscale of the SF-36 verses population norms when comorbid major depressive disorder and age were controlled. Similarly, Ouimette and colleagues[51] found that patients with PTSD scored significantly lower on the SF-36 general health and physical function domains than patients without PTSD when comorbid major depressive disorder was controlled.

In 2004, Schnurr and Green[52] presented evidence that traumatic exposure leads to poor physical and mental health as well as impaired psychosocial functioning. They ascertained that by addressing the physical health consequences of traumatic exposure in treatment, the burden on individuals and society may be reduced.

Mueser and colleagues[53] examined the relationship between PTSD symptoms, health, QOL, and work outcomes in 176 clients with severe mental illness participating in a 2-year randomized controlled trial of three vocational rehabilitation programs. The overall rate of PTSD among the group was 16% and these individuals reported more severe psychiatric symptoms, worse reported health, lower self-esteem, and lower subjective QOL. Subjects with PTSD also had worse employment outcomes with lower rates of work, fewer hours worked, and lower wages earned.

Holbrook and colleagues[54] studied 401 trauma patients aged 12-19 years diagnosed with injuries (excluding severe brain or spinal cord injury) in which QOL was measured using the Quality of Well-Being scale. PTSD was significantly and strongly correlated with female gender, older age, lower socioeconomic status, drug and alcohol abuse, and adolescent behavioral problems. An ongoing diagnosis of PTSD was associated with marked QOL deficits throughout the 24-month follow-up interval.

d'Ardenne and colleagues[55] used the Manchester Short Assessment of Quality of Life (MANSA) to study subjective QOL in 117 patients with PTSD in a specialized clinic in East London, United Kingdom. When scores were compared with nonclinical subjects and other groups of subjects with mental disorders, PTSD patients had lower scores in 7 of the 8 single subjective QOL domains (e.g., life in general, finances, social relations, leisure activities, personal safety, living situation, family relationships, and mental health) with the exception being the living situation domain.

Treatment outcome studies

Malik and colleagues[56] evaluated QOL for 16 patients with PTSD by administering the SF-36 as part of a 12-week, double-blind, placebo-controlled trial of fluoxetine. At baseline, subjects with PTSD reported greater impairment than subjects with major depression and OCD on several of the domains (bodily pain, general health, vitality, social functioning, emotional role limitations, and mental health). Twelve weeks of fluoxetine treatment caused significant improvement in those domains compared to placebo-treated groups.

Rapaport and colleagues[57] studied PTSD and QOL in 359 subjects across a 64-week sertraline treatment trial. Assessments included the Q-LES-Q, SF-36, and the social and occupational functioning items of the Clinician Administered PTSD scale (CAPS-2). QOL was significantly impaired at baseline for all subjects. Those receiving sertraline treatment demonstrated statistical and clinically significant improvement on all measurement domains during a one-year treatment period.

Even though there is still little evidence as to the effect of PTSD, both epidemiological and clinical studies have reported that individuals with PTSD suffer from impaired functioning and QOL in a multitude of domains including mental health, physical health, employment, and social functioning with evidence suggesting that patients may also be at a greater risk for drug and alcohol problems and violence. Initial pharmacologic trials suggest that effective treatment improves both short-term and longer-term QOL and overall functioning.

Generalized Social Phobia

Social phobia is defined as marked, persistent, and excessive fear of most social situations in which a person is exposed to unfamiliar people or to possible scrutiny by others. Any exposure provokes anxiety; therefore, the situations are avoided or sometimes endured with intense distress[22]. It is a common and debilitating psychiatric disorder that is highly comorbid with other psychiatric conditions. Primary social phobia has its onset in early to late adolescence; however secondary social phobia may occur later in life, usually as the result of some other mental disorder[58]. Lifetime prevalence of social phobia in the U.S. is 12.1%[1] and 6.7% in European countries[59] with younger individuals representing the highest rates in both samples. Studies support that generalized social phobia is associated with significant educational underachievement, increased financial dependency, decreased work productivity, social impairment, and poorer quality of life[60,61].

Community samples

Stein and Kean[62] used Ontario Health Survey (OHS) results to examine the relationship between subjective life satisfaction and social phobia using the Quality of Well-Being Scale and found that patients with current and/or lifetime social phobia were dissatisfied with their main activity, family life, friends, leisure activities, and income level. Depressive comorbidity contributed only modestly to these findings. The findings of work impairment are consistent with the Epidemiologic Catchment Area (ECA) study that reported the rate of financial dependency among subjects with uncomplicated social phobia to be 22.3%[63].

Magee and colleagues[64] analyzed data from the National Comorbidity Survey (NCS) and reported that the presence of social phobia was negatively correlated with education and income levels. Social phobia was also associated with decreased levels of social support. Approximately half of the subjects with social phobia reported at least one negative outcome (i.e., significant role impairment, professional help seeking, and use of medication more than once) at some point in their lives as a result of the disorder.

Clinical research samples

Schneier and colleagues[14] administered the Liebowitz Self-Rated Disability Scale (LSRDS) to 32 outpatients with generalized social phobia and found they exhibited more impairment in most areas of functioning, especially work, education, and relationships than the normal volunteer comparison group. These findings were extended in a study comparing patients with social phobia and herpes conducted by Wittchen and Beloch[65] employing both the SF-36 and the Work Productivity And Impairment (WPAI) questionnaire. They reported that subjects with social phobia had significantly lower scores on the SF-36 subscales of vitality, general health, mental health, role limitations due to emotional health and social functioning than subjects with herpes. Subjects with social phobia had lower education levels, diminished work productivity, and fewer romantic involvements than subjects with herpes. The social phobia cohort also had significantly diminished work productivity, greater rates of unemployment, missed more hours of work, and had worse work performance.

Antony and colleagues[66] used the Illness Intrusiveness Rating Scale (IIRS) to measure the extent to which anxiety disorders interfere with domains of functioning in patients with panic disorder, obsessive-compulsive disorder, and social phobia. Subjects with social phobia reported greater impairment on the IIRS in areas of social relationships, self-expression and self-improvement than subjects with either panic disorder or OCD. In a second study, Wittchen and colleagues[67] attempted to investigate QOL, work productivity, and social impairment across the spectrum of generalized social anxiety disorders. They compared and contrasted 65 subjects with "pure" generalized social phobia, 51 subjects with social phobia and another psychiatric comorbidity, and 34 patients with sub threshold social phobia with normal controls. Subjects with comorbidities reported the most profound reductions in QOL as measured by the SF-36. They also found that social phobia (pure, comorbid, and sub threshold) adversely affected most areas of life including education, career, and romantic relationships. The work productivity of subjects with all forms of social phobia was also significantly diminished by subjects with social phobia reporting higher rates of unemployment, missed more hours of work, and reduced work performance. Thus, the entire spectrum of subjects with social phobia had significant impairment with a monotonic gradient of impairment related to the severity of illness.

There have been two longitudinal follow-up studies of patients with generalized social phobia. An eight-year longitudinal study of 176 patients with social phobia by Yonkers and colleagues[68] found that only 38% of women and 32% of men experienced a complete remission during the study period. Women were slightly more functionally impaired than men particularly in terms of household functioning. In a subsequent study, 33 patients with social phobia who completed a SF-36 rated themselves as significantly more impaired in social functioning and having worse general mental health than the general population[69].

Treatment outcome studies

Safren et al.[70] (1996) employed the Quality of Life Inventory (QOLI) to examine the perception patients with social phobia had about the quality of their lives. QOL was inversely correlated with measures of social phobia severity, functional impairment, and depression. Cognitive-behavioral group therapy caused a clinically and statistically significant improvement in QOL scores. Although the number of pharmacotherapy studies investigating QOL outcomes is limited, Stein and colleagues[71] performed a 12-week, double-blind, randomized placebo-controlled trial comparing the contrasting fluvoxamine and placebo. Patients taking active medication had significantly greater improvement on the work functioning, family life, and home functioning items of the Sheehan Disability Scale (SDS) compared to placebo-treated subjects.

In summary, community studies suggest that patients with social phobia experience severe disruptions in their financial stability, social activities, and support. However, clinical studies have reported significant and repetitious findings of reduced QOL and impairment in work/career functioning and satisfaction, educational goals and success, and involvement in social and romantic relationships. Initial treatment trials (pharmacologic and psychotherapeutic interventions) report demonstrable improvements in QOL, work, and family functioning.

AN OVERVIEW OF QUALITY OF LIFE FINDINGS AND IMPLICATIONS FOR FUTURE RESEARCH

In general, QOL research in psychiatry is still in its infancy. As part of any field that is emerging, there are a variety of challenges that must be overcome. One of the greatest challenges is developing a consensus about what the definition of QOL impairment is. Investigators are attempting to create more of a consensus about how we define QOL. However, there still needs to be considerable work performed in this area. One major problem is the lack of a concise and agreed upon theoretical construct that should be subsumed in a definition of QOL. As would be expected, this means that rating scales used to assess QOL are difficult to compare and contrast across studies. Although these scales may be measuring similar constructs, they frequently are not completely overlapping. Thus, there is a need to attempt to harmonize the definition and the scales used to assess QOL in anxiety disorders.

Another challenge faced by psychiatry at this time is the need to establish normal ranges of QOL and functional impairment associated with different anxiety disorders. Most of the extant research has come from either epidemiological surveys or from rarified clinical populations. There are very few studies investigating QOL and functional impairment in true clinical populations where one could compare and contrast different anxiety disorders simultaneously. Furthermore, there have been few attempts to fully investigate the range of impairment that might be associated with different levels of severity and the impact of comorbidity (single or multiple) associated with these psychiatric syndromes.

Another area that requires further research is the interaction between disorders, QOL (an individual's perception), and objective measurements of dysfunction, work, and productivity. Although it is recognized that these concepts are interrelated, there are no studies that determine the extent of overlap or explore potential interactions that may occur across these domains of illness. Future work attempting to understand this complex interplay is necessary.

QOL measurement needs to be more highly integrated into both psychotherapy and pharmacotherapy treatment trials. Effective treatment must go beyond a decrease in signs and symptoms of a disorder and should take into account how treatments can impact QOL and functioning of individuals. Unfortunately, at this time, most studies consider QOL as an afterthought rather than as an integral measure in treatment trials. The incorporation of well-validated measures of QOL as one of the primary outcomes for future treatment trials is imperative to fully appreciate the benefits of available therapies for anxiety disorders. Another possible use of QOL and functional impairment is to help determine the entry criteria of subjects with anxiety disorders in research studies. Anxiety disorder studies have notoriously high placebo response rates. Preliminary analyses of data by Rapaport and colleagues suggest that subjects with the greatest levels of functional impairment and QOL impairment at baseline are least likely to respond to placebo in pharmacological trials[72].

The DSM-IV-TR and ICD-9-CM require inclusion of QOL or function impairment as part of the definition of anxiety disorders. At this time, it has been difficult to develop any consensus about what level of impairment is necessary to help define the disorder. Studies investigating the range of impairment may allow us to determine some standard functional impairment levels that help define anxiety disorders.

In summary, our current research in QOL demonstrates significant impairments associated with most disorders. This lays a foundation that is crucial for significantly more thoughtful investigation.

REFERENCES

1. Kessler RC, Berglund P, Demler O, et al. Lifetime prevalence and age-of-onset distributions of DSM-IV disorders in the national comorbidity survey replication. Arch Gen Psychiatry 2005;62:593–602
2. Greenberg PE, Sisitsky T, Kessler RC, et al. The economic burden of anxiety disorders in the 1990s. J Clin Psychiatry 1999;60(7):427–435
3. Stein MB, Roy-Byrne PP, Craske, MG, et al. Functional impact and health utility of anxiety disorders in primary care outpatients. Med Care 2005;43(12):1164–1170
4. Rapaport MH, Clary C, Fayyad R, et al. Quality of life impairment in depressive and anxiety disorders. Am J Psychiatry 2005;162:1171–1178
5. World Health Organization. World Health Organization constitution: basic documents.1948; Geneva: World Health Organization
6. Kaplan RM, Anderson JP. The general health policy model: an integrated approach. Quality of Life and Pharmacoeconomics in Clinical Trials. 1996; Lippincott-Raven Publishers, Philadelphia, PA
7. Mendlowicz MV, Stein MB. Quality of life in individuals with anxiety disorders. Am J Psychiatry 2000;157(5):669–682
8. Patrick DL, Erickson P. What constitutes quality of life? Concepts and dimensions. Clin Nutr 1988;7:53–63
9. Ware JE, Sherbourne CD. The MOS 36-itme short-form health survey (SF-36). I: conceptual framework and item selection. Med Care 1992;30:473–483
10. Devins GM, Binik YM, Hutchinson TA, et al. The emotional impact of end-stage renal disease: importance of patients' perception of intrusiveness and control. Int J Psychiatry Med 1983;13:327–343
11. Endicott J, Nee J, Harrison W, et al. Quality of life enjoyment and satisfaction questionnaire: a new measure. Psychopharmacol Bull 1993;29:321–326
12. Weissman MM, Bothwell S. Assessment of social adjustment by patient self-report. Arch Gen Psychiatry 1976;33:1111–1115
13. Sheehan DV. The anxiety disease. 1986; Bantam Books, New York
14. Schneier FR, Heckelman LR, Garfinkel R, et al. Functional impairment in social phobia. J Clin Psychiatry 1994;55:322–331
15. Frisch MB, Cornell J, Villanueva M, et al. Clinical validation of the quality of life inventory: a measure of life satisfaction for use in treatment planning and outcome assessment. Psychological Assessment 1992;4:92–101
16. McDowell I, Newell C. Measuring health: a guide to rating scales and questionnaires. 1987; Oxford University Press, New York
17. Oliver JPJ, Huxley PJ, Priebe S, et al. Measuring the quality of life of severely mentally ill people using the lancashire quality of life profile. Soc Psychiatry Psychiatr Epidemiol 1997; 32:76–83
18. Spitzer RL, Kroenke K, Linzer M, et al. Health-related quality of life in primary care patients with mental disorders. JAMA 1995;274(19):1511–1517
19. Ormel J, VonKorff M, Ustun TB, et al. Common mental disorders and disability across cultures: results from the WHO collaborative study on psychological problems in general health care. JAMA 1994;272:1741–1748
20. Karno M, Golding JM, Sorenson SB, et al. The epidemiology of obsessive-compulsive disorder in five U.S. communities. Arch Gen Psychiatry 1988;45(12):1094–1099
21. Mogotsi M, Kaminer D, & Stein DJ. Quality of life in the anxiety disorders. Harvard Rev Psychiatry 2000;8:273–282
22. American Psychiatric Association: Diagnostic and Statistical Manual of Mental Disorders, Fourth Edition, Text Revision. 2000; Washington, DC
23. Massion AO, Warshaw MG, Keller MB. Quality of life and psychiatric comorbidity in panic disorder and generalized anxiety disorder. Am J Psychiatry 1993;150:600–607

24. Kessler RC, Du Pont RL, Berglund P, et al. Impairment in pure and comorbid generalized anxiety disorder and major depression at 12 months in two national surveys. Am J Psychiatry 1999;156:1915–1923
25. Wittchen HU, Carter RM, Pfister H, et al. Disabilities and quality of life in pure and comorbid generalized anxiety disorder and major depression in a national survey. Int Clin Psychopharmacol 2000;15(6):319–328
26. Stein MB, Heimberg RC. Well-being and life satisfaction in generalized anxiety disorder: comparison to major depressive disorder in a community sample. J Affect Disord 2004; 79: 161–166
27. Bourland SL, Stanley MA, Snyder AG, et al. Quality of life in older adults with generalized anxiety disorder. Aging and Mental Health 2000;4(4):315–323
28. Jones GN, Ames SC, Jeffries SK, et al. Utilization of medical services and quality of life among low-income patients with generalized anxiety disorder attending primary care clinics. Int J Psychiatry Med 2001;31(2):183–198
29. Wetherell JL, Thorp SR, Patterson TL, et al. Quality of life in geriatric generalized anxiety disorder: a preliminary investigation. J Psychiatr Res 2004;38(3):305–312
30. Stanley MA, Beck JG, Novy DM, et al. Cognitive-behavioral treatment of late-life generalized anxiety disorder. J Consult Clin Psychol 2003;71(2):309–319
31. Davidson JRT, Bose A, Wang Q. Safety and efficacy of escitalopram in the long-term treatment of generalized anxiety disorder. J Clin Psychiatry 2005;66(11):1441–1446
32. Dahl AA, Ravindran A, Allgulander C, et al. Sertraline in generalized anxiety disorder: efficacy in treating the psychic and somatic anxiety factors. Acta Psychiatrica Scandinavica 2005;111(6): 429–435
33. Blank S, Lenze EJ, Mulsant BH, et al. Outcomes of late-life anxiety disorders during 32 weeks of citalopram treatment. J Clin Psychiatry 2006;67(3):468–472
34. Markowitz JS, Weissman MM, Ouellette R, et al. Quality of life in panic disorder. Arch Gen Psychiatry 1989;46:984–922
35. Katerndahl DA, Realini JP. Quality of life and panic-related work disability in subjects with infrequent panic and panic disorder. J Clin Psychiatry 1997;58(4):153–158
36. Alonso J, Angermeyer MC, Bernert S, et al. Disability and quality of life impact of mental disrders in Europe: results from the European Study of the Epidemiology of Mental Disorders (ESEMeD) project. Acta Psychiatrica Scandinavica, Supplementum 2004;420:38–46
37. Sherbourne CD, Wells KB, Judd LL. Functioning and well-being of patients with panic disorder. Am J Psychiatry 1996;153:213–218
38. Hollifield M, Katon W, Skipper B, et al. Panic disorder and quality of life: variables predictive of functional impairment. Am J Psychiatry 1997;154(6):766–772
39. Candilis PJ, McLean RY, Otto MW, et al. Quality of life in patients with panic disorder. J Nerv Ment Dis 1999;187:429–434
40. Rubin HC, Rapaport MH, Levin B, et al. Quality of well being in panic disorder: the assessment of psychiatric and general disability. J Affect Disord 2000;57:217–221
41. Gregor KL, Zvolensky MJ, Yartz AR. Perceived health among individuals with panic disorder: associations with affective vulnerability and psychiatric disability. J Nerv Ment Dis 2005;193(10): 697–699
42. Telch MJ, Schmidt NB, Jaimez TL, et al. Impact of cognitive-behavioral treatment on quality of life in panic disorder patients. J Consult Clin Psychol 1995;63(5):823–830
43. Rollman BL, Belnap BH, Mazumdar S, et al. A randomized trial to improve the quality of treatment for panic and generalized anxiety disorders in primary care. Arch Gen Psychiatry 2005;62:1332–1341
44. Jacobs RJ, Davidson JR, Gupta S, et al. The effects of clonazepam on quality of life and work productivity in panic disorder. Am J Managed Care 1997;3(8):1187–1196
45. Rapaport MH, Pollack M, Wolkow R, et al. Is placebo response the same as drug response in panic disorder? Am J Psychiatry 2000;157:1014–1016

46. Davidson JR. Trauma: the impact of posttraumatic stress disorder. J Psychopharmacol 2000;14: S5–S12
47. Zatzik DF, Marmar CR, Weiss DS, et al. Posttraumatic stress disorder and functioning and quality of life outcomes in a nationally representative sample of male Vietnam veterans. Am J Psychiatry. 1997a;154(12):1690–1695
48. Zatzick DF, Weiss DS, Marmar CR, et al. Posttraumatic stress disorder and functioning and quality of life outcomes in female Vietnam veterans. Mil Med 1997b;162:661–665
49. Warshaw MG, Fierman E, Pratt L, et al. Quality of life and dissociation in anxiety disorder patients with histories of trauma or PTSD. Am J Psychiatry 1993;150:1512–1516
50. Zayfert C, Dums AR, Ferguson RJ, et al. Health functioning impairments associated with posttraumatic stress disorder, anxiety disorders, and depression. J Nerv Ment Dis 2002;190:233–240
51. Ouimette P, Cronkite R, Henson BR, et al. Posttraumatic stress disorder and health status among female and male medical patients. J Trauma Stress 2004;17:1–9
52. Schnurr PP, Green BL. Understanding relationships among trauma, post-traumatic stress disorder, and health outcomes. Adv Mind Body Med 2004;20(1):18–29
53. Muesser KT, Essock SM, Haines M, et al. Posttraumatic stress disorder, supported employment, and outcomes in people with severe mental illness. CNS Spectrums 2004;9(12):913–925
54. Holbrook TL, Hoyt DB, Coimbra R, et al. Long-term posttraumatic stress disorder persists after major trauma in adolescents: new data on risk factors and functional outcome. J Trauma 2005;58:764–771
55. d'Ardenne P, Capuzzo N, Fakhoury WKH, et al. Subjective quality of life and posttraumatic stress disorder. J Nerv Ment Dis 2005;193(1):62–65
56. Malik ML, Connor KM, Sutherland SM, et al. Quality of life and posttraumatic stress disorder: a pilot study assessing changes in SF-36 scores before and after treatment in a placebo-controlled trial of fluoxetine. J Trauma Stress 1999;12:387–393
57. Rapaport MH, Endicott J, Clary C. Posttraumatic stress disorder and quality of life: results across 64 weeks of sertraline treatment. J Clin Psychiatry 2002;63(1):59–65
58. Wittchen HU, Fehm L. Epidemiology, patterns of comorbidity, and associated disabilities of social phobia. Psychiatr Clin North Am 2001;24(4):617–641
59. Fehm L, Pelissolo A, Furmark T, et al. Size and burden of social phobia in Europe. Eur Neuropsychopharmacol 2005;15(4):453–462
60. Lipsitz JD, Schneier FR. Social phobia. Epidemiology and cost of illness. Pharmacoeconomics 2000;18(1):23–32
61. Walker JR, Kjernisted KD. Fear: the impact and treatment of social phobia. J Psychopharmacol 2000;14(2 Suppl 1):S13–S23
62. Stein MB, Kean YM. Disability and quality of life in social phobia: epidemiological findings. Am J Psychiatry 2000;157:1606–1613
63. Schneier FR, Johnson J, Hornig CD, et al. Social phobia: comorbidity and morbidity in an epidemiologic sample. Arch Gen Psychiatry 1992;49:282–288
64. Magee WJ, Eaton WW, Wittchen HU, et al. Agoraphobia, simple phobia, and social phobia in the National Comborbidity Survey. Arch Gen Psychiatry 1996;53:159–168
65. Wittchen HU, Beloch E. The impact of social phobia on quality of life. Int Clin Psychopharmacol 1996;11:S15–S23
66. Antony MM, Roth D, Swinson RP, et al. Illness intrusiveness in individuals with panic disorder, obsessive-compulsive disorder, or social phobia. J Nerv Ment Dis 1998;186(5):311–315
67. Wittchen HU, Fuetsch M, Sonntag H, et al. Disability and quality of life in pure and comorbid social phobia: findings from a controlled study. Eur Psychiatry 2000;15(1):46–58
68. Yonkers KA, Dyck IR, Keller MB. An eight-year longitudinal comparison of clinical course and characteristics of social phobia among men and women. Psychiatr Serv 2001;52(5): 637–643
69. Simon NM, Otto MW, Korbly NB, et al. Quality of life in social anxiety disorder compared with panic disorder and the general population. Psychiatr Serv 2002;53(6):714–718

70. Safren SA, Heimberg RG, Brown EJ, et al. Quality of life in social phobia. Depress Anxiety 1996;4(3):126–133
71. Stein MB, Fyer AJ, Davidson JRT, et al. Fluvoxamine treatment of social phobia (social anxiety disorder): a double-blind, placebo-controlled study. Am J Psychiatry 1999;156:756–760
72. Rapaport MH. Enhancing Precision in Clinical Trials: Using Quality of Life to Differentiate True Drug Response from Placebo Response. 2001; NCDEU Presentation

CHAPTER 15

QUALITY OF LIFE IN OBSESSIVE-COMPULSIVE DISORDER

JULIO BOBES, M.-P. GARCÍA-PORTILLA, MARIA-TERESA BASCARÁN, PILAR-ALEJANDRA SÁIZ, MARIA-TERESA BOBES-BASCARÁN AND MANUEL BOUSOÑO

Department of Psychiatry, University of Oviedo, Spain

Abstract: Obsessive-compulsive disorder (OCD) is a severe mental disorder with a lifetime prevalence of 1.6% in the US population, although the identified obsessive-compulsive spectrum may affect up to 10% of the US population. Furthermore, OCD is a chronic, profoundly disabling illness that impacts negatively on the academic, occupational, social and family patients' functioning as well as on their families' lives. Indeed, OCD is tenth in the World Bank's and WHO's ten leading causes of disability ranking.

A growing number of treatments have been recently shown to be useful for OCD symptoms and, to a less degree, for the disabilities that the disorder involves.

In spite of all the above mentioned benefits and disabilities, there are relatively few reports on quality of life in patients with OCD and on the relationship between patients' quality of life and clinical variables, particularly about the effect of available treatments on quality of life.

Since the first published paper on the topic, there has been a general agreement that moderate and severe OCD show lower quality of life level than that of the general population, other mentally ill patients (depressed and heroin dependent patients) and that of patients with chronic medical conditions (such as diabetes type II and kidney transplant patients). Areas that have been found to be the most impaired are social functioning and role limitations due to emotional problems.

In this chapter we will discuss in depth recent findings, controversies and concerns regarding quality of life of obsessive-compulsive disorder patients and its treatment, as well as provide future direction and areas of interest

Keywords: Quality of life, Obsessive-compulsive disorder, Anxiety disorders

INTRODUCTION

Nowadays, Obsessive-Compulsive Disorder (OCD) is included in the group of anxiety disorders and is defined as recurrent obsessions or compulsions, recognized by the patient as excessive or unreasonable, that are severe enough to be time

M.S. Ritsner and A.G. Awad (eds.), Quality of Life Impairment in Schizophrenia, Mood and Anxiety Disorders, 293–303.

consuming or cause marked distress or significant impairment[1]. Its course can be defined as a chronic waxing and waning of symptoms. The lifetime prevalence of OCD in the National Comorbidity Survey Replication was 1.6% (SE 0.3)[2], and the twelve-month prevalence was 1.0% (SE 0.3)[3]. In this last study OCD was found to be the anxiety disorder with the highest percentage of serious classifications: 50.6% (SD 12.4) of the cases were classified as serious, 34.8% (SD 14.1) as moderate, and 14.6% (SD 5.7) as mild.

Obsessive-compulsive disorder is recognized by clinicians as a chronic and disabling illness that impacts negatively on the academic, occupational, social, and family function of patients[4,5]. This impact carries over onto their families, friends and society[4]. Indeed, surveys conducted by WHO have demonstrated that OCD is one of the five mental disorders –along with unipolar depression, schizophrenia, alcohol use, and bipolar disorder- that cause the highest disability in the world, as also it ranks high among the group[6].

OCD shows a high degree of comorbidity, particularly with other anxiety disorders and depression. This high comorbidity implies a significant negative impact on disability and quality of life of these patients.

ISSUES IN ASSESSING QUALITY OF LIFE IN OCD

Quality of life is defined by the World Health Organization[7] as the "Individuals' perceptions of their position in life in the context of the culture and value systems in which they live, and in relation to their goals, expectations, standards, and concerns". In the field of Medicine, health-related quality of life is a multidimensional construct reflecting an individual's global physical and mental well-being[8]. A more detailed definition is that of Wenger and Furberg: "those attributes valued by patients, including: their resultant comfort or sense of well-being; the extent to which they were able to maintain reasonable physical, emotional, and intellectual function; and the degree to which they retain their ability to participate in valued activities within the family, in the workplace, and in the community"[9]. This concept is of special relevance in the field of chronic disorders, since these often have a negative effect upon functioning and health-related quality of life. In this sense, OCD patients have signalled striking influences of OCD on their academic, occupational and social functioning and accordingly a general influence on their quality of life[10].

In recent years quality of life has been included as a reliable intermediate outcome in psychiatry along with management of side-effects and subjective response to drugs[11]. Quality of life assessment is based on the principle of applying medical care and interventions bearing in mind the principle of autonomy of the patients, which necessarily includes taking their opinions into account both in diagnostic evaluations and prognoses, as well as for therapeutic interventions and medical care management[11]. In the case of the obsessive-compulsive disorder the International OCD Conference 2000 recognized the need for standardizing the research of quality of life[12].

QUALITY OF LIFE IN OCD

Compared with other psychiatric disorders, quality of life research in anxiety disorders in general and in OCD in particular is at the dawn. However, as Medlowicz and Stein [13] stated it is expected that a better understanding of the impact of these disorders on the quality of life will increase public awareness of anxiety disorders as serious mental disorders worthy of further investment in research, prevention and treatment.

The majority of studies, including the most recent studies, did not considerer some factors of great relevance for quality of life such as length of illness, number of symptoms, illness subtype, insight, etc. In addition, methodological problems exist in relation to the small sample size, the lack of specific quality of life instruments for OCD, matched comparative populations, and the shortness of the follow-up period, if any.

Impact of OCD on Quality of Life

Since the paper of Koran et al in 1996 [14] several studies have investigated the impact of OCD on the quality of life of these patients (Table 1). The majority of them used the SF-36 for assessing quality of life and compared their results with the general population norms. Results can be summarized as follows:

1. Quality of life of OCD patients is worse than that of general population.
2. The most impaired areas are social functioning, role emotional and mental health.
3. Physical areas appears to be affected to a less extent.
4. Gender does not affect the quality of life of OCD patients. This is a striking fact taking into account that in general population and in affective disorders women report lower levels of quality of life than men. In contrast, women with schizophrenia tend to report higher levels than that of men.
5. The severity of the disorder negatively impacts on the quality of life of these patients, especially the severity of obsessions.
6. The presence of depressive symptomatology further affects their quality of life.
7. There are equivocal findings when comparing quality of life of OCD patients to that of other psychiatric disorders; some studies demonstrated that OCD patients show worse levels of quality of life than schizophrenic patients whereas other studies demonstrated the opposite. The same occurs when compared to depressive patients.
8. Studies analyzing the differential effects of OCD available treatments on the quality of life of these patients virtually do not exist.

Following, are in depth descritpin of the different OCD studies published examining their findings, controversies and concerns regarding quality of life and OCD treatment.

Eisen et al [15] studied 197 consecutive adult individuals, seeking treatment and recruited to be part of a naturalistic study on the course of OCD. The mean age was 40.6 years (SD 12.9), 56.9% were women, 35.5% were single and 42.6% had

Table 1. General studies on quality of life in OCD

1st Author, year	Country	Patients	QoL instruments	Results
Eisen, 2006	USA	197	Q-LES-Q, SF-36	Compared with community norms, OCD patients showed worse scores on all the Q-LES-Q scales and on all the SF-36 scales but PF
Rodriguez-Salgado, 2006	Spain	64	SF-36	Compared to Spanish SF-36 population norms, OCD patients showed worse scores in all scales except in PH and BP
Akdede, 2005	Turkey	23 OCD 22 healthy subjects	WHOQOL-BREF-TR	Compared to healthy subjects patients scored worse in psychological and social domains
Moritz, 2005	Germany	79 32 healthy controls	SF-36	Compared to healthy controls OCD patients scored lower in all SF-36 scales. Compared to the healthy subsample of the German SF-36 norm population z scores were greater than −2 on the V, SF, RE, and MH scales
Rapapport, 2005	USA	521	Q-LES-Q	Compared to community norms OCD had lower scores, and 26% had severe impairment in QoL
Bystritsky, 2001	USA	31	QoLI	Compared to schizophrenic OCD patients reported worse quality of life
Bobes, 2001	Spain	36 in maintenance treatment	SF-36	Compared to Spanish SF-36 population norms, OCD patients showed worse scores on GH, VT, SF, RE, and MH Compared to schizophrenic patients OCD patients reported better QoL in PF, RP, and GH Compared to depressed patients OCD patients showed worse scores on BP, V, SF, RE, and MH
Koran, 96	USA	60 medication-free outpatiens	SF-36	Compared to USA SF-36 population norms, OCD patients scored lower on the SF and RE Compared to depressive patients OCD patients reported higher QoL in all SF-36 domains but RE

OCD: Obsessive-Compulsive Disorder; Q-LES-Q: Quality of Life Enjoyment and Satisfaction Questionnaire; QoL: Quality of Life; QoLI: Lehman's Quality of Life Interview; SF-36: Medical Outcomes Survey 36-Item Short-Form Health Survey; WHOQOL-BREF-TR: WHO Quality of Life Scale – Brief Form in Turkish graduated from college. Only 4% of the sample came from minorities. At the time of intake, 84% of the subjects were taking selective serotonin reuptake inhibitors. In the study they assessed (1) psychopathology by means of the Yale-Brown Obsessive-Compulsive Scale (Y-BOCS) for severity of the disorder, the Modified Hamilton Rating Scale for Depression (MHRSD), and the Brown Assessment of Beliefs Scale (BABS) for delusional thinking; (2) quality of life was assessed using the Quality of Life Enjoyment and Satisfaction Questionnaire (Q-LES-Q) and the Medical Outcomes Survey 36-Item Short-Form Health Survey (SF-36); and (3) psychosocial functioning was evaluated employing the Social and Occupational Functioning Assessment Scale (SOFAS) and the Range of Impaired Functioning Tool (LIFE-RIFT). [Please reference all the scales]

The mean Y-BOCS total score was 21.41 (SD 7.8), the mean MHRSD score was 11.44 (8.7), and the mean BABS score was 7.23 (SD 4.9). In relation to the Q-LES-Q they found that compared with community norms (n = 89) mean scores for OCD patients were significantly worse on all the summary scales [general (short-form), physical health, emotional well-being, household, leisure, social, and work/school]. The pattern with the SF-36 was similar; compared to US population (n = 2474) OCD patients scored significantly lower on all scales except on physical functioning (role physical, bodily pain, general health, vitality, social functioning, role emotional, and mental health). There was a negative and significant correlation between the severity of the disorder and the quality of life –SF-36 mental health, role emotional and social functioning, and Q-LES-Q general (short-form)-, although the correlation coefficients were low to moderate (0.28 to 0.49). Obsessions showed slightly higher correlations than compulsions. In addition, they found a significant decline in quality of life occurred if Y-BOCS total score was 20 or greater. Scores on the delusional thinking scale to a less degree correlated with quality of life. In the hierarchical regression using the Q-LES-Q as the dependent variable, only the marital status, the Y-BOCS obsessions score and the MHRDS were significantly associated with Q-LES-Q scores and accounted for 42% of the variance. Missing was information concerning the impact of depressive symptoms on quality of life since the authors employed the MHDRS and they did not report the results in the paper (only reported MHDRS data for the regression results).

Rodriguez-Salgado et al[16] studied 64 adult outpatients under psychopharmacological and/or psychotherapeutic treatment. Patients were interviewed by a clinical psychiatrist with the Mini International Psychiatric Interview (MINI), the Y-BOCS and the Hamilton Rating Scale for Depression (HRSD), and the SF-36 was completed . The mean age was 36.4 years, 54.7% were female, 57.8% were married, 46.9% had a university degree, and 40.6% were unemployed. Almost 50% had another psychiatric disorder (18.8% major depression, 4.7% dysthymia, 9.4% anxiety disorders, 14.1% substance abuse, 10.9% other and 14.1% more than one comorbid disorder), 73.4% were suffering from an acute medical disease at the time of the study and 39% from a chronic medical disease.

The mean Y-BOCS scores were 22.6 (SD 6.2) for the global OCD severity, 11.3 (SD 3) for obsessions and 12.3 (SD 4.3) for compulsions, and the mean HRSD score was 9.4 (5.6). In relation to quality of life, compared to scores in the Spanish general population (n = 9151) OCD patients scored significantly worse on all SF-36 scales except on physical functioning and bodily pain. The Y-BOCS global scores correlated negatively with all SF-36 scales although, as in the study of Eisen et al[15], the correlation coefficients were low to moderate (0.3 to 0.5). Again, obsessions had greater negative impact on quality of life than compulsions. Psychiatric comorbidity affected to a greater extent patients' quality of life (bodily pain, general health, vitality, social functioning and mental health) than the medical comorbidity (general health and social functioning). Finally, as expected, unemployed patients scored significantly lower in the role physical scale than employed patients. The multiple regression analysis demonstrated that depressive symptomatology affected to a greater extent the quality of life of OCD patients (physical functioning, bodily pain, general health, vitality, and mental health) than the OCD symptomatology itself (general health and vitality).

Kivircik Akdede et al[17] studied 23 OCD outpatients under treatment and 22 healthy controls (without self or family history of psychiatric disorders). Patients with a HDRS>15 were excluded. The mean length of illness was 10.7 years (SD 9.6). Severity of OCD was assessed with the Y-BOCS, depressive symptomatology with the HDRS, overvalued ideas with the Overvalued Ideas Scale (OIS), and quality of life using the WHO Quality of Life Scale-Brief Form (WHOQOL-BREF). Patients' mean age was 32.1 (SD 11.4), 73.9% were female, and 39.1% had a university degree. No significant statistical differences were observed between patients and healthy controls across all these variables.

Patients reported significantly worse quality of life than controls in the psychological and social domains. The global severity of the disorder negatively affected the WHOQOL-BREF psychological domain. Obsessions negatively affected the physical, psychological and environmental domains whereas compulsions only caused negative impact on the psychological domain. Overvalued ideas did not affect quality of life. Unfortunately authors did not report results on the influence of depressive symptoms on quality of life.

To our knowledge this is the first study that reported the influence of cognitive functioning on OCD patients' quality of life. The worse they performed in working memory, attention, mental flexibility and motor ability the lower they scored in the psychological domain (Auditory Consonant Trigram Test, Digit Span, and Trail Making Test). In addition, the social domain was negatively affected by low performance in attention test (Digit Span).

Moritz et al[18] included 79 patients with a diagnosis of primary OCD and 32 healthy controls. Among patients, 37% also fulfilled criteria for major depression or dysthymia and 13% for another anxiety disorder. All but 7 patients were treated with cognitive-behavioural therapy, 41% received antidepressants and 9% received antipsychotics. Evaluations were made at baseline and at the end of treatment (approximately 10 weeks), and they used the Y-BOCS, the short form of the Hamburg Obsessional Compulsive Inventory (HOCI), the HDRS, and the Beck Depression Inventory (BDI). In addition patients fulfilled the SF-36.

At baseline, patients compared to healthy controls showed significantly decreased mean scores in all SF-36 scales. Furthermore, z scores on the physical role, general health, vitality, social functioning, emotional role and mental health were between 2 and 4 SD lower than those of the healthy subsample of the German SF-36 norm population. In contrast with other studies physical scales were also affected. The authors conclude that this may be due to the fact that they have excluded somatic ill patients from the German normative sample, avoiding this bias. The areas most affected in OCD patients were vitality, social functioning and mental health (scores below the 25th percentile of the normative sample).

Depression, both self and clinician rated, showed the strongest associations with SF-36 scales specially those referred to mental health. Contrary to other investigations, Moritz et al found that compulsions (Y-BOCS and HOCI compulsions scores) were stronger correlated with quality of life than obsessions.

When the OCD sample was split in two subsamples according to the type of compulsions –"washers" (n = 42) and "checkers" (n = 39)- washers showed poorer quality of life on physical functioning, general health, social functioning and emotional role than nonwashers. Checkers reported worse quality of life on emotional role and mental health than noncheckers.

Rapaport et al[19] reported baseline quality of life data from 521 OCD patients drawn from several multicenter trials investigating the efficacy of sertraline in anxiety disorders. The QoL instrument employed was the Quality of Life Enjoyment and Satisfaction Questionnaire (Q-LES-Q). The mean age was 38.6 years (SD 11.8), 49% were female and 64% were employed. OCD patients had lower mean Q-LES-Q percentage scores than the community normative percentage score (83% of the maximum score of 70) (community sample n = 67; mean age 32.4 years, 65.8% women). Furthermore, 26% of the patients had severe impairment in quality of life defined as 2 or more standard deviations below the community norm. In addition relative to normative population OCD patients showed lower scores across all of the Q-LES-Q domains. Compared to post-traumatic stress disorder and affective disorders, OCD showed more impairment on the social, family, leisure, ability to function and vision domains. When examining the impact of specific symptoms on the quality of life they found that only 1.4% of the variance in Q-LES-Q scores was explained by them.

Masellis et al[20] conducted a study with the aim to determine the specific impact of obsessions, compulsions and comorbid depression on the quality of life of OCD patients. They recruited 43 consecutively referred OCD patients that were assessed using the Y-BOCS, the BDI and the Illness Intrusiveness Rating Scale (IIRS). The IIRS measures objective and subjective interference of symptoms across 13 life domains considered important to quality of life. The total score ranges from 13 to 91; the greater the score the higher the intrusiveness.

The mean age was 34.9 years (SD 8), 58% were female; 37.3% were married and 39.5% had completed college or university. The mean age at onset of the illness was 14.4 years (SD 9), the Y-BOCS scores were 19.7 (SD 8.4), 9.9 (SD 4.5) and 9.7 (SD 4.2) for total, obsessions and compulsions respectively. The mean BDI score was 16.7 (SD 12) and the mean IIRS 45.7 (SD 17.7). In the analysis they found that demographics and age at onset of the illness did not predict quality of life (illness intrusiveness scores). In contrast depression and obsessions severity significantly predicted illness intrusiveness; although depression accounted for greater variance than obsessions. As in the majority of such studies on the issue, compulsions did not predict quality of life at all.

Bystritsky et al[21] studied 31 OCD and 68 schizophrenic outpatients who participated in two day treatment programs. The mean ages were 32.6 and 36.2 years for OCD and schizophrenic patients respectively. OCD psychopathology was assessed using the Y-BOCS, the HDRS and the Hamilton Anxiety Rating Scale (HARS). Quality of life was evaluated employing the Independent Living Skills Survey (ILSS) and the Quality of Life Interview (QoLI). Evaluations were made at the intake and at the time of completion participation in the programs (6-week or 6-month).

At intake patients with OCD scored similar to schizophrenic patients in all social and independent living skills except for health maintenance where the OCD patients scored significantly higher. At the time of discharge from the programs, OCD patients scored significantly higher than schizophrenics in six out of the 10 life domains. However, the social and independent living skills of OCD patients in spite of having improved remained less than satisfactory. The global QoL at baseline and at the end of treatment, schizophrenic patients scored significantly higher than OCD patients (4.97 and 4.95 versus 3.70 and 4.34). However, OCD patients significantly improved their quality of life after the treatment period whereas no significant improvement was obtained in the schizophrenic group. These surprising data may be due to the fact that OCD patients preserve a good insight into their functional impairments and limitations.

Bobes et al[22] evaluated the quality of life of 36 consecutive OCD outpatients under maintenance treatment with different serotonergic medications using the SF-36. In addition patients were evaluated with the Y-BOCS, the HDRS and the World Health Organization Short Disability Assessment Schedule (WHO DAS-S). Mean age was 34 years (SD 11.4) and 55.6% were male. The mean total Y-BOCS score was 23.6 (SD 9), and the mean scores for obsessions and compulsions were 12.1 (SD4.8) and 11.5 (SD 5.8) respectively. The mean HDRS was 14.7 (SD 8.1) and 44.8% obtained scores > 17. Compared to Spanish population norms (n = 9151; mean age: 45.2; 48.2% male) z scores of OCD patients felt below 2 SD on social functioning, and between 1 and 2 on role emotional and mental health. Contrary to the study of Bystritsky et al[21] OCD patients compared to schizophrenic patients did not show significant differences in the mental health subscales and reported better quality of life in the scales related to physical health, that is, physical functioning, role physical and general health. In addition, compared to depressive patients (n = 729; mean age 47.3, 28.3% male) OCD patients showed lower scores on the following SF-36 scales: bodily pain, vitality, social functioning, role emotional and mental health.

Koran et al[14] included 60 OCD outpatients who were participating in pharmacological treatment trials. Patients were evaluated using the Y-BOCS and the HDRS and they completed the SF-36. The mean age was 40.1 years (SD 10.6), 43% were female and 50% were married. They found that mean scores of the OCD patients were lower than those of the US population (n = 2474; 57% female) mainly in mental health (z score −1.25), role emotional (z score −0.86), vitality (z score −0.79) and social functioning (z score −0.64). The severity of the disorder only correlated significantly with the social functioning SF-36 scale (r = −.39). Compared to depressive patients OCD patients reported higher quality of life on all life domains except role emotional. This observation is opposite to that found by Bobes et al[22].

Effects of OCD Treatments on Quality of Life

Even though data about quality of life of OCD patients are scarce, data regarding the effect of different OCD treatments are practically nonexistent.

The first study was developed by Bystritsky et al[23] and included 30 consecutive OCD patients admitted to the UCLA partial hospitalization program. These patients were treated with a combination of cognitive-behavioural therapy, medication and psychosocial interventions over 6 weeks. The mean age was 34 years and 33% were female. The severity of the OCD was rated with the Y-BOCS, the HDRS and the HARS. Quality of life was assessed using the QoLI.

Between admission and discharge significant improvements were found in both objective and subjective measures for the following 3 life domains: activities, health and social. In addition, significant improvements were also observed in the QoLI subjective measures for family, living situation, life and safety domains. However, no changes were found in significant objective measures as work, disability, employment, and family support. Psychopathology did not significantly affect the quality of life of these patients; only the total Y-BOCS score significantly correlated with the global QoLI score (r = .42).

Moritz et al[18] in their previously described study analyzed the impact of response to treatment on quality of life. They found that OCD patients, contrary to healthy controls, showed improvement in all SF-36 scales except for role physical and bodily pain. At discharge only the vitality scale was significantly different between responders and non-responders. In addition when patients were split in two groups according to the median QoL change (median = 44 points change) neither demographic nor psychopathological baseline variables differentiate between responders and non responders.

Koran et al[24] reported quality of life data from a multicentre 80-week clinical trial with sertraline conducted in US. A total of 649 were included in the single-blind phase of the trial (weeks 1-52) and 223 in the double-blind phase (weeks 53-80; sertraline arm included 109 patients and placebo arm 114). Psychopathology was evaluated using the Y-BOCS and the Clinical Global Impression scales (CGI). Patients fulfilled the Quality of Life Enjoyment and Satisfaction Questionnaire Q-LES-Q.

At baseline, the mean age was 38.6 years (SD 11.9), the mean length of the illness was 21.4 years (SD 12.4) and the mean scores on the symptom measures were Y-BOCS: 26.1 and CGI-S: 4.8 points. During the single-blind phase the symptomatic improvement observed was associated with improvement in quality of life (mean pre-treatment Q-LES-Q score: 60 versus end of week 52 score: 78.2). During the double-blind phase sertraline was significantly superior to placebo (sertraline endpoint Q-LES-Q score: 77.3 versus placebo endpoint 75.5).

CONCLUSIONS

In summary, it can be concluded that there is a growing interest for assessing quality of life in OCD patients as demonstrated by the increasing number of papers published. However much more work must be done in order to improve the methodology employed.

In general, and in spite of the limitations of the studies reviewed, it can be said that OCD negatively affects the quality of life of patients, and such impact is greater with comorbid depression. (The clarify effect of gender on the quality of life seems to be neutralized by the illness itself.)

Unfortunately, we don't have enough data on the effect of the different treatments on quality of life that can help clinicians in making decisions about the choice of treatment approach.

REFERENCES

1. American Psychiatric Association. Diagnostic and Statistical Manual of Mental Disorders (4th ed.). DSM-IV. Washington, DC: American Psychiatric Association. 2004.
2. Kessler RC, Berglund P, Demler O, Jin R, Walters EE. Lifetime prevalence and age-of-onset distributions of DSM-IV disorders in the National Comorbidity Survey Replication. Arch Gen Psychiatry 2005; 62: 593–602.
3. Kessler RC, Chiu WT, Demler O, Walters EE. Prevalence, severity, and comorbidity of 12-month DSM-IV disorders in the National Comorbidity Survey Replication. Arch Gen Psychiatry 2005; 62: 617–627.
4. Hollander E, Kwon JH, Stein DJ, Broatch J, Rowland CT, Himelein CA. Obsessive-compulsive and spectrum disroders: overview and quality of life issues. J Clin Psychiatry 1996; 57: (Suppl 8): 3–6.
5. Stein DJ, Roberts M, Hollander E, Rowland C, Serebro P. Quality of life and pharmaco-economic aspects of obsessive-compulsive disorder. S Afr Med J 1996; 36: 1579–1585.
6. Organización Mundial de la Salud. Informe sobre la Salud en el Mundo 2001. Salud mental: nuevos conocimientos, nuevas esperanzas. Ginebra: Organización Mundial de la Salud. 2001.
7. The WHOQOL Group. The World Health Organization Quality of Life Assessment (the WHOQOL): position paper from the World Health Organisation. Soc Sci Med. 1995; 41: 1403–1409.
8. Ware JE. SF-36 Health Survey: Manual and interpretation guide. Boston: Health Institute, New England Medical Center. 1993.
9. Wenger NK, Furberg CD. Cardiovascular disorders. In: Spilker B (Ed.) Quality of Life Assessment in Clinical Trials. New York: Raven Press, 1990; 335–45.
10. Sorensen CB, Kirkeby L, Thomsen PH. Quality of life with OCD. A self-reported survey among members of the Danish OCD Association. Nord J Psychiatry 2004; 58: 231–236.

11. Bobes J. Current status of quality of life assessment in schizophrenic patients. Eur Arch Psychaitry Clin Neurosc 2001; 251(Suppl 2): II/38–42.
12. Stein DJ, Allen A, Bobes J, Eisen JL, Figuera ML, Iikura Y et al. Quality of life in obsessive-compulsive disorder. CNS Spectrums 2000; 5(6 Suppl 4): 37–39.
13. Mendlowicz MV, Stein MB. Quality of life in individuals with anxiety disorders. Am J Psychiatry 2000; 157: 669–682.
14. Koran LM, Thienemann ML, Davenport R. Quality of life for patients with obsessive-compulsive disorder. Am J Psychiatry 1996; 153: 783–788.
15. Eisen JL, Mancebo MA, Pinto A, Coles ME, Pagano ME, Stout R, Rasmussen SA. Impact of obsessive-compulsive disorder on quality of life. Compr Psychiatry 2006 ; 47 : 270–275.
16. Rodriguez-Salgado B, Dolengevich-Segal H, Arrojo-Romero M, Castelli-Candia P, Navio-Acosta M, Perez-Rodriguez MM, Saiz Ruiz J, Baca-Garcia E. Perceived quality of life in obsessive-compulsive disorder: Related factors. BMC Psychiatry 2006; 6:20.
17. Kivircik Akdede BB, Alptekin K, Akvardar Y, Kitis A. Quality of life in patients with obsessive-compulsive disorder: Relations with cognitive functions and clinical symptoms. Turkish Journal of Psychiatry 2005: 16: 13–19.
18. Moritz S, Rufer M, Fricke S, Karow A, Morfeld M, Jelinek L, Jacobsen D. Quality of life in obsessive-compulsive disorder before and after treatment. Compr Psychiatry 2005; 46: 453–459.
19. Rapaport MH, Clary C, Fayyad R, Endicott J. Quality of life impairment in depressive and anxiety disorders. Am J Psychiatry 2005; 162: 1171–1178.
20. Masellis M, Rector NA, Richter MA. Quality of life in OCD: Differential impact of obsessions, compulsions, and depression comorbidity. Can J Psychiatry 2003; 48: 72–77.
21. Bystritsky A, Liberman RP, Hwang S, Wallace CJ, Vapnik T, Maindment K, Saxena S. Social functioning and quality of life comparisons between obsessive-compulsive and schizophrenic disorders. Depress Anx 2001; 14: 214–218.
22. Bobes J, González MP, Bascaran MT, Arango C, Sáiz PA, Bousoño M. Quality of life and disability in patients with obsessive-compulsive disorder. Eur Psychiatry 2001; 16: 239–45.
23. Bystritsky A, Saxena S, Maidment K, Vapnik T, Tarlow G, Rosen R. Quality of life changes among patients with obsessive compulsive disorder in a partial hospitalization program. Psychiatr Services 1999; 50: 412–414.
24. Koran LM, Hackett E, Rubin A, Wolkow R, Robinson D. Efficacy of sertraline in the long-term treatment of obsessive-compulsive disorder. Am J Psychaitry 2002; 159: 88–95.

PART III

TREATMENT AND REHABILITATION ISSUES

CHAPTER 16

ANTIPSYCHOTIC MEDICATIONS, SCHIZOPHRENIA AND THE ISSUE OF QUALITY OF LIFE

A. GEORGE AWAD[1] AND LAKSHMI N.P. VORUGANTI[2]

[1]*Department of Psychiatry and the Institute of Medical Science University of Toronto, Chief of Psychiatry, Humber River Regional Hospital, Canada*
[2]*McMaster University, Hamilton, ON, Canada*

Abstract: Schizophrenia as a long term disabling illness generally runs a chronic course with acute psychotic relapses that frequently requires hospitalizations. Antipsychotic medications have emerged as the cornerstone in treatment of the disorder in addition to other important interventions such as rehabilitation, psychosocial and economic support. Unfortunately antipsychotic medications has presented a number of significant limitations in terms of inconveniencing or serious side effects as well as there inability to improve certain symptoms at different stages of the illness. 30% to 50% of patients on antipsychotic medications develop serious dislike to medication as a result of becoming dysphoric, a situation that lead to compromised adherence to medications and subsequent relapse and compromised quality of life.

Reviewing published studies involving both new and old antipsychotic medications reveal a good deal of methodological limitations that make it difficult to reach any definitive conclusion about the impact of antipsychotic medications on quality of life. However, the weight of the evidence so far can only suggest a trend favouring the new antipsychotics in terms of their positive impact on quality of life. On the other hand it is clear that medications by themselves can not directly improve quality of life but certainly can improve the potential for patients to benefit from other important interventions such as rehabilitation and psychosocial support that can have direct impact on quality of life. A number of methodological and conceptual issues are proposed to improve the quality of quality of life studies. Finally, though quality of life by itself is an important outcome it is time for the field to look beyond quality of life measurements. In addition to being an outcome, quality of life improvement can be also construed as mediator of other important outcomes such as decreased relapse rate, re-hospitalization and medical resources utilization. Similarly, improvement in quality of life can enhance strategies for improving adherence to medications and other therapeutic regimens as well as better and longer tenure in the community

Keywords: Schizophrenia, Antipsychotics, Quality of life, Conceptual models, Quality of life scales, Dopamine, Neurobiological correlates

M.S. Ritsner and A.G. Awad (eds.), Quality of Life Impairment in Schizophrenia, Mood and Anxiety Disorders, 307–319.

INTRODUCTION

Schizophrenia is a severe disabling psychiatric disorder which generally runs a long term course with acute psychotic relapses that may frequently require hospitalization. The clinical picture varies according to the stage of the illness and generally includes a wide range of psychotic symptoms, such as delusions, hallucinations and conceptual disorganization. The disorder also includes a number of negative and deficit symptoms such as blunted affective responses, emotional and social withdrawal, and lack of spontaneity as well as neurocognitive deficits. In addition to the personal sufferings of the patients and the families, schizophrenia extols a major economic impact in terms of long term psychiatric and psychosocial services required but also in terms of lost productivity and long term disability[1]. Over the past five decades with the introduction of the first antipsychotic Chlorpromazine in the 1950's the pharmacological approach has become the cornerstone of clinical management of schizophrenia in addition to psychosocial and economic support as well as rehabilitation programs. Up until the early 1980's a wide range of antipsychotic medications similar to Chlorpromazine has been introduced and all has been potent dopamine D^2 receptor antagonists. It is not surprising then that these "first generation antipsychotics" possessed a wide range of side effects notably among them, significant extrapyramidal symptoms and irreversible neurological side effects such as tardive dyskinesia which proved inconveniencing for patients. Similarly many of these first generation antipsychotics were not well tolerated subjectively by at least 30% to 50% and ultimately lead to medication refusal and compromised compliance behaviour resulting in increased risk for hospitalization and medical resources utilization[2–5]. In addition there seems to be at least 30% of persons with Schizophrenia who do not adequately respond to such medications.

In the late 1980's and the subsequent decade a number of new "second generation antipsychotics" pushed their way to clinical practice as an alternative to replace the old medications. This new group included Amisulpride, Remoxipride (withdrawn from use), Risperidone, Olanzapine, Sertindole, Quetiapine, Zotepine, Ziprasidone and more recently Aripiprazole. Similarly Clozapine which was in use in the 1970's but was withdrawn because of serious hematological side effects has been reintroduced for treatment resistant patients who failed to respond adequately to other antipsychotics. Though a number of these "second generation" antipsychotics has different pharmacological properties all of them continued to possess to a variable degree some dopamine D^2 blocking properties. The superiority of the new antipsychotics compared to the old medications has been the subject of extensive research and ongoing debate[6–9]. In general there seems to be an agreement that the new and old antipsychotics are comparable in terms of the efficacy in the treatment of positive symptoms. There is also general agreement that overall the second generation antipsychotics have in general a more favourable side effect profile and improved subjective tolerability compared to the first generation antipsychotics. On the other hand the new antipsychotics have brought with them a host of endocrine and metabolic side effects as well as excessive weight gain that introduced clinical limitations as well as increased cost to manage such potentially serious side effects over and above the initial high acquisition cost of the new medications themselves[10–13].

THE AIMS OF ANTIPSYCHOTICS MEDICATIONS

In a previous publication we defined the aims of antipsychotics medications similar to any medications used for the long term treatment of chronic illnesses such as Schizophrenia to include[2]:

1. Efficacy without adverse effects,
2. Improved quality of life and subjective tolerability,
3. Positive long term outcomes, and
4. Cost effectiveness

Though this chapter concerns itself with the impact of antipsychotic medications on quality of life, such issue can only be explored in the context of a number of other important issues such as symptoms improvement, medication side effects, psychosocial management and long term outcomes.

A CONCEPTUAL MODEL FOR QUALITY OF LIFE

Though there exist a number of definitions and conceptualization of quality of life there has been no conceptual model specific to the impact of medications on quality of life. Such specific models can allow for experimentally testing the role of various factors and its relative contribution to the variations in outcome. Such model can be of value in clinical trials testing of new antipsychotic medications.

In 1997 we developed and reported on a clinically intuitive conceptual and integrative model specific for quality of life of persons on antipsychotic medications[14]. According to this model quality of life is defined as the outcome of dynamic interaction between three major primary determinants: the level of psychotic symptoms, medication side effects and the level of psychosocial performance. A number of secondary factors have also been identified as contributing to the construct of quality of life: personality characteristics, premorbid adjustment, value and attitudes towards health, illness and medications; resources, their availability and adequacy. The primary determinants and the second order factors as modulators are integrated in a circular model which underscores the inter-relatedness of the primary determinants and also emphasizes the multidimensional nature of the construct of quality of life. The model though clinically intuitive in its development has allowed to experimentally test the interrelationship of its components as well as estimate their relative contribution. An analyses of data from a sample of stable chronic patients with schizophrenia attending medication clinic, demonstrated that psychotic symptoms and the subjective distress caused by specific side effects of medication such as akathesia and neuroleptic dysphoria account for nearly half of the variation in quality of life in such symptomatic but stable schizophrenia population[14]. Surprisingly in our initial study, the contribution of psychosocial performance was minimal which may have to do with methodological issues such as the sensitivity of measures for psychosocial performance and possibly may also relate to the specific population understudy. Such stable but symptomatic chronic population tend to be irregular in attending their rehabilitation programs and more focused on medications. However, subsequent studies in different populations have documented the contribution of psychosocial performance and its impact on quality of life [15–17].

Our conceptual model as the first of its kind has managed to focus attention on the role of symptoms and medications and their impact on quality of life evidenced by a large number of subsequent publications [18–23]. Similarly the model has attracted attention to the need for developing conceptual models specific to particular clinical situations or specific population. Amplification of the model and development of new models has broadened applicability of quality of life constructs to other social or vocational interventions by elaborating the balance of positive protective and negative limiting factors that impact on quality of life [24,25].

THE IMPACT OF ANTIPSYCHOTIC MEDICATIONS ON QUALITY OF LIFE

In a recent and extensive review about the impact of new second generation antipsychotics on quality of life we have previously provided detailed review of published reports related to individual antipsychotics [26]. Though there is increased interest in exploring the impact of antipsychotic medications on quality of life as evidenced by the increase in the number of publications, one can not reach a clear and definitive conclusions about their impact on quality of life [26,27]. Many of the studies are uncontrolled and suffer from methodological limitations. Frequently the inclusion of quality of life assessments in clinical trials seems to be an after thought. Many of the studies are short term lasting only a few weeks with no long-term follow up. The use of several measurement scales based upon different theoretical constructs seems to limit any reliable comparative analysis. Some of the instruments used are of unknown psychometric properties and maybe inappropriate for use in the schizophrenia population or are not sensitive enough to detect small changes in quality of life as expected in such relatively short term trials [28]. Similarly the different time framework for assessment of quality of life in different studies complicates any meaningful interpretation [29]. A review of the commonly used instruments for measuring quality of life has been previously provided by us and others [28].

Overall a great number of studies dealt with comparisons of the impact of the new and old antipsychotics on quality of life. Fewer studies have attempted to compare the impact of quality of life among individual new antipsychotics. Generally speaking most of the published literature has dealt with **Haloperidol** as a representative of the old antipsychotics, **Risperidone, Olanzapine** and **Clozapine** are among the new antipsychotics that have been more extensively studied. Though few recent literature started to emerge about Quetiapine, Ziprasidone and Aripiprazole, there is not enough data yet to allow meaningful comparisons.

Several uncontrolled non-comparative cross-sectional, mostly post marketing studies reported improvement in quality of life under **Risperidone** treatment[29–32] while a number of comparative studies of Risperidone versus Haloperidol[9,33–35] or Olanzapine[9,31,36–39], many of them suffer from methodological limitations. The majority of data are obtained from industry initiated multi-center clinical trials.

For Olanzapine a similar picture evolves as a number of studies reported better quality of life on Olanzapine compared to Haloperidol[31–35], or Risperidone[36–38]. Other studies failed to show any difference between Olanzapine and Risperidone,[9,31,39]. A recently reported study deserves some comments as a case study for the current state of research[40]. The study has a well controlled double blind long-term design comparing the impact of either Haloperidol or Olanzapine on quality of life. This study represents a relatively much better design yet at the end the results were confounded by a number of limitations[41]. The use of SF-36, a well validated and commonly used health status index does not seem to be the most appropriate instrument in schizophrenia, particularly that it is mostly skewed towards physical symptoms which are not relevant to the schizophrenia population. It is possible that the instrument was not sensitive enough to pick small differences between the two treatments or alternatively the use of relatively lower dosages of haloperidol than what was in previous studies may have eliminated any advantages related to Olanzapine. Another critical issue is the use of a diagnostically heterogeneous population that included schizophrenia, schizoaffective and brief psychosis. It is possible that the presence of affective states may confer better course or prognosis in the schizoaffective group.

Clozapine has been reintroduced to clinical practice in the late 1980's with a main indication for use in schizophrenic patients unresponsive to other antipsychotics. It then follows that most of the studies with Clozapine has included a rather chronic and more symptomatic population who have not responded to a variety of treatments before. A number of studies have reported significant improvements in quality of life on Clozapine[42–46]. Yet a number of other studies failed to find a difference comparing Clozapine to first generation antipsychotics[31–47].

The literature about the impact of Quetiapine on quality of life seems to be limited with only few studies pointing to a more favourable trend toward improvement in quality of life compared to first generation antipsychotic but comparable to that of Risperidone and Olanzapine[9,31,48].

With all these methodological limitations the following synthesis can be provided:

1. In spite of the methodological limitations the balance of evidence points to just a trend towards more favourable impact of second generation antipsychotics on quality of life compared to first generation antipsychotics[26,27].
2. There is need for well designed controlled and long term comparative studies not only between old and new antipsychotics but also between various new antipsychotics in a head to head design.

3. Since the majority of the studies reported are industry related, there is need for more independent studies.

QUALITY OF LIFE AND PHARMACOECONOMICS OF ANTIPSYCHOTICS

Interest in the pharmacoeconomics of antipsychotic medications followed the introduction of a number of new second generation antipsychotics; all of them have higher acquisition costs compared to the old first generation antipsychotics. Attempts to justify the higher acquisition costs of the new antipsychotics focused on demonstration of accrued benefits that can offset the higher cost. In that context, (estimating) the impact of new antipsychotics on quality of life has emerged as an important component in the study of the pharmacoeconomics of schizophrenia. The potential impact of treatment on a number of specific domains of quality of life such as use of leisure time, satisfying relationships, feeling of satisfaction and wellbeing, , engaging in a productive role within the family or the society have become important considerations. Most of the recent pharmacoeconomic studies of new antipsychotics have mainly focused on cost minimization by demonstrating a reduction in cost of certain components of care notably decrease in re-hospitalization or hospital days[42–52]. Unfortunately the majority of these studies suffered from methodological limitations and the overall clinical experiences do not support projected reduction in hospitalization rates nor significant savings from the use of the new antipsychotics. Though our conclusions from reviewing the literature and from our own experiences point to a modest favourable trend of new antipsychotics on quality of life, likely related to improved side effect profile and better subjective tolerability, it is clear that medications alone are not enough to substantially raise the level of quality of life. Other interventions are required in addition to medications. Thus in reality improving the quality of life in patient with schizophrenia on new antipsychotics should result in higher costs at least initially as a result of the cost of other interventions such as rehabilitation programs, economic and psychosocial support[26].

A recent interesting approach in cost infectiveness analysis is the concept of cost utility which has the advantage of combining both cost and quality of life considerations. Cost utility approach proved successful as applied to a number of medical conditions such as rheumatoid arthritis, chronic pain, cancer....etc., but its applicability to schizophrenia proved to be a challenge in view of the complex clinical picture. In a recent study,[53,54] we reported on the feasibility of applying utility analysis in patients with schizophrenia. We demonstrated that the majority of patients with schizophrenia have experienced considerable difficulty in completing a number of utility measurements specifically; standard Gamble and Magnitude Estimation which accordingly we had to drop from our analysis. On the other hand our data demonstrate good convergent validity as well as significant inter-correlations between utility measures for current health states and quality of life. Although the concept of cost utility analysis is well grounded in scientific and economic theories and seems to be promising, it has several limitations in the schizophrenia population. There are only a few reported studies so far and the approach requires refinement and reappraisal. Further discussion of cost utility analysis is presented in chapter 20 of this book byDernovsek et al,[55]. Other significant reviews about the pharmacoeconomic of antipsychotics are published elsewhere[56–58].

THE NEUROBIOLOGY OF EMOTIONAL AND SUBJECTIVE RESPONSES

Quality of life as a subjective construct requires a number of prerequisite in order for the person to reach an accurate assessment of their own subjective state of well-being, satisfaction ... etc. There needs to be a degree of cognitive intactness not only in terms of some basic insight but also their ability to interact effectively with their environment. A number of studies and reviews have correlated cognitive ability with objective measures of functioning such as vocational potential, job performance or educational tasks[59–61] while other researchers failed to confirm any strong relationship between cognitive impairment and subjective constructs such as quality of life or feeling of satisfaction and wellbeing[62–64]. It is plausible that subjective constructs are basically different from objective constructs and possibly being mediated through different mechanisms. Another prerequisite is the presence of some basic emotional and affective responsivity[65]. At a clinical basic level, significant mood changes can alter the ability of the person to judge their own internal and external reality. In this context and running the risk of taking a reductionist approach, what happens in the brain in terms of the basic neurobiological processes involved in cognition and elaboration of emotions ultimately can shape to a large part the person's ability to judge their internal subjective state and its reliable interpretation. Raising the issue of neurobiological correlates with subjective constructs like feelings of wellbeing or quality of life is a rather new and recent development since historically the field has largely focused on psychosocial aspects of quality of life as a psychosocial construct.

A good deal of basic animal and clinical data over the past three decades have assigned a significant role to the dopamine pathways in the brain and its role in rewards/punishment as well as pleasurable behaviour[66,67]. On the other hand dopamine is central to the pharmacological actions of antipsychotics since all of them are dopamine antagonists. Recent neuroimaging studies including ours and others have broadened knowledge about the role of dopamine in the genesis of such subjective constructs as neuroleptic dysphoria in humans[68–70]. This issue and related psycho-neurobiological issues will be dealt with in great detail in a subsequent chapter by Voruganti and Awad[71]. (chapter 2)

ISSUES THAT REQUIRE FUTURE RESOLUTION

1. Researchers need to clearly define the concept of quality of life as applied in their particular studies. There may not be a single definition of quality of life that covers all clinical situations as well as different stages of the illness. It is

incumbent then on researchers to clearly describe what they meant by quality of life as applied to their studies. It is also incumbent on journal editors to require such explicit definition and description of the concept of quality of life as applied in various studies. Such valuable suggestion by the late Alvin Feinstein still awaits broad implementations by reviewers and journal editors[72].

2. In exploring the impact of antipsychotic medications on quality of life it is imperative to choose the right measurement instruments. Such instruments need not only to be psychometrically sound but they must also have adequate sensitivity to detect the expected small changes on quality of life in medication clinical trials, particularly that the majority of such clinical trials with antipsychotic are of relatively short term nature.
3. The time course of quality of life assessment is critical. No significant changes even on the best medications are expected in few weeks[31]. Similarly it has to be understood that medications by themselves have no direct impact on improving quality of life. Medications optimally can make it possible to benefit from other interventions such as psychosocial manipulations or rehabilitation programs which are more effective in achieving a higher level of quality of life. This seems to be a significant issue overlooked in multicenter trials with different centers having different approaches and different support programs. Without knowing what patients have received in different centers, it is difficult to know from where improvement or deterioration in quality of life is coming[73].
4. The objective/subjective dichotomy needs to be resolved and reconciled. The majority of stable schizophrenic patients are able to reliably judge their level of satisfaction with their quality of life[21,74]. Historically there has been reluctance among clinicians to take patient's self reports as reliable expression of their inner feelings. Yet the paradox is that psychiatric diagnosis is mostly based on what patients tell their doctors without any means of objectively verifying their reported subjective experiences, such as hallucinations or delusions.
   On the other hand there are certain clinical situations such as in severe acute psychotic episodes or in those deteriorated chronic patients who are severely impaired in their cognitive functioning to reliably assess and express their subjective experiences. In the purest definition, quality of life is a subjective construct. Yet we believe that in schizophrenia it may be wise and practical to include subjective self reports as well as some objective measures. To preserve the subjective nature of quality of life we propose that the concept of quality of life be limited to subjective self reports. On the other hand we propose that important objective measure such as handling aspects of life situation: housing, dealing with finances and emergencies. . . etc. can serve as confirmatory objective measures but need to be labeled differently as measures of standards of living[75,76].
5. There is need for developing conceptual models appropriate to the clinical situation, the population under study and the stage of the illness. We subscribe to the notion that there is nothing more practical than having theoretical models. New conceptual models enhance our understanding of the issues contributing to

the concept of quality of life itself as well as having practical utility in terms of recognition of the multidimensional of aspects of quality of life. Clear conceptual models can enhance the development of more appropriate measurement instruments, based on such conceptual constructs.

6. As indicated before, though the important interventions that can be instrumental in improving quality of life are mostly psychosocial in nature, yet the evolving neurobiological science of emotion and subjective responses requires increased attention. The new research technology in terms of brain imaging particularly in the area of medication effects may prove helpful in the future in developing the right medications that may impact positively on quality of life and other subjective emotional experiences.
7. Though improvement in quality of life in itself can be construed as an outcome of treatment it needs also to be viewed as a possible mediator for other outcomes such as better compliance with medications or therapeutic regimens, better level of satisfaction and psychosocial adjustment. In essence the field has to move beyond simply measuring quality of life and explore the role of improved quality of life in achieving other outcomes.

CONCLUSIONS

Over the last five decades since antipsychotic medications has evolved as the cornerstone in the management of psychotic illness such as schizophrenia, exploring the impact of these medications on quality of life seems to be relevant and important. Antipsychotic medications by virtue of their impact on psychotic symptoms as well as their potential for inconveniencing and sometimes serious side effects can influence quality of life positively or negatively.

Methodologically many of the studies exploring the impact of antipsychotic medications on quality of life have been deficient and there is need for rigorous improvement in the design and methodology of such studies as well as expanding our understanding of the concept of quality of life on medications and the factors impacting on it.

Overall in spite of the methodological limitations there seems to be just a trend favouring the new second generation compared to first generation antipsychotics in achieving better quality of life. However it needs to be clearly understood that medications alone are not enough for significant improvement in quality of life unless they are coupled with adequate psychosocial support and rehabilitation efforts. It is time to look beyond quality of life as being an outcome by itself, but also possibly as being a mediator of other important outcomes such as improved adherence with therapeutic regimens, better tenure in the community and reduced risk of relapse.

Finally though the concept of quality of life has been traditionally and historically defined as a psychosocial construct, new and recent brain research has expanded our knowledge about the basic neurobiological processes involved in elaboration of emotions, the genesis of subjective responses and cognitive ability; all are important issues in understanding subjective constructs such as quality of life, satisfaction and well-being.

BIBLIOGRAPHY

1. Davies L.M., Drummond M.F. The Economic Burden of Schizophrenia. Psychiat. Bull 1990; 14: 522–555.
2. Awad A.G., Voruganti L.N.P., Heselgrave R.J. The Aim of Antipsychotic Medications: What Are They and Are They being Achieved? CNS Drugs 1995; 4: 8–16.
3. Arana G.W. An Overview of Side Effects Caused by Typical Antipsychotics. J. Clin Psychiatry; 2000; 61 (Suppl 8): 5–11.
4. Awad A.G. Antipsychotic Medications: Compliance and Attitudes Towards Treatment. Curr. Opin. Psychiatry. 2004; 17: 75–80.
5. Awad A.G., Masty V., McDonnell D. The Link Between Drug Attitudes, Compliance Behaviour and Resource Use Among Individual with Schizophrenia. Poster Presented at the 150 2nd American Psychiatric Association Meeting 1999; (Washington, D.C.).
6. Kerwin R.W., Taylor D. New Antipsychotics: A Review of their Current Status and Clinical Practice. CNS Drugs 1996; 6: 71–82.
7. Awad A.G., Voruganti L.N.P. New Antipsychotics, Compliance, Quality of Life, and Subjective Tolerability – Are Patients Better Off? Can J Psychiatry 2004; 49: 297–302.
8. Lieberman J., Stroup T.S., McEvoy J.P., Swartz M.S., Rosenheck R.A., Perkins D.O., Keefe, R., Davis C., Severe J., Hsiao J., and the Catie Investigators. Effectiveness of Antipsychotic Drugs in Patients with Chronic Schizophrenia. The New England J. MED; 353: 1209–1223.
9. Voruganti L.N.P., Cortese L., Owyeumi L., Kotteda V., Cernovsky Z., Zirul S., Awad A.G. Switching from Conventional to Novel Antipsychotic Drugs: Results of a Prospective Naturalistic Study. Schizophr. Res 2002; 57: 201–208.
10. Wirshing D.A., Spellberg P.J., Erhart S.M., Marder S.R., Wirshing W.C. Novel Antipsychotic and New Onset Diabetes. Biol. Psychiatry 1998; 44: 778–783.
11. Newcomer J.W. Second Generation (atypical) Antipsychotic and Metabolic Effects: A Comprehensive literature Review. CNS Drugs 2005; 19: 1–93.
12. De Nayer, DeHert M., Scheem A., Van Gaal L., Peuskins J., Conference Report: Belgain Consensus on Metabolic Problems Associated with Second Generation Antipsychotics. Int J. Psychiatry Clin Pract. 2005; 9: 130–137.
13. Rothbard A., Murrin M.R., Jordon N., Kuno E., McFarland B., Stroup T.S., Morrissey J.P., Stiles P., Boothroyd R., Merwin E., Shern D. Effects of Antipsychotic Medication on Psychiatric Service Utilization and Cost. J Ment Health Policy Econ 2005; 8:83–93.
14. Awad A.G. Voruganti L.N.P., Heselgrave R.J., Preliminary Validation of a Conceptual Model to Assess Quality of Life in Schizophrenia. Quality of Life Res, 1997; 6: 21–26.
15. Lambert M., Naber D. Current Issues in Schizophrenia: Overview of Patient Acceptability, Functioning Capacity and Quality of Life. CNS Drugs 2004; 18: 5–17.
16. Norman R.M., Malla A.K., McLean T., The Relationship of Symptoms and Level of Functioning in Schizophrenia to General Wellbeing and the Quality of Life Scale. Acta Psychiatr Scand 2000, 102: 303–309.
17. Barowne S, Clarke M, Gervin M., Waddington J.L., Larkin C., O'Callaghan B., Determinants of Quality of Life at First Presentation with Schizophrenia. Br. J. Psychiatry 2000; 176: 173–176.
18. Karow A., Naber D., Subjective Wellbeing and Quality of Life under Atypical Antipsychotic Treatment. Psychopharmacology 2002; 162: 3–10.
19. Ritsner M., Kurs R., Impact of Antipsychotic agents and their Side Effects on the Quality of Life in Schizophrenia. Expert Rev. Pharmacoeconomics Outcomes. Res. 2002; 2: 89–98.
20. Awad A.G., Antipsychotic Medications in Schizophrenia: How Satisfied are our Patients? In Hellewell J. Editor: Clear Perspectives 1999, 2: 1–6. London, Shire Hall International.
21. Voruganti L.N.P., Heselgrave R.J., Awad A.G., Seeman M., Quality of Life Measurement in Schizophrenia: Reconciling the question for subjectivity with the question of Reliability.
22. Kilian R., Angermeyer M.C., The Effect of Antipsychotic Treatment on Quality of Life of Schizophrenic patients under naturalistic Treatment conditions: An application of random effect

regression models and propensity scores in an observational prospective trail. Quality of Life Res 2005; 14: 1275–1289.

23. Awad A.G., Lapierre Y.D., Augus C., Rylander A., Quality of Life and Response of Negative Symptoms in Schizophrenia to Haloperidol and the Atypical Antipsychotic Remoxipride J. Psychiatry Neurosci. 1997; 22: 244–248.
24. Ritsner M, Kurs R., Gibel A., Hirshchmann S., Shinkarenko E., Ratner Y., Predictors of Quality of Life in Major Psychosis: A Naturalistic Follow up Study. J. Clin Psychiatry 2003; 64: 308–315.
25. Ritsner M., Ben-Avi I., Ponizovsky A., Timinsky I Bistrov E., Modai I., Quality of Life and Coping with Schizophrenic Symptoms. Quality Life Res. 2003; 12: 1–9.
26. Awad A.G., Voruganti L.N.P., Impact of Atypical Antipsychotics on Quality of Life in Patients with Schizophrenia. CNS Drugs 2004; 18: 877–893.
27. Corrigan P.W., Reinke R., Landsberger S.A., Charate A., Toombs G.A. The Effect of Atypical Antipsychotic Medications on Psychosocial Outcome Schizophr Res 2003; 63: 91–101.
28. Awad A.G. Voruganti L.N.P., Heselgrave R.J., Measuring Quality of Life in Patients with Schizophrenia. Pharmacoeconomics 1997; 11: 32–47.
29. Jeste D.V., Klausner M, Brecher M., Clyde C., Jone R. Clinical Evaluation of Risperidone in the Treatment of Schizophrenia: A 10 week Open Label Muticentre Trial. Psychopharmacology 1997; 131: 239–247.
30. Barcia D., Ayuso J.L, herraiz M.L. Calidad de vida en Pacientes Esquizofrenicos Tratados con Risperidona. An Psiquiatr 1996; 12: 403–412.
31. Voruganti L.N.P., Cortese L., Oyewumi L.K., Cernovsky Z., Zirul S., Awad A.G. Comparative Evaluation of Conventional and New Antipsychotic Drugs with reference to their subjective Tolerability, Side Effect, Profile and Impact on Quality of Life. Schizophr. Res. 2000; 43: 135–145.
32. Bobes J., Gutierrez M., Gibert J. et al. Quality of Life in Schizophrenia: Long Term Follow-up in 362 Chronic Spanish Schizophrenic Outpatients undergoing Risperidone Maintenance Treatment. Eur Psychiatry 1998; 13: 158–163.
33. Franz M., Lis S., Pluddemann K., et al. Conventional vs. Atypical Neuroleptic: Subjective Quality of Life in Schizophrenia Patients. Br J Psychiatry 1997; 170: 422–425.
34. Hamilton S.H., Revicki D.A., Edgell E.T., et al. Clinical and Economic Outcomes of Olanzapine compared with Haloperidol for Schizophrenia: Results from a Randomized Clinical Trial. Pharmacoeconomics 1999; 15: 469–480.
35. Revicki D.A., Ginduso L.A., Hamilton S.H., et al. Olanzapine vs. Haloperidol in the Treatment of Schizophrenia and Other Psychotic Disorders: Quality of Life and Clinical Outcomes of a Randomized Clinical Trial. Qual Life Res 1999; 8: 417–426.
36. Tran P.V., Hamilton S.H., Kunz A.J., et al. Double-Blind Comparison of Olanzapine vs. Risperidone in the Treatment of Schizophrenia and Other Psychotic Disorders. J Clin Psychopharmacology 1997; 17: 407–418.
37. Ho B.C., Miller D., Nopoulos P., et al. A Comparative Effectiveness Study of Risperidone and Olanzapine in the Treatment of Schizophrenia. J Clin Psychiatry 1999; 60: 658–663.
38. Naber D., Moritz S., Lambert M., et al. Improvement of Schizophrenic Patients' Subjective well-being under Atypical Antipsychotic Drugs. Schizophr Res 2001; 50: 79–88.
39. Tempier R., Pawliuk N. Influence of Novel and Conventional Antipsychotic Medication on Subjective Quality of Life. J Psychiatry Neuroscience 2001; 26: 131–136.
40. Strakowksi S., Johnson J., DeBello M., Hamer R., Green A., Tohen M., Lieberman J., Glick I., Patel J., Quality of Life during Treatment with Haloperidol or Olanzapine in the year following a first psychotic Episode. Schizophr Res 2005; 78: 161–169.
41. Awad A.G. First Episode Psychosis: Olanzapine and Haloperidol provide similar improvement in Quality of Life and Social Functioning; Commentary Evidence-Based Mental Health 2006; 9:47.
42. Meltzer H.Y., Burnett S., Bastani B., et al. Effects of Six months of -Clozapine Treatment on the Quality of Life of Chronic Schizophrenic Patients. Hosp Community Psychiatry 1990; 41: 892–897.
43. Meltzer H.Y., Okayli G., Reduction of Suicidality during Clozapine Treatment of Neuroleptic-resistant Schizophrenia: Impact on Risk Benefit Assessment. Am J Psychiatry 1995; 152: 183–190.

44. Naber D. A self-rating to Measure Subjective Effects of Neuroleptic Drugs, Relationships to Objective Psychopathology, Quality of Life, Compliance and other Clinical Variables. Int Clin Psychopharmacology 1995; 10 Suppl. 3: 133–138.
45. Rosenheck R., Cramer J., Xu W., et al. A Comparison of Clozapine and Haloperidol in Hospitalized Patients with Refractory Schizophrenia. N Engl J Med 1997; 337: 809–815.
46. Bellack A.S., Schooler N.R., Marder S.R., Kane J.M., Brown C.H., Young Y. Do Clozapine and Risperidone Affect Social Competence and Problem Solving? Am J Psychiatry 2004; 16: 364–367.
47. Essock S.M., Hargreaves W.A., Covell N.H. et al. Clozapine's Effectiveness for Patients in State Hospitals: Results from a Randomized Trial. Psychopharmacology Bull 1996; 32: 683–697.
48. Hellewell J.S. Kalai A.H., Langham S.J. et al. Patient Satisfaction and Acceptability of Long-Term Treatment with Quetiapine. Int J Psychiatry Clin Pract 1999; 3: 105–113.
49. Addington D.E., Jons B., Bloom D. Reduction of Hospital Days in Chronic Schizophrenic Patients treated with Risperidone: A Retrospective Study. Clin Ther 1993; 15: 917–926.
50. Albright P.S., Livingston S., Keegan D.L. Reduction of Health Care Resource Utilization and Cost following the use of Risperidone for patient with Schizophrenia previously treated with Standard Antipsychotic Therapy: A Retrospective Analysis using the Saskatchewan Health Linkage database. Clin Drug Invest 1996; 11: 289–299.
51. Reid W.H., Mason M., Toprac M. Saving in hospital bed days related to treatment with Clozapine. Hosp. Community Psychiatry 1994; 45: 261–263.
52. Dickson R.A. Hospital Days in Clozapine Treated Patients. Can J. Psychiatry 1998; 43: 945–948.
53. Awad A.G., Voruganti L.N.P. Cost-Utility Analysis in Schizophrenia. J. Clin Psychiatry 1999; 60 suppl. 3: 22–26.
54. Voruganti L.N.P., Awad A.G., Oyewumi L.K. Assessing Health Utilities in Schizophrenia: A feasibility Study. Pharmacoeconomics 2000; 17: 273–286.
55. Dernovsek M.Z., Prevolnik-Rupel V., Tavcar R. Cost Utility Analysis Research and Practical Applications (Chapter ____).
56. Hargreaves W.A., Shumway M. Pharmacoeconomics of Antipsychotic Drug Therapy. J. Clin Psychiatry 1996; 57: suppl. 9: 66–76.
57. Revicki D. Pharmacoeconomic Studies of Atypical Antipsychotic Drugs for the Treatment of Schizophrenia. Schizophr. Res 1999; 35: 101–109.
58. Zito J.M. Pharmacoeconomics of the New Antipsychotics for the Treatment of Schizophrenia. Psychiatric Clin. North A.M. 1998; 21: 181–202.
59. Green M.F., What are the Functional Consequences of Neurocognitive Deficits in Schizophrenia? Am J Psychiatry 1996; 153: 321–330.
60. Keefe R., Poe M., Waker T., Kang J., Harvey P.H. The Schizophrenia Cognition Rating Scale: An interview based Assessment and its Relationship to Cognition, real-work functioning and functional capacity. Am J Psychiatry 2006; 163: 426–432.
61. Velligan D.I., Mahurin R.K., Diamond P.L., Hazelton B.C., Eckert S.L., Miller A.L. The Functional Significance of Symptomatology and Cognitive Function in Schizophrenia. Schizophr Res 1997; 25: 21–31.
62. Heselgrave R.J., Awad A.G. Voruganti L.N.P., The Influence of Neurocognitive deficits and symptoms on Quality of Life in Schizophrenia. J. Psychiatry Neuroscience 1997; 22: 235–243.
63. Moritz s., Ferahli S., Naber D. Memory and Attention Performance in Psychiatric patients: Lack of correspondence between Clinician-rated and patient-rated functioning with neuropsychological test results. J. Int Neuropsychol Soc. 2004; 10: 623–633.
64. Bowie C., Reichenberg A., Patterson T., Heaton R., Harvey P. Determinants of Real-World Functional Performance in Schizophrenia Subjects: Correlations with Cognition, Functional Capacity and Symptoms. Am J Psychiatry 2006; 163: 418–425.
65. Hooker C., Park S. Emotion processing and its relationship to Social Functioning in Schizophrenic Patients. Psychiatry Res 2002; 112: 41–50.
66. Wise R.A. Addictive Drugs and Brain Stimulation Reward. Ann Rev Neurosci. 1996; 19: 319–340.
67. Fibiger H.C. Neurobiology of Depression – focus on Dopamine. Adv Biochem Psychopharmacology 1995; 49: 1–7.

68. Voruganti L.N.P., Slomka P., Zabel P., Costa G., So A., Mattar A., Awad A.G. Subjective Effects of AMPT induced Dopamine Depletion in Schizophrenia. The Correlation between D2 binding ration and Dysphoric Responses. Neuropsychopharmacology 2001; 25: 642–650.
69. de Hann L, Lavalaye J., Linszen D., Dingemans P.M.A.J., Booij J. Subjective Experiences and Striatal Dopamine D2 Receptor Occupancy in Patients with Schizophrenia stabilized by Olanzapine or Risperidol. Am J Psychiatry 2000; 157: 1019–1020.
70. Voruganti L.N.P., Awad A.G. Subjective and Behavioural Consequences of Striatal Dopamine Depletion in Schizophrenia-findings from an in VIVO SPECTS study. Schiz Res (in press)
71. Voruganti L.N.P., Awad A.G. Role of Dopamine in Pleasure, Reward and Subjective responses as Important aspects of Quality of Life. (Chapter 2)
72. Gill T.M., Feinstein A.R. A Critical Appraisal of the Quality of Quality of Life Measurements. JAMA 1994; 272: 619–626.
73. Awad A.G. Voruganti L.N.P. Intervention Research in Psychosis: Issues related to the Assessment of Quality of Life. Schizophr Bull 2000; 26: 557–564.
74. Awad A.G., Voruganti L.N.P. The Subjective/objective dichotomy in Schizophrenia: Relevance to nosology, research and management. In: Gaebel W., editor. Zukunftsperspktiven in Psychiatries Und Psychotherapie. Darmstadt: Steinkopff Verlag, 2002: 21–27.
75. Skantze K., Malm U., Dencker S.J., May P.R.A., Corrigan P. Comparison of Quality of life with standard of living in Schizophrenic Out-patients. Br J. Psychiatry 1992; 161: 797–801.
76. Skantze K. Subjective Quality of Life and Standard of living: A 10-year Follow up of Out-patients with Schizophrenia. Acta Psychiatr Scand. 1998; 98: 390–399.

CHAPTER 17

QUALITY OF LIFE OUTCOMES OF ECT

PETER B. ROSENQUIST AND W. VAUGHN MCCALL
Department of Psychiatry and Behavioral Medicine, Wake Forest University School of Medicine, Medical Center Blvd., USA

Abstract: Electroconvulsive therapy (ECT) continues to stand the test of time. There is a growing demand for patient reported outcome data including measures of HRQL, patient satisfaction and utility. ECT is efficacious for a number of psychiatric conditions but it is often reserved for the most ill patients because compared to drugs or psychotherapy it has a higher cost per unit of treatment, requires anesthesia, and because of a continuing stigma. As well, ECT has a particular set of side effects including the potential for memory impairment, and cardiac death. Mental illness produces profound deficits in HRQL and loss of HRQL is linear to severity of illness, and may therefore be part of decision to refer for ECT. At baseline studies have shown that ECT patients have reduced HRQL and functioning. ECT treatment improves HRQL and functioning in a lasting way for patients with major depression and schizophrenia. HRQL improvements are related to ECT's effect on mood whereas functional improvements are more closely related to cognition

INTRODUCTION: SOMETHING TO PROVE

In his book *Electroshock: Restoring the Mind*, Max Fink begins a chapter devoted to the Patient's Experience of ECT as follows: "The public's images of electroshock too often reflect practices that were discarded more than 40 years ago. The picture of a pleading patient being dragged to a treatment room, where he is forcibly administered electric currents as his jaw clenches, his back arches, his body shakes, all while he is held down by burly attendants may be dramatic but it is wholly false. Patients are not coerced into treatment. They may be anxious, but they come willingly to the treatment room. They have been told why the treatment is recommended and have given their consent."[1]

Hollywood's contribution to the public perception of electroconvulsive therapy through such movies as *One Flew Over the Cuckoo's Nest* aside, ECT has for many years stood alone as the sole procedure approved for the treatment of mental disorders, making it a likely target for the antipsychiatry lobby. Such organizations as the "Committee for Truth in Psychiatry" (backed by the Church of Scientology) promulgate anecdotal reports of those allegedly harmed by the procedure. Few

M.S. Ritsner and A.G. Awad (eds.), Quality of Life Impairment in Schizophrenia, Mood and Anxiety Disorders, 321–331.

come forward to counter the tide of misinformation with the exception of a handful of such famous personages as talk show host Dick Cavett, who wrote of his series of treatments in 1980, "In my case, ECT was miraculous".[2] The attitude of patients and their families toward the treatment and perceived stigma by society influence the choice for ECT, and color as well the perception of its benefits.[3] ECT is seen often as the treatment of last resort, and similar in that way to say, chemotherapy for the treatment of cancer there is a real fear that the cure may be worse than the disease. This has implications in terms of the Distress/Protection/ Vulnerability (DPV) model of HRQL elaborated in this volume. To the distress-producing factors associated with a severe mental disorder at the point of considering a course of ECT, one must add the distress of perceived loss of autonomy and the question of what must be faced if the treatment does not succeed. Therefore, it is imperative that the evidence base solidly supports the continued use of ECT, and the demonstration of benefits to HRQL are central to the argument that while this treatment may in fact be more costly, monetarily and in terms of bearing the burden of side effects and an enduring stigma against it's use, it is *worth it*.

When England's National Institute for Clinical Excellence (NICE) issued its recent guidance concerning ECT, the Health Technology Assessment underpinning the guidelines summarized the evidence base concerning health related quality of life (HRQL) and ECT: "There are no trials exploring the impact of ECT on quality of life. This had important implications for the cost effectiveness modelling within the NICE review."[4] One may surmise that the perceived dearth of evidence regarding HRQL at the time the NICE review was conducted in December of 2001 contributed to the limitation of ECT within the British National Health Service to cases of proven treatment resistance and where the depressive or manic episodes are characterized by catatonia or are potentially life threatening.[5] The search strategy employed in the NICE review also exposed a general lack of studies addressing the comparative effectiveness of treatment for depressed patients, particularly those that have attempted to estimate the quality of life of patients suffering from depression. However, there are a growing number of studies measuring HRQL for patients receiving ECT and this chapter is intended to summarize and explore what these studies say about the HRQL of patients selected to receive this treatment of apparent last resort, and how ECT affects HRQL outcomes. But first we turn to the literature on the efficacy and side effects of ECT for various conditions, as this has framed to some extent the questions concerning HRQL.

ECT EFFICACY AND TOLERABILITY

Typical of the literature examining the efficacy of psychotherapy or pharmacotherapy, the majority of ECT clinical trials have employed clinician-rated symptom and global assessment measures, though patient self-report instruments frequently have been included as well. Studies using these outcomes support the use of ECT in major depression, bipolar disorder, and schizophrenia, but the strongest evidence exists for depression for which there are both sham-controlled studies and

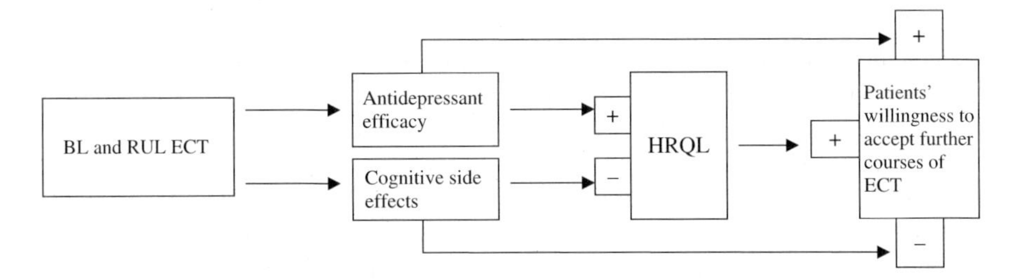

Figure 1. Relationship between mood, memory, HRQL, and patient preferences

comparisons with pharmacotherapy alone.[5–7] A number of characteristics of the ECT method influence the response rates and also the side effect burden. These include the electrode placement and ECT stimulus type and intensity. The majority of ECT are administered with right unilateral (RUL) and bilateral (BL) placements and while both placements have been shown to be beneficial, RUL ECT is more sensitive to underdosing.[8–11]

Certain patient characteristics appear to predict a better response to ECT, as compared to pharmacotherapy alone. These include psychotic depressive states, and those accompanied by catatonic features and acute suicidality.[6] While chronicity and treatment resistance, defined as a failure to respond to adequate antidepressant medication trials, also diminishes the likelihood of response to ECT, a significant proportion of patients will respond.[12] The durability of the improvements brought on by a course of ECT is also of concern to the field, with the majority of relapses among ECT responders occurring within the first six months. In one longitudinal study, the rates of relapse at six month follow up were 68% for those who had failed an adequate trial of antidepressant during the episode of depression vs. 36% of patients who had not received an adequate trial prior to receiving ECT.[11]

In addition to the potential of autobiographical memory deficit, possible side effects of the ECT treatment include confusion, cardiovascular problems and an increased risk of falls.[13] ECT requires undergoing a general anesthetic, with its attendant risks. Returning to the DPV model of HRQL, one might hypothesize for patients receiving ECT improvements in HRQL through a reduction of distress associated with symptoms and functional impairments, and an increased distress or vulnerability brought about by possible cognitive and other side effects (Figure 1). For treatment resistant patients the likelihood of relapse is higher and hopeless and helpless cognitions more likely with the passage of time, potentially dampening HRQL directly, and also willingness to undergo further courses of ECT.[14]

BASELINE HRQL DEFICITS IN ECT PATIENTS

Depression leads to HRQL deficits across many domains and treatment settings that equal or exceed that from other chronic diseases.[15–17] For those whose depression is very severe, there are likely to be significant role limitations, and chronic functional impairment limiting social and occupational attainment.[16] Existing research has documented that HRQL is influenced not only by the severity of the depression but also advancing age of the patient.[18] The severity of initial symptoms may also play a role in treatment outcomes for those who receive treatment. Shmuely and colleagues documented improvements in HRQL among geriatric patients three months after discharge from an inpatient psychiatric unit.[19] Over a third of this population received ECT. However, they noted that the severity of depression at the time of admission and physical health were predictive of failure to improve in role and social functioning as well as body pain. In the clinical setting, this has prognostic implications, and highlights the importance of controlling for a myriad of factors in studies of ECT outcome.

ECT AND HRQL

There is growing evidence that ECT contributes to improved function and HRQL in patients with major depression and to a lesser extent, schizophrenia. Table 1 lists eight publications that have incorporated measures of HRQL. Given the severity of illness for which ECT is indicated none of these studies were designed with a sham control and three compared ECT with medication. Five studies examined outcomes beyond one month post-ECT course. Not included in this list are studies referencing ECT related changes in the Global Assessment of Functioning Scale (GAF). As a measure of HRQL, the GAF is limited because it is a clinician-rated scale but it does specifically reference patient functioning in addition to symptoms.[20] In personality disordered patients, the GAF correlated with SF-36 vitality and social functioning scores[21] and in patients with schizophrenia was found to be significantly correlated with several measures of quality of life, including the SF-36[22] and QLS[23]. Therefore we will briefly discuss studies measuring GAF outcomes of ECT treatment. Thienhaus et al. reported on a case series of 6 older patients (mean age 71 ± 5 years) who received a course of ECT followed by continuation ECT and found that GAF scores increased from 53 ± 11 to 68 ± 11. Statistical testing was not reported for the change in GAF scores. Brodaty et al surveyed 81 inpatients before and within two weeks of completion of ECT index course and found that global assessment of functioning scores improved from pre-ECT to post-ECT (pre-ECT: mean = 36.2, SD = 15.1; post-ECT: mean = 64.9, SD = 16.5; $t = -11.3$, $p < 0.001$).[24] In long term follow-up (mean = 109.9 weeks, SD = 31.5) gains in the GAF were sustained. Dew et al documented a similar magnitude of improvement from admission to discharge GAF in 78 patients following an acute course of ECT.[25] Suzuki et al demonstrated a dramatic improvement in GAF scores for patients with schizophrenia treated with ECT. The total GAF score increased significantly (from 10.7 ± 8.4 just before ECT to 61.3 ± 17.0 1 week after ECT; $T = 0.00$, $Z = 2.934$, $P = 0.003$). While these studies provide evidence for ECT's beneficial effects on global functioning, patient-reported measures of HRQL will undoubtedly carry more weight in future analyses of utility and cost-effectiveness.

Table 1. Studies of ECT and HRQL

Author/Year	Population/ Methods	HRQL Measure *[Observation period]*	Results
Chanpattana 2001	Refractory Schizophrenia Maintenance ECT vs. flupenthixol vs. combination. (N = 46)	QLS, SOFAS *[One year follow up of ECT responders]*	93% relatpse rate for maintenance ECT alone, flupenthixol alone 40% relapse rate for combination
Casey 1996	40 depressed patients with and without personality disorder	SFS *[Baseline/ @6 week intervals for 6 months]*	80% improvement in SFS scores from baseline following ECT which persisted 6 months post-ECT
Fisher 2004	Depressed inpatients Comparison of ECT (N = 129) vs No ECT (N = 843)	HoNOS, SF-36 *[Admission/Discharge]*	Equal improvement on the HoNOS and SF-36 scales. ECT patients had lower baseline and SF-36 vitality subscale lowered baseline and
Shmuely 2001	Elderly depressed inpatients (N= 100)	MOS-6 *[Admission/Discharge/3 months post discharge]*	Treatment with ECT not predictive of outcome in stepwise regression of MOS items
McCall et al 2001	Depressed inpatients Comparison of ECT (N=31) vs No ECT (N=59)	BASIS-32 (DLRF and RSO), IADL Scale *[Baseline, post-ECT, 12 month follow up]*	Lower baseline and greater improvement @ 12 months on DLRF and IADLS for ECT treated group.
McCall 2004	Depressed inpatients N=77 (40 RUL 37 BL)	BASIS-32 and IADL, Cognitive measures Hamilton Depression, Beck Depression *[2 weeks and 4 weeks]*	Improvements in IADL scale and the DLRF subscale of BASIS-32
Rosenquist 2006	Depressed inpatients N=77	Patient preference for future ECT; BASIS-32 and IADL, Cognitive measures, HRDS, BDI *[Baseline, 4 weeks]*	HRDS baseline and change scores, IADL baseline predict patient choice for future ECT ($R2 = 0.34$, $P < 0.0001$). Quality-of-life variables and measures of change in cognition not significant in the model.
McCall 2006	Depressed inpatients N = 283	SF-36 *[Baseline/ Immediate Post-ECT / 6 months]*	Low baseline SF-36 87% improved Post ECT 78% remained improved at 6 months

BASIS-32:Behavior and Symptom Identification Scale[35]
DLRF Daily Role Functioning RSO Relationship to Self and Others
IADL: Instrumental Activities of Daily Scale (Personal Self-Maintenance Scale)[36]
SF-36: Medical Outcome Study 36 item short form survey.[33]
MOS-6: Medical Outcome Study 6 item general health survey[33]
HoNOS: Health of the Nation Outcome Scales[32]
QLS: Quality of Life Scale[23]
SOFAS Social and Occupational Functioning Assessment Scale[20]
HRSD: Hamilton Rating Scale for Depression[28]
BDI: Beck Depression Inventory[38]
SFS: Social Functioning Schedule[27]

There is a need to further characterize global improvements from ECT in terms of specific domains of HRQL, and to consider the contribution of comorbid conditions. Casey et al.[26] assessed the change in social functioning in 40 depressed patients with and without personality disorder following a course of ECT using the Social Functioning Schedule (SFS)[27], which includes measures of employment, household chores, finance, self care, relationships, and recreation. In this study, the SFS scores were assessed prior to onset of ECT and every 6 weeks up to 6 months. Statistical testing on the difference between pre and post ECT SFS scores was not presented, but the numerical change in scores was impressive, showing an approximately 80% improvement in SFS scores from baseline following ECT which persisted 6 months post-ECT. Patients with comorbid personality disorder receiving ECT had a less dramatic acute improvement in SFS scores, but appeared to "catch up" with non-personality disordered patients over time and there were no significant differences in readmission rate for the two groups. Hamilton Depression Scale (HRSD)[28] scores were highly correlated with SFS scores during the follow up period (0.69 at discharge and 0.84 at 6 months). This high degree of correlation suggests that for social functioning at least, short-term response rates using HRSD scores may be a valid HRQL proxy for purely depressed patients and less so for those with personality disorder. However, functional recovery predictably may lag behind symptomatic recovery for severely ill patients.[29–31]

Several studies have focused on the relative benefits in terms of HRQL for ECT for patients with major depression versus those treated with medication alone. Fisher et al compared a series of depressed hospitalized patients receiving ECT with those who did not using Health of the Nation Outcome Scales (HoNOS)[32] and the SF-36[33] assesed at admission and discharge. The HoNOs is a clinician rated 12 item scale that includes ratings of problems with psychiatric symptoms as well as cognition and problems with making supportive social relationships; problems with activities of daily living; accommodation difficulties; and problems with occupation and activities. Both groups improved equally on the HoNOS and SF-36 scales. The only difference between groups on HRQL measures was found on the SF-36 vitality subscale which was lower at admission ($F = 5.80$, $df = 1970$, $p < .05$) and showed a more rubust improvement ($F = 14.18$, $df = 1970$, $p < 0.001$) for the group receiving ECT.[34] Thirty seven percent of the Shmuely et al elderly sample received ECT during hospitalization and though the baseline and post treatment improvements in HRQL scores were not reported separately from those receiving medications, ECT did not emerge as a significant predictor of outcome in a stepwise logistic regression.[19]

Further comparative data are provided by McCall et al who contrasted 31 depressed inpatients treated with ECT with 59 depressed inpatients treated with medications.[31] HRQL was assessed with the daily-living and role functioning (DLRF) and relationship to self and others (RSO) subscales of the BASIS-32[35], while functional capacity was assessed with an instrumental activities of daily living (IADL) scale from the Personal Self-Maintenance Scale.[36] Adequacy of medication treatment was determined by comparison with the Prudic scale.[37] Patients receiving

ECT were older than the medication-treated patients, but the 2 groups were not different in depression severity as measured by Beck Depression Inventory (BDI)[38] and HRSD. ECT patients reported more impairment in terms of activities on the IADL scale and more problems on the DLRF scale. At 12 month follow up, the degree of improvement in depression scores, DLRF improvement, and IADL improvement was significantly greater in the ECT group than the medication group.[31] The lower baseline functioning for ECT treated patients in this and other studies[18] provides evidence that the guidance regarding the use of ECT for more severe depression put forth in the NICE report is already being followed in practice. Further study is needed to elucidate whether the HRQL benefits of ECT in comparison with medication treatment apply equally to patients who are less ill at baseline.

ECT SIDE EFFECTS AND HRQL

Can ECT treatment actually be detrimental to overall quality of life? Side effect burden, such as ECT's effect upon cognition could possibly contribute to HRQL thereby diminishing the net benefit of the treatment. Complicating this equation are the cognitive deficits associated with depression which can independently lead to decrements in HRQL and function.[39,40] In order to determine the potential effects upon HRQL of cognitive changes associated with ECT treatment, McCall et al assessed depressed ECT patients with the BASIS-32 and IADL scales, before ECT and at 2 weeks and 4 weeks after ECT along with a battery of memory and mood instruments. There were no significant differences between the groups receiving right unilateral or bilateral electrode placement in any of these variables, so the results were pooled for analysis. At baseline, prior to ECT, participants were severely depressed, had minimal cognitive dysfunction and had significant deficits in quality of life and functional status. Significant improvement in mood variables as well as the IADL scale and the DLRF subscale of the BASIS-32 were seen at 2 and 4 weeks, with $\geq$80% of patients reporting improvement on each HRQL and function measure.[41] Thus, in patients with major depression, ECT is associated with early improvement in functioning and in HRQL. The sample showed improvement in measures of cognition as well, except for the autobiographical memory test, a test revealing only memory loss, not improvement. Furthermore, change in IADL at 2 weeks was most closely related to change in cognition, not mood. This is consistent with the finding that baseline IADL function in patients treated with ECT is related to the MMSE score but not to mood.[42] Therefore, improvement in HRQL was related to mood, whereas improvement in IADLs were related to improvement in global cognition.

Do patients attribute these improvements to their ECT treatment? Patients in this same study were asked if they would be likely to repeat a course of ECT based upon their experience and a majority were so disposed.[43] Neither baseline nor treatment-associated changes in cognitive or HRQL variables predicted a positive attitude towards ECT. Rather, the significant predictors were lower baseline HRSD and IADL scores, and higher percentage improvement in HRSD.[14]

Brodaty and colleagues found that the prevalence and total burden of 20 common ECT related side effects[44] (muscle aches, constipation, headache, confusion, etc.) in 81 elderly patients with major depression were unchanged after ECT even as HRSD and GAF scores improved.[45] Side effect burden was noted to be related to depression level before ($r = 0.38$, $P < 0.005$) and after ($r = 0.57$, $P < 0.001$) ECT. Improvement in depression correlated with reduction in side effect burden ($r = 0.46$, $P = 0.001$). However, even after controlling for change in HRSD scores, there was a significant decline in side effect burden following ECT. Furthermore, patients' scores on neuropsychological measures did not appear to change after ECT or between pre-ECT and follow-up. Taken together, the limited evidence would seem to suggest a more complex relationship between cognition, mood and ECT and the related HRQL outcomes.[46] Simultaneous assessment of non-cognitive side effects and HRQL and by comparison with pharmacotherapy may shed further light on how patients treated with ECT perceive their improvement.

DURABILITY OF HRQL OUTCOMES

The time course and durability of ECT related improvement in HRQL are not well characterized and an important focus for future research. McCall et al used the SF-36 as a generic measure of HRQL in a community sample.[47] In this study 283 patients were tested before ECT, immediately after, and again at 6 months. Baseline SF-36 scores were low, consistent with inpatient samples.[48] Eighty-seven percent of patients had improvement in the SF-36 score immediately after ECT, and 78% still showed positive change at 6 months. This is surprising given rates of remission of only 45.9%, based upon minimum 60% reduction in HRSD and post ECT score of less than 10. Even more so was the finding that patients who did not respond to ECT or who subsequently relapsed overall still showed improved SF-36 scores compared to baseline. Those who sustained remission or were late remitters at 24 weeks post treatment showed a greater differentiation in SF-36 change scores than those who relapsed or never remitted, with that greatest separation between groups seen for vitality, social functioning, role-emotional and mental health subscales. This pattern of results is intriguing and coupled with the observation that changes from baseline global HRQL and HRSD scores shared only 21.2% of common variance at the post ECT assessment, there is reason to hope that in time HRQL measures will tell us something new about remission and recovery from major depression.

The problem of how to manage patients with chronic, recurrent and treatment resistant psychotic and affective states remains. The long term effects of HRQL in patients receiving maintenance ECT has not been studied in mood disorders. However, Chanpattana and Kramer examined HRQL outcomes for 46 patients with refractory schizophrenia who had responded to an index course of ECT as evidenced by marked reductions in the BPRS scores, and substantial increases in QLS[23], SOFAS[49], GAF, and MMSE[50] scores. The patients were then followed prospectively over one year, comparing the efficacy of three continuation treatments after acute ECT: flupenthixol alone, continuation ECT alone, and continuation ECT and flupenthixol combined. Whereas 14 of 15 (93%) patients suffered relapse in both the group treated with flupenthixol alone and the group treated by continuation ECT alone, only 6 of 15 (40%) patients suffered relapse in the group treated with continuation ECT and flupenthixol in combination. Though relatively small and lacking blinding and controls, this study suggests that for ECT responsive patients with schizophrenia, HRQL improvements can best be sustained by continuation ECT combined with neuroleptics.

DIRECTIONS FOR FUTURE RESEARCH

More research is needed to evaluate the short and longer term effects of ECT upon HRQL. Future studies of ECT should include established measures of efficacy along with patient-reported satisfaction and global and disease specific HRQL instruments to permit an increased understanding of the interplay between these variables in ECT patients. The mechanisms whereby ECT exerts its clinical effects have not been determined, but as this work progresses, it may shed light on the neurophysiology of HRQL outcomes. It will be helpful as well to include HRQL measures in future studies comparing ECT with antidepressant medication and emergent brain stimulation treatments such as vagus nerve stimulation and transcranial magnetic stimulation.

REFERENCES

1. Fink M. Electroshock: Restoring the Mind. New York: Oxford University Press; 1999
2. Pettinati HM, Tamburello TA, Ruetsch CR, et al. Patient attitudes toward electroconvulsive therapy. Psychopharmacol Bull 1994;30: 471–475
3. Dowman J, Patel A, Rajput K. Electroconvulsive therapy: attitudes and misconceptions. J ECT 2005;21: 84–87
4. Greenhalgh J, Knight C, Hind D, et al. Clinical and cost-effectiveness of electroconvulsive therapy for depressive illness, schizophrenia, catatonia and mania: systematic reviews and economic modelling studies. Health Technol Assess 2005;9: 1-iv
5. National Institute for Clinical Excellence. Technology Appraisal Guidance 59: Guidance on the use of electroconvulsive therapy. London: National Institute for Clinical Excellence; 2005:
6. Abrams R. Electroconvulsive Therapy. 3rd Edition ed. New York, NY: Oxford University Press; 1997
7. Pagnin D, de Q, V, Pini S, et al. Efficacy of ECT in depression: a meta-analytic review. J ECT 2004;20: 13–20
8. Chanpattana W, Chakrabhand ML, Sackeim HA, et al. Continuation ECT in treatment-resistant schizophrenia: a controlled study. J ECT 1999;15: 178–192
9. Sackeim HA, Decina P, Kanzler M, et al. Effects of electrode placement on the efficacy of titrated, low-dose ECT. Am J Psychiatry 1987;144: 1449–1455
10. Sackeim HA, Prudic J, Devanand DP, et al. Effects of stimulus intensity and electrode placement on the efficacy and cognitive effects of electroconvulsive therapy. N Engl J Med 1993;328: 839–846
11. Sackeim HA, Prudic J, Devanand DP, et al. A prospective, randomized, double-blind comparison of bilateral and right unilateral electroconvulsive therapy at different stimulus intensities. Arch Gen Psychiatry 2000;57: 425–434

12. Dombrovski AY, Mulsant BH, Haskett RF, et al. Predictors of remission after electroconvulsive therapy in unipolar major depression. J Clin Psychiatry 2005;66: 1043–1049
13. Rosenquist PB, McCall WV. Electroconvulsive Therapy. In: Aminoff M, Darroff R, eds. Encyclopedia of the Neurological Sciences. San Diego: Academic Press; 2003:
14. Rosenquist PB, Dunn A, Rapp S, et al. What predicts patients' expressed likelihood of choosing electroconvulsive therapy as a future treatment option? J ECT 2006;22: 33–37
15. Wells KB, Stewart A, Hays RD, et al. The functioning and well-being of depressed patients. Results from the Medical Outcomes Study. JAMA 1989;262: 914–919
16. Goldney RD, Fisher LJ, Wilson DH, et al. Major depression and its associated morbidity and quality of life in a random, representative Australian community sample. Aust N Z J Psychiatry 2000;34: 1022–1029
17. Stewart AL, Sherbourne CD, Wells KB, et al. Do depressed patients in different treatment settings have different levels of well-being and functioning? J Consult Clin Psychol 1993;61: 849–857
18. McCall WV, Cohen W, Reboussin B, et al. Pretreatment differences in specific symptoms and quality of life among depressed inpatients who do and do not receive electroconvulsive therapy: a hypothesis regarding why the elderly are more likely to receive ECT. J ECT 1999;15: 193–201
19. Shmuely Y, Baumgarten M, Rovner B, et al. Predictors of improvement in health-related quality of life among elderly patients with depression. Int Psychogeriatr 2001;13: 63–73
20. Goldman HH, Skodol AE, Lave TR. Revising axis V for DSM-IV: a review of measures of social functioning. Am J Psychiatry 1992;149: 1148–1156
21. Narud K, Mykletun A, Dahl AA. Quality of life in patients with personality disorders seen at an ordinary psychiatric outpatient clinic. BMC Psychiatry 2005;5: 10
22. Reine G, Simeoni MC, Auquier P, et al. Assessing health-related quality of life in patients suffering from schizophrenia: a comparison of instruments. Eur Psychiatry 2005;20: 510–519
23. Heinrichs DW, Hanlon TE, Carpenter WT, Jr. The Quality of Life Scale: an instrument for rating the schizophrenic deficit syndrome. Schizophr Bull 1984;10: 388–398
24. Brodaty H, Berle D, Hickie I, et al. Perceptions of outcome from electroconvulsive therapy by depressed patients and psychiatrists. Aust N Z J Psychiatry 2003;37: 196–199
25. Dew RE, Kimball JN, Rosenquist PB, et al. Seizure length and clinical outcomes in electroconvulsive therapy using methohexital or thiopental. 21 ed 2005:16–18
26. Casey P, Meagher D, Butler E. Personality, functioning, and recovery from major depression. J Nerv Ment Dis 1996;184: 240–245
27. Remington M, Tyrer P. The Social Functioning Schedule: A brief semi-structured interview. 40 ed 1979:151–157
28. Hamilton M. A rating scale for depression. J Neurol Neurosurg Psychiatry 1960;23: 56–62
29. Blazer DG. Severe episode of depression in late life: the long road to recovery. Am J Psychiatry 1996;153: 1620–1623
30. Mintz J, Mintz LI, Arruda MJ, et al. Treatments of depression and the functional capacity to work. Arch Gen Psychiatry 1992;49: 761–768
31. McCall WV, Reboussin BA, Cohen W, et al. Electroconvulsive therapy is associated with superior symptomatic and functional change in depressed patients after psychiatric hospitalization. J Affect Disord 2001;63: 17–25
32. Wing JK, Beevor AS, Curtis RH, et al. Health of the Nation Outcome Scales (HoNOS). Research and development. Br J Psychiatry 1998;172: 11–18
33. Ware JE, Jr., Sherbourne CD. The MOS 36-item short-form health survey (SF-36). I. Conceptual framework and item selection. Med Care 1992;30: 473–483
34. Fisher LJ, Goldney RD, Furze PF, et al. Electroconvulsive Therapy, Depression, and Cognitive Outcomes. Journal of ECT 2004;20: 174–178
35. Eisen SV, Dill DL, Grob MC. Reliability and validity of a brief patient-report instrument for psychiatric outcome evaluation. Hosp Community Psychiatry 1994;45: 242–247
36. Lawton MP, Brody EM. Assessment of older people: self-maintaining and instrumental activities of daily living. Gerontologist 1969;9: 179–186

37. Prudic J, Haskett RF, Mulsant B, et al. Resistance to antidepressant medications and short-term clinical response to ECT. Am J Psychiatry 1996;153: 985–992
38. Beck AT, Steer RA, Garbin MG. Psychometric Properties of the Beck Depression Inventory: Twenty-Five Years of Evaluation. Clinical Psychology Review 1988;8: 77–100
39. McCall V, Dunn A. Cognitive deficits are associated with functional impairment in severely depressed patients. Psychiatry Res 2003;121: 179–184
40. Alexopoulos GS, Vrontou C, Kakuma T, et al. Disability in geriatric depression. Am J Psychiatry 1996;153: 877–885
41. McCall WV, Dunn A, Rosenquist PB. Quality of life and function after ECT. British Journal of Psychiatry 2004; 405–409
42. McCall WV, Dunn AG. Cognitive deficits are associated with functional impairment in severely depressed patients. Psychiatry Res 2003;121: 179–184
43. Iodice AJ, Dunn AG, Rosenquist P, et al. Stability over time of patients' attitudes toward ECT. Psychiatry Res 2003;117: 89–91
44. Sackeim HA, Ross FR, Hopkins N, et al. Subjective Side Effects Acutely Following ECT: Associations with Treatment Modality and Clinical Response. Convuls Ther 1987;3: 100–110
45. Brodaty H, Berle D, Hickie I, et al. "Side effects" of ECT are mainly depressive phenomena and are independent of age. J Affect Disord 2001;66: 237–245
46. Rami-Gonzalez L, Bernardo M, Boget T, et al. Subtypes of memory dysfunction associated with ECT: characteristics and neurobiological bases. J ECT 2001;17: 129–135
47. McCall W, Prudic J, Olfson M, et al. Health-related quality of life following ECT in a large community sample. 2006:
48. Oslin DW, Streim J, Katz IR, et al. Change in disability follows inpatient treatment for late life depression. J Am Geriatr Soc 2000;48: 357–362
49. Goldman HH, Skodol AE, Lave TR. Revising axis V for DSM-IV: a review of measures of social functioning. Am J Psychiatry 1992;149: 1148–1156
50. Folstein MF, Folstein SE, McHugh PR. "Mini-mental state". A practical method for grading the cognitive state of patients for the clinician. J Psychiatr Res 1975;12: 189–198

CHAPTER 18

QUALITY OF LIFE IN MENTAL HEALTH SERVICES

SHERRILL EVANS

Centre for Carework Research, Department of Applied Social Sciences, University of Wales, Swansea

Abstract: This chapter describes the nature of current community mental health services and the concept of Quality of Life (QOL) as it relates to them. It also provides an understanding of the nature, scale and content of the current evidence base for QOL research in mental health service settings. A search strategy for isolating the most relevant sources and types of published material is included, and the best quality evidence relating to the QOL of people in receipt of community mental health care is reviewed. Finally, QOL measurement and outcomes for people with severe mental illnesses (SMI) in receipt of community based care are compared with those for people with more common mental disorders (CMD) such as anxiety and depression, and mentally healthy general population groups, to provide a broad understanding of the impact of illness on life-quality

Keywords: Quality of Life, Community Mental Health Services, Severe Mental Illness, Outcomes, Measurement

BACKGROUND

Mental health problems are widespread and common. They contribute 12% of the Global Burden of Disease, second only to infectious disorders (23%), and are a bigger burden than AIDS, TB and malaria combined (10%)[1]. Mental disorders are disabling and are the cause of huge economic and social costs to individuals and societies. They affect the employment and productivity of the person with the disorder, and also of the family/caregiver. Mental disorders lead to high health service utilization, and high rates of utilization of other formally delivered services including social services, housing, education, and in some cases the criminal justice system. Most middle and low-income countries devote less than 1% of their health expenditure to mental health. Consequently mental health policies, legislation, community care facilities, and treatments for people with mental illness are not given the priority they deserve. In some of the major high income western countries there has been a call for the scarce resources to be targeted at people with the most severe and complex mental health problems, and in many instances this aim

M.S. Ritsner and A.G. Awad (eds.), Quality of Life Impairment in Schizophrenia, Mood and Anxiety Disorders, 333–353.

has been incorporated into mental health policies and strategies[2,3]. Recent evidence suggests that these policies have had the desired effect of concentrating community service resources on the people with the most complex and severe disorders in some countries such as the UK[4], but not so convincingly elsewhere[5,6].

The growing recognition of mental ill health as a major social problem has encouraged policy makers, health professionals and other stakeholders to look for new solutions. One major example was the European Ministerial Conference on Mental Health, sponsored by the World Health Organisation[7]. In response to the WHO initiative, the European Commission issued a 'green paper' on mental health in Europe[8], and like policymakers elsewhere in the world[9–12] recognized that a major fundamental objective is to improve the quality of life of people with mental health problems, and that high quality community mental health services ought to be organized to focus on this outcome[13] (para 5.0). The EU green Paper also called for more research into the quality of life consequences of mental ill health in order to improve current practice[8] (para 6.3).

Quality of Life in Mental Health Services

The current literature about QOL is extensive and diverse, encompassing the varied conceptualisations of well-being, satisfaction and happiness and incorporating generic, social and economic indicators as well as health-related and disease-specific approaches to measurement. The terms health-status, functioning, health-related quality of life (HRQOL) and QOL are often used interchangeably, but each has subtle and important differences in respect of their dimensionality, perspective and scope[14].

- Health-status refers specifically to the state of physical and mental health and often incorporates the patient's perspective of these attributes.
- Functioning relates to one's capacity to perform everyday activities associated with daily living (e.g. cleaning and cooking) and independence (e.g. personal care) and to engage in social activities.
- HRQOL concentrates on the effects of illness on physical, psychological and social aspects of life.
- QOL extends its focus beyond the health domain and the effects of illness to incorporate a wide range of human experiences. A distinguishing characteristic of QOL measures is that they incorporate user values, judgements and individual preferences[15].

In health, the interest in QOL stemmed from a desire to evaluate the impact of clinical interventions and to assess the relative merits of different health systems using patient-centred indicators rather than more traditional outcome measures (e.g. morbidity, mortality or number of patients treated)[16]. QOL models and measures often focus on those aspects of life-quality that can be attributed directly to illness and treatments, and as such are best described as health-related quality of life[17]. While some HRQOL measures include global perceptions of functioning or well-being[18], they tend to be limited to aspects of physical, mental and social functioning affected by health-status (e.g. SF-36)[19]. Leidy et al[20] suggested that while these domains of functioning are essential they should be regarded as a minimum requirement of HRQOL assessment.

QOL Measurement in Mental Health Service Settings

In the community mental health services, HRQOL measures such as these are often considered not fit for purpose, because of their specific focus. More comprehensive, multi-dimensional indicators of QOL are required that recognise that health is not necessarily a dominant feature of life-quality[21], and which capture the multi-dimensionality of life and the interventions provided by community mental health services, such as those aimed at improving other aspects of life e.g. work, home and family life, social life etc. Using HRQOL measures to evaluate services that address the wider needs of patients with chronic illness may miss positive effects, because the instruments do not capture non-health dimensions[22].

For this reason, QOL in community mental health settings tends to focus on more socio-scientific approaches to measurement, defining QOL in more generic terms, regarding it as a multi-dimensional concept that incorporates all aspects of life, and referring to the sense of well-being and satisfaction experienced by people[21,23], under their current life conditions[24]. Concept mapping exercises have demonstrated a considerable level of agreement between researchers and research participants, about the domains of life-quality that should be included[21,25,26]. Nevertheless, patients appeared to focus more on standard of living and lifestyle, whereas psychiatrists concentrated on illness and the absence of handicaps and disabilities, and emphasised the importance of professional help and self-help[26].

Psycho-social indicators have been favoured over psychological indicators of life-quality that focus on an individual's psychological well-being and provide a summary of a person's sense of overall happiness and positive mental health[27], because the latter do not elicit a sufficient understanding of other life experiences and conditions, to be considered generic QOL measures. The favoured psychosocial methods emphasise an individual's perceived or subjective QOL and focus on an integrated model of social and cognitive or behavioural factors. Andrews & Withey[21] identified two components of subjective well-being: judgements about life satisfaction, which have been described as a set of evaluations made by an individual for each major life domain[27], and indicators of positive and negative affect[28]. This conceptualisation of QOL recognises the multi-dimensionality of the concept, and takes for granted the basic essentials of life[23] while accommodating the concerns of community, family, social and economic life[22]. Typically, objective and subjective QOL are assessed in a range of life-domains to provide a comprehensive understanding of the component parts of an individual's life (e.g. work, financial and living situation, family and social life and health), in addition to a subjective measure of life overall.

This conceptualisation of QOL has been adopted widely in surveys of the general population[29–31] and in health services research[22,24,32,33], following the seminal works of Hadley Cantril[34], Andrews & Withey[21] and Campbell et al[23]. This approach has the advantage of allowing comparison of outcomes across different patient groups and with general populations, but does so at the cost of failing to include disease-specific elements that may be considered crucial aspects of patient outcome. Nevertheless, generic QOL measures of this type are often considered to be the best available[35] because they provide the most direct assessment of people's experiences and perceptions of life. Debate continues however, about the specific features of this type of QOL measurement, such as the relative benefits of objective and subjective indicators, domain-specific and global measures.

Modelling QOL in Mental Health Service Settings

Several models of QOL have been developed that apply to mental health service samples, but many do not specify the nature or direction of causality, instead simply describing associations between certain variables and QOL[36,37].

Most causal, psychosocial models derive from Cantril's model of QOL[34], which focused on the association between satisfaction, needs and aspirations. The association between these variables was a common feature of the models formulated by Campbell et al[23] and Andrews & Withey[21] in their national surveys of the American population, which provided the theoretical base from which most QOL research in the mental health field derives[38]. One of the most influential models, developed by Anthony Lehman[24] suggested that the experience of 'general' well-being was a product of personal characteristics, objective life-circumstances in a variety of domains and satisfaction with life in those domains. Although it was not explicit in the original publication[24], Angermeyer & Kilian[37] suggested that Lehman's model assumed QOL was dependent upon the degree of compliance between an individual's life-circumstances and their wants, needs and wishes.

Subsequent models have tended to reflect the theoretical backgrounds of their modifiers[39–41], as is the case with the model of QOL presented in Figure 1[42], which underpins the comparisons of severe mental illness, common mental disorder and

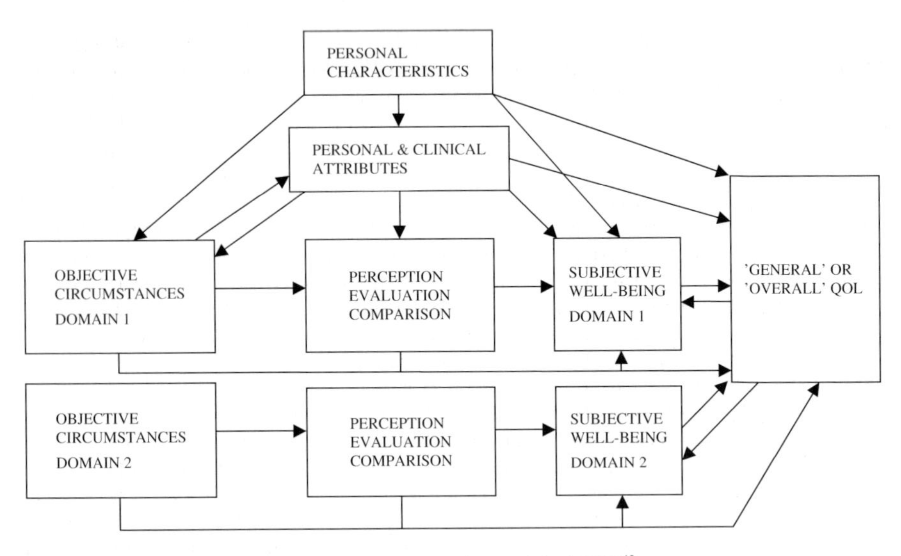

Figure 1. An alternative model of QOL[42]

healthy general population groups, presented towards the end of this chapter. The advantages of this 'alternative' model are that it distinguishes QOL from mental health or affect, and happiness, and is not complicated unnecessarily by attempts to accommodate domain importance (given that weighting measures on the basis of domain importance is unnecessary and undesirable)[43]. In addition, it has the capacity to control for the effects of adaptation and to explore the relevance of some theories developed recently on the basis of analytical models (e.g. top-down, bottom-up, bi-directional theories and homeostatic theories.

QUALITY OF LIFE IN COMMUNITY MENTAL HEALTH SERVICES – THE LITERATURE REVIEWED

Given the remit of 'outcomes in mental health services', this review (based on the strategy outlined[1]) focuses mainly on SMI samples, as services are increasingly targeted on this group, and as far as possible on longitudinal studies, in which the

---

[1] The challenges posed by the volume and nature of the QOL literature, were overcome by adopting a strategic and targeted approach to the review. Systematic searches of Medline, EMBASE, Psychinfo, IBSS, SSCI, ASSIA, Sociological Abstracts, Social Services Abstracts, Cinahl, Premedline, SIGLE (to provide coverage of the 'grey' literature), Cochrane Database of Systematic Reviews (CDSR), Cochrane Controlled Trials Register (CCTR) and the Campbell Collaboration's Social, Psychological, Educational & Criminological Trials Register (C2-SPECTR) were supplemented using the web-site search engine Copernic, to maximise retrieval of journal articles, books, dissertation and conference abstracts covering a range of perspectives. As a consequence, the potential for publication bias was reduced. Supplementary 'hand searching' of the two journals 'Quality of Life Research' and the 'Journal of Happiness Studies' and 'snowball searching' of references included in publications identified by electronic methods helped to facilitate comprehensive retrieval.
Pre-determined inclusion criteria were incorporated in order to increase the homogeneity of studies and minimise potential selection bias. The criteria used in this review were based on four core elements namely study design, participants, intervention and outcome measures. Evidence relating to QOL outcomes was provided by studies of the type described in the first three of the five classifications of evidence outlined by Bandolier (1995), namely, systematic reviews, randomised controlled trials (RCTs) and well-designed intervention studies (with or without randomisation). Study participants needed to have a severe mental illness (including those with psychotic disorders); be an adult of working age (18–65); be residing independently or semi-independently in the community; and should not be learning disabled or abusing substances (unless the study reported dual-diagnosis separately). Community based clinical and social interventions that reported at individual, group and/or service level (e.g. housing schemes) were included. Studies using generic measures of QOL that were validated and reliable, and reporting individual level and/or aggregated data were included in the review. Studies that were limited to HRQOL, health-status or disease-specific measures, or which reported the development of new measures without presenting any outcome data, were not included.
Publications subject to review were classified using a quality rating system based on an approach recommended by the Cochrane Collaboration Depression, Anxiety & Neurosis group (CCDAN; *www.iop.kcl.ac.uk/iop/ccdan.index.htm*). This process helped to highlight the best quality relevant evidence and isolate any publications that might be included in a meta-analytic review. The procedure also pinpointed studies of inferior quality that were of questionable value because the reliability and validity of measures or findings, was not established. Using objective indicators the classification took account of the clarity of reporting and adequacy of methods in relation to 16 independent features.

impact of clinical and/or social interventions is assessed. Nevertheless, due to the paucity of good quality data of this type the results of cross-sectional studies are included in descriptive reports of objective circumstances and subjective life-quality; comparisons with other population samples such as mentally healthy participants, people with common mental disorders and other disadvantaged groups in the general population are also provided in order to highlight the ways in which severe mental illness can impact on objective life opportunities and quality. Throughout, the results of multivariate analyses are given precedence, as they portray the relationship between quality of life (as the dependent variable) and other potentially explanatory variables, while taking account of the role of other related factors.

This search strategy demonstrated that a large proportion of QOL publications are either conceptual and data free (38%) or report data that are not directly relevant to today's community mental health services (50%) e.g. some papers related to hospital discharge and deinstitutionalisation, others related to non-mental health samples, focusing instead on depression and QOL in people with physical health problems, and for many QOL was not the primary outcome. Unfortunately, the nature of evidence is often of poor (45%) or dubious (36%) quality, when categorised according to the quality criteria outlined in the search strategy in endnote[1]). Theoretically driven research is rare, as studies are frequently undertaken by clinicians to inform operational practice rather than to advance theoretical understanding. Relatively few studies are based on large samples that were selected randomly, with most utilising convenience samples that were not subject to any inclusion criteria. In order to enhance the reliability and validity of the literature review that follows focuses on papers of good quality (i.e. those using random samples, validated measures and appropriate analyses, and achieving reasonable response rates etc – see endnote[1]).

The vast majority of relevant studies were cross-sectional, and only a quarter had a longitudinal design. While the diagnostic characteristics of samples varied between studies the main focus was on people with schizophrenia, who constituted between 37% and 91% of SMI samples. Only a small proportion (4%-16%) of people with SMI were currently married[24,44,45], or living as married (12%-17%)[46,47], which is substantially lower than in 'representative household samples' of the general population (68%-70%)[21,30,48,49].

Objective Life-circumstances of People with SMI

The majority of SMI respondents lived independently (76%-81%)[46,47], many living in their own homes (93%) and without family (74%)[45]. 78% of an American community based sample lived in private homes and on average had been resident for one year[50]. Nevertheless, 15% of one UK sample was living in supported housing[47] and 26% of another sample had lived in a supervised residence within the past two years; within the same time period 5% had been homeless and 10% had been in jail[46]. The median length of stay in hostels was 30 months[24] and the mean stay was 40 months, but all length of stay data were skewed towards longer periods of time (SD = 41.8). This mean length of stay was lower than that for inpatients (53 months (SD 96.1)), but higher than for residents in supervised homes (29 months (SD 31.9)[44]. Cultural differences were evident however, with the majority of participants in Italian and Brazilian studies, living with their relatives[51,52].

The proportion of people with SMI who were working tended to be very low[45,52–54]. Employment rates ranged from 13% - 15% in some studies, rising to 35% in an Italian sample[51], but even the latter figure was considerably lower than those for general population samples (57%-70%)[48,55]. When working, very few people with SMI (3%) worked full-time; in a study of working and non-working groups, workers were on average employed for 18.5 hours per week (range eight to 40 hours)[50]. Although workers in an Italian study worked for an average of 29 hours per week the number of hours worked varied considerably between individuals (4-55 hours)[51]. Most people worked in community settings (71%), but 24% worked in sheltered environments and a further 6% worked in both settings; 83% of those working in community settings were in supported employment and only six workers were in open employment (12%). A greater proportion of men ($n = 33$; $\chi^2 = 4.949$, $df = 1$, $p < 0.05$) and younger people (aged 36 c.f. 40; $t = 2.59$, $df = 90$, $p < 0.01$) were employed[50].

Unsurprisingly, incomes levels tended to be low[52], but varied between countries and obviously over time. Two-thirds of participants in a 1982 study[53] received less than $60 per month for personal expenses not including board and lodge, equating to £32 per month at current exchange rates. The median monthly income in a Swedish study conducted in 1999 was 7,400 Swedish kroner[45] (£542), whereas in Italy in 2005 average incomes were €671 per month (£452), but ranging from €7.75 (£5.22) – €774.70 (£522).[51] As one would expect, incomes were higher for workers ($786 (£416) c.f. $665 (£352); $t = 1.98$, $df = 90$, $p < 0.05$) than for non-workers, in an American study conducted in 1996[50].

Most people reported weekly (39%) or daily contact (31%) with family but for a small number contact with relatives was annual or less (n = 11, 9%). More than half had a close friend (58%) or someone that they could rely on (56%) and 43% had visited a friend recently[45].

Finally, Lehman et al[53] found that a considerable proportion of the SMI sample (33%) had been robbed or assaulted in the last year either inside or outside of their home. It is difficult to assess how this figure compared with the local general population in 1982, but the figure is considerably higher than rates of assault (1.2%) and robbery (0.6%) in the general population of America, as reported in the 7th United Nations Survey of Crime Trends and Criminal Justice Systems.[56]

Subjective Life-quality in SMI Samples

Subjective QOL ratings were moderate at 'general' and domain levels but rarely reached the range for satisfaction (five to seven) on the delighted-terrible scale[45–47,53,57–60]. Domain scores were usually in the range four to five and tended to be highest in relation to leisure activities and social relationships, but as in studies of the general population satisfaction with family relationships was also high[53,57,58]. Ratings for work and finance were very low[46,47,52,53,57,58,60], as were the proportions of respondents reporting satisfaction in these domains (42% & 54% respectively)[45].

Factors Relating to Life-quality in SMI Samples

While personal & objective characteristics were not a major influence on subjective evaluations of 'general' QOL, their influence on domain life-quality tends to be greater[54]. Lehman[24,57] demonstrated that all objective variables made a contribution to the explanation of 'general' life-quality, with the major contributors being total medical care used, being a victim of crime, intimacy of social relations, frequency of family contact, and in the latter paper, hours worked per week and seeking employment. Similarly having a disability, living with family, living in supervised accommodation, being in work or being homeless in the last two years, were influential in explaining 'overall' life-quality[46]. Having access to larger social network groups also made a major positive contribution to ratings of 'global' and more noticeably, domain-specific life-quality (i.e. social)[59]. The contribution made by demographic and diagnostic variables (4–9%) to the explanation of 'general' life-quality in board and care, supervised residence and inpatient samples was overshadowed when objective variables were included in the model (14%–27%)[44], but objective social variables were related less consistently to 'overall' life-quality than clinical and subjectively reported needs[46].

Having the opportunity to make improvements to certain aspects of one's life[54] and being empowered to make one's own life choices[61] appear to be important factors in explaining subjective life-quality in people with SMI, and may mediate the relationship between objective life-circumstances and subjective life-quality at a domain level. For example, it is easy to imagine a situation where an individual lives in a pleasant block of flats, with friendly neighbours, in an area with all the amenities that one could want, but that individual giving a low rating of QOL in the living situation domain, because if they were able or had the choice, they would like a house or flat with a garden that they can enjoy with their dog, or live closer to their friends and family.

Subjective variables (e.g. QOL, needs and value based indicators such as valuing work) seem to contribute most to the explanation of 'overall' and domain life-quality[24,50,57]. For example, Lehman[24,57] found that introducing subjective variables into a regression model of 'general' QOL increased the amount of variance explained considerably (e.g. from $R^2 = 0.25$ to $R^2 = 0.66$). This model highlighted the importance of domain satisfactions with social life, leisure, health and finances. The dominant influence of subjective QOL indicators was observed consistently across outpatient (board & care – 57%, supervised residence – 49%) and inpatient samples (40%), and models of QOL appeared to be comparable with those for general populations, in which subjective variables explained between 42% and 61% of the variance in life-quality[21,23]. 'Global' satisfaction with services was also found to be significantly associated with 'global' life-quality[59] and with QOL in the social domain, whereas subjectively rated 'unmet need' variables appeared to be inversely related to QOL[46,59].

Depression and anxiety were negatively related to evaluations of 'general' and domain life-quality[24,52,59], but it is clear that domain and 'global' QOL scales differ reliably from mental health scales for depression and anxiety, indicating that QOL and these two aspects of mental health are different constructs[24]. Controlling for mental health does not change the relationship between objective and subjective variables or those between domain and 'general' life-quality except in health. Other clinical indicators of psychiatric morbidity (CPRS), depression (MADRS) and negative symptoms (SANS) were related to life-quality, as were symptom severity and course type ('episodic' or 'continuous'), although the coefficients for these clinical variables were sometimes smaller than those for objective life conditions[46].

Comparisons with Other Community Based Samples

Objective life-circumstances

A study comparing the QOL of people with severe mental illness (SMI) in receipt of community mental health services, with people with common mental disorders and a healthy general population living in the community[42,54] indicated that the three health-status groups differed in terms of demography and baseline life-circumstances. The SMI group was disadvantaged compared to one or both other groups, in all domains except illness/disability. There was a downward trend in mean scores and percentage values from the healthy population to the CMD and SMI groups for many outwardly positive aspects of life (e.g. employment, marriage, income, social contact), and a similar upward trend from healthy population to SMI groups for some more negative aspects of life (e.g. benefit receipt, crime victimisation). Overall, CMD group members appeared more similar to the healthy population than to the SMI group, with the striking exception of victimisation, and presence of depressive symptoms. The CMD group differed markedly from both other groups in its reporting of long-standing illness/disability. As one might expect, the magnitude of mental health-status effects was greatest for depression, income and benefit receipt, having a close friend and employment status variables.

The differential effects of health-status on income might be more pronounced in the UK than in other European countries. Pukrop et al[62] found that although most people in his entire sample had an income of less than DM1500 (German marks) per month, the proportion receiving incomes in this range was highest in the general population (51%) and lower but similar in the depression (40%) and schizophrenia (41%) samples. Unfortunately, statistical significance was not presented and the results were affected undoubtedly by the high proportion of students included in the general population group.

Perceived life-opportunities also varied between SMI, CMD and healthy population groups at baseline, tending to be most restricted in the CMD group. The SMI group was significantly less restricted than the CMD group in safety and living situation, and both groups were significantly more restricted than the healthy population in the health domain; opportunities were not more or less restricted in the SMI group than in the other groups for any other domains. Similarly sized restricted opportunity effects were observed in all domains[42,54].

Subjective life-quality

Studies comparing different population groups confirm many of the findings reported in single general population or SMI studies. Nevertheless, whereas moderate satisfaction scores are reported by SMI samples, high ratings of QOL are made by general populations; SMI samples report statistically significant lower scores in most domains. The starkest differences were observed in family, safety and finance domains where the proportions reporting satisfaction (scores ≥ 5 on the delighted-terrible scale) in the SMI sample were considerably lower than those reported for the general population (family 57% cf 91%; safety 49% cf 72%; finance 34% c.f. 81%)[21,44,57]. The SMI sample also reported statistically significantly lower scores for life in 'general' and QOL in some domains than other socially disadvantaged groups within the general population, such as people from low socio-economic groups, minority ethnic backgrounds and single parents. The SMI sample were never more satisfied than members of the other groups but were equally pleased with their living situation, leisure, job and health. Between-group differences were most notable in the social domain where people with SMI scored lower than all other groups; less significant contrasts were found in personal safety and life in 'general'. In finances and family, SMI group scores were similar to those for unmarried parents, but for life in 'general' SMI group scores were more similar to those for black people.

Direct comparisons between SMI, CMD and healthy population groups[42,54] suggested that personal and lifestyle differences were accompanied by variations in subjective QOL ratings. For most aspects of life-quality there was a downward trend in ratings between healthy population, CMD and SMI groups (reflecting inversely the objective life-conditions of the three groups). Repeated measures regression analyses confirmed that these group effects were stable over time and were not confounded with other explanatory variables. For most aspects of life-quality (except life 'overall'), SMI group ratings were significantly lower than those for the healthy population, but their ratings for finance were significantly higher than the healthy population group. This result seems to be counter-intuitive given the low income levels of the SMI group, but it is likely to be explained by the high benefit take up in the SMI groups relative to the other health-status groups, brought about by the systematic and thorough assessment of benefits provided by the occupational therapists in the mental health service concerned[58]. This process may have produced increases in income of up to £3,100 per annum for those receiving a full benefits assessment[63], given that the average benefits underpayment for people with SMI can average £45 per week[64]. Under these circumstances, SMI members may have expressed higher rates of satisfaction, either because of increases in their income, or due to the knowledge that they were receiving their full benefit entitlement.

In the same study[42,54], CMD group membership was associated with statistically significantly lower scores (compared to the healthy population) in all aspects of life-quality. B coefficients demonstrated that for some aspects of life-quality (e.g. 'general' QOL, finance and safety) the magnitude of the effect of SMI group membership was similar to that of CMD group membership, in that QOL ratings were lowered by similar amounts compared to the healthy population. In other respects the effects of SMI and CMD group membership differed considerably; having a severe mental illness seemed to have a stronger impact on general health, family and living situation, whereas having a common mental disorder had a stronger association with mental health and life 'overall'.

Finally, a study comparing people with major depression with those with other depressions and no depression suggested that the major depression group scored statistically significantly lower than the other groups in most dimensions of life-quality, although differences between the major and other depression groups in the 'physical senses' dimension were not statistically significant[65]. Nevertheless, the findings also demonstrated that people with major and other forms of depression remained able to distinguish between life-domains when rating their subjective QOL.

Modelling life-quality in different mental health-status groups

The earliest attempts to compare models of life-quality between general population and SMI samples were presented by Lehman in 1983[57], although relative performance was not tested directly in models of the three data-sets[21,23,57]. These simple comparisons suggested that the same hierarchical regression models performed well in SMI and general population samples. For example, personal characteristics explained 4% of the variance in 'general' life-quality in the SMI sample compared to 11% in Andrews & Withey's survey[21] and 5% in Campbell et al's work[23]. Slightly more variance was explained in the SMI sample (23%)[57] than in the general population (18%)[23]by a combination of personal and objective indicators. In the final model, which included personal, objective and subjective indicators the amount of variance explained in the SMI sample (58%)[57] was similar or higher than in the general population (42% and 61% respectively)[21,23].

A comparative model of HRQOL in general population, depression and schizophrenia groups also demonstrated similarities in the construction of subjective QOL, in different population groups, suggesting that features of depression (e.g. waking up depressed, feelings of hopelessness and worthlessness) were closely related to 'general' life-quality in the general population sample and to a lesser extent (fewer items) in the depression sample[62]. The impact on 'general' QOL of health, vitality, psychosocial, and material factors was secondary to depression in the schizophrenia and depression groups. In the general population sample the impact of health and material factors was only peripheral and was overtaken by emotional and spare time elements, which had a secondary impact on 'general' life-quality. Nevertheless, comparison between study groups suggested that the model was best applied in the general population, and while the model structure was validly replicated in the depressed sample the model was less valid in the schizophrenia sample. It is possible that the balance between health-related and disease-specific (n = 6) rather than generic indicators (n = 2) of life-quality may have contributed to these findings.

The results of a more recent study that examined models of life-quality across three mental health-status groups (ie SMI, CMD and healthy population), by controlling for the interaction between mental health-status group and other explanatory variables[42,54], bears some similarities to Pukrop's findings but appears to contradict Lehman's assertion that QOL models are similar in SMI and general population samples. This comparative model indicated that QOL was explained differently in the three study groups (i.e. associations between QOL and objective circumstances, presence of depressive symptoms and time often varied significantly between groups). Although a number of explanatory variables were associated with most aspects of life-quality, their relative contributions varied between domains and between groups.

Separate regression models for each health-status group showed that most objective lifestyle indicators were not statistically significantly associated with subjective life-quality in the SMI group, but that statistically significant associations between life-conditions and subjective QOL were observed more frequently in CMD and healthy population groups. Similarly, opportunity variables contributed to the explanation of all aspects of life-quality, in all health-status groups, but the magnitude of the association was often greater in the SMI group, although a smaller number of statistically significant associations were observed between opportunity and 'general' and 'overall' QOL in the SMI group, than in CMD and healthy population groups. Finally, depressive symptoms contributed significantly to the explanation of most QOL models in all study groups, but the magnitude of the effect tended to be greatest in the SMI group. Depression and other health-related variables often contributed more to models of 'general', 'overall' and health, especially in the SMI group.

Changes in Life-quality

The only way to examine QOL outcomes for SMI groups is to evaluate changes in QOL over time. There are some concerns within the mental health services about the responsiveness of QOL measures to changes brought about by treatment, despite evidence demonstrating responsiveness in general population studies[30,31,66], and in studies of SMI samples[58,67–73]. While accepting that the magnitude, extent (i.e. numbers affected by change) and breadth (number of aspects of QOL) of changes have been small in some studies[51,69,74–78], and that others have failed to demonstrate any QOL changes over time[47,79–83], this chapter aims to demonstrate that concerns about responsiveness are unfounded, often being based on unreasonable expectations of the service intervention.

In fact, relatively few studies examined changes in life-quality and many of those that did are of questionable value due to the quality of their designs, the lack of any intervention other than being in receipt of community mental health services, or their inability to demonstrate improvements in any outcomes, including objective life-circumstances and subjective QOL. Often, the assumption appears to be that as health improves, so does quality of life, but this naïve premise fails to recognise the complexity of the QOL concept and its association with mental ill health, which may result from factors relating to life-events and life-circumstances, and which also impacts greatly on one's capacity to participate in or enjoy other aspects of life such as work, income, family and social life etc.

Factors associated with QOL changes

Hospital treatment alone does not necessarily effect changes in subjective QOL[81,82], whereas improving clinical conditions and effecting a reduction in patient-rated unmet needs in the social domain has been associated with an increase in subjective QOL over 4 years; interestingly however, changes in staff-rated unmet needs did not show any association with changes in subjective life-quality[84]. Slade et al[85] also demonstrated that QOL improvements can be attained by addressing unmet needs, but given the inverse relationship between needs and QOL variables, one wonders whether needs assessment based on standardised measures such as the Camberwell Assessment of Need[86] or one of its derivatives, provides any added value over and above those provided by the indicators of objective life-conditions, life-opportunities and domain QOL ratings included in many measures of life-quality.

Case-management provides a facility to broker a variety of services and interventions that might impact on aspects of life other than health, and as such might be expected to deliver QOL improvements. Two studies evaluating the impact of wider community based interventions in the form of intensive and standard services provided evidence about QOL changes over time that was contradictory, although their findings about the impact of intensive interventions were somewhat complementary. While Huxley et al[58] reported statistically significant changes in 'overall' QOL for their whole sample, and in standard and intensive case-management groups in a two year period, Taylor et al[47] reported no moderate effects ($ES \geq 0.40$) in either service over the same period. Similarly, statistically significant improvements were observed in all life-domains except family, in intensive and standard conditions[58], even when Simes test adjustments for multiple outcomes testing were applied, whereas Taylor et al[47] found no changes over time when the same adjustments were made. Even so, Huxley et al[58] found that the magnitude of the intensive case-management effect varied by study site, for global and domain-specific indicators of life-quality. Improvements in QOL were not restricted to people with any particular diagnosis, or to people whose depression was less severe, however.

Many of the regression analyses undertaken within mental health service contexts do not comply with the theoretical models that have been developed, e.g. many ignore objective indicators of life-circumstances in favour of needs, service satisfaction and clinical factors. To date, no studies have identified, which variables predict improvements in objective life-conditions, and although being female, unmarried, of older age, less educated and more disabled predicted worsening objective QOL over time, the amount of variance explained was small. A model of QOL change reported by Ruggeri et al[51] suggested that greater clinician-rated anxiety and depressive symptoms had a negative effect on satisfaction with health and general well-being, and that psychological status, self-esteem and satisfaction with service were important predictors of subjective QOL in most life-domains. Ritsner's work[87] appeared to confirm the role of self-construct variables, but did not fully support evidence relating to the impact of changing psychological status. An inverse relationship was observed between changing levels of distress, paranoid symptoms, side-effects and insight, and changes in some aspects of QOL; reductions in distress, paranoid symptoms and side-effects were associated with QOL improvements, whereas increased insight was associated with deteriorations in QOL. Increased self-esteem and self-efficacy were also associated with QOL improvements, as were positive changes in social support from others in relations to satisfaction with leisure activities. Changes in other psychological symptoms, coping styles, support from family and friends, and length of follow-up, were not associated with QOL changes in any domain. Finally, an alternative regression model presented by Huxley et al[58], which conforms closely to the original models of life-quality presented by Lehman[24,57], indicated that the major contributors to changes in 'overall' life were changes in depression, site of intervention and objective deterioration in social life. Site was also an important factor in producing positive changes in satisfactions with work, finances, living situation and family. The importance of the site of intervention demonstrates the importance of considering any site-specific interventions that exist, either when conducting single-site or multiple-site studies. For example, in Huxley et al's study[58] the site that achieved the greatest breadth and magnitude of QOL changes, was also the one that provided comprehensive assessments of benefits entitlement, which might be expected to lead to improvements in many aspects of life-quality, if incomes increased as a result. The evidence supports this assertion somewhat, given that changes to objective life-conditions were influential in most domains, including work, finance, safety, family and social aspects of life, where improved circumstances had a positive effect and deteriorating circumstances had a negative effect on life-quality. Depression made an important contribution to explanations of variance in all domains except finances and family, as reduced depression scores (i.e. improvement) were associated with increases in QOL. The influence of other health-related variables was limited and confined largely to the health domain. These results are compatible with those reported for general populations[30,31].

Finally, a recent study undertaken by the author[42], had the advantage of examining models of QOL change in SMI, CMD and healthy population groups, in the way advocated recently by researchers in the field. The results suggested that statistically significant changes in subjective QOL can be observed in all mental health-status groups, although the nature and magnitude of changes may differ between groups, depending on the nature of the intervention and whether it is applied at an individual or system level. For example, in this study the magnitude of change was greatest in the SMI group where moderate and large effects were observed in most domains, as a result of an individual level case-management intervention. In contrast, QOL deteriorated over time in many domains in the healthy general population group, which during the time period was subject to the disruption associated with a long-term, urban regeneration initiative.

Changes in objective life-conditions were a major contributor to the explanation of most aspects of domain life-quality, but contributed slightly less to the explanation of changes in 'general' and 'overall' life-quality. This result is understandable when one considers the myriad of factors that merit consideration when evaluating 'how one feels about life in general', but highlights the importance of measuring and reporting domain-specific as well as global life-quality. Improved life conditions were associated with increases in QOL, whereas deteriorating situations were associated with reduced QOL scores. Objective change indicators were particularly influential in domain models of work, family and social life, in which objective changes were associated more highly with changes in subjective QOL, than any other variable.

Changes in depression status were also significantly associated with changes in most aspects of life-quality, but not in safety, family or social domains. Nevertheless, changing depression status did not dominate or make a major contribution to the explanation of any QOL change model, and the relationship was independent of study group. As one would expect, becoming depressed was associated with declining QOL and stopping being depressed was associated with improved QOL. The magnitude of the depression effect was greatest for 'general' life-quality.

The model appeared to demonstrate that the explanation of QOL change varied for different aspects of life-quality. Change models for 'general' and 'overall' life-quality and QOL in work, finance, general health and mental health were similar in some respects, in that they tended to be associated with changes in domain-specific objective circumstances, changes in depression status and mental health-status group. Models of living situation, safety, family and social life also shared similarities in that few variables were associated with changes in these aspects of subjective QOL. Some differences were observed between health-status groups, however. Decreases in income were associated with deteriorating QOL in finance in the general population, but not in the other groups, whereas income increases were associated with improving domain QOL in the CMD and SMI groups (thereby sustaining the argument presented earlier in relation benefits assessments). The association between objective change and subjective changes in 'general' and 'overall' life-quality were limited largely to general population and CMD groups, and to a relatively small number of objective change indicators.

To provide an overview, studies that demonstrated QOL change tended to involve a specific intervention. Nevertheless, the results presented by Evans[42] and Huxley et al[58] contradict those of other studies of case-management interventions, which failed to demonstrate QOL changes[47,79,88,89]. This might be because other studies have failed to deliver effective or sufficiently targeted interventions, or to generate improvements in other outcomes such as symptom severity, psychopathology, functioning or needs[79,83,89,90]. It could be argued therefore that concerns about QOL measurement would be even greater if subjective QOL changes had been observed under these circumstances, which they were not. Unfortunately, the comparable stability of other change measures, in the same studies, has not deterred authors from concluding (without any corroborating evidence) that it is the poor responsiveness of QOL measures, or even adaptation to life's circumstances that has contributed to the lack of subjective change in their studies[47,69,72,81].

Until recently there was no evidence to sustain or contradict arguments about the role of 'adaptation' and 'response shift' in the measurement of subjective QOL within SMI or other health-status groups. Recent work by the author[42,91], provides such evidence in relation to three population groups: people with SMI, CMD and a healthy general population, thereby allowing an examination of the extent and influence of adaptation (defined in terms of resignation and aspirations), and a comparison of its influence in different study groups.

The results suggested that adaptation occurs infrequently - only 32% of the time, on average, across all health-status groups and life-domains, indicating that most people respond appropriately and do not become resigned to their objective circumstances or have unfulfilled aspirations. In the SMI group, the average domain figure for no adaptation was 67%, compared to 65% in the CMD group and 71% in the healthy population group. On average, respondents had fewer than one aspiration or resignation, and while the SMI group experienced resignation in a greater number of domains than both other groups, the CMD group had more aspirations than the other groups. Importantly, although significant differences were observed between the different health-status groups, the pattern of results in the 'unwell' (i.e. SMI and CMD) groups was not vastly different from the healthy population group. The main differences lay in physical health, mental health and living situation, where resignation reached its highest level in the SMI group. Secondly, the pattern of results show similarities between groups in domains other than health, supporting the contention that adaptation effects extend beyond the health domain, although it appears not to affect most people, even in health-related QOL domains.

Even when it occurs, the magnitude of the adaptation effect is small in most domains. In all models, adaptation variables tended to have less of an effect on follow-up QOL than other variables such as objective changes, changes in depression and baseline QOL ratings[42,91]. Resignation seemed to be more closely related to global QOL ratings than to domain-specific ratings of life-quality (with which it was associated in models of living situation and mental health only). Aspiration appeared more consistently and made a larger contribution in all models of general, overall and domain life-quality. Estimates of the net effect of aspirations/resignations on follow-up QOL suggested an overall reduction in most subjective QOL ratings, which is important as in the mental health services, the assumption has tended to be that QOL scores are higher (not lower) than one might expect and remain stable over time, because of the dominant effect of resignation. Moreover, ratings for family and health might be reduced by a greater amount than in other aspects of life-quality, because aspirations and resignation were both associated with a reduction in scores. In contrast, 'general' QOL scores are unlikely to be affected by adaptation because the reduction brought about by aspirations, would be offset by the increase arising from resignation[91].

Models including the interaction between adaptation and health-status groups suggested that the association with QOL varied by group. Individual group models indicated that adaptation variables were less associated with general and overall QOL and changes in these aspects of life-quality, in the SMI group, than in the other health-status groups. Resignation was less strongly associated with domain QOL and domain QOL changes, and aspirations were more strongly associated with domain QOL and domain QOL changes in the SMI group, than in the other two groups[42]. The effect of aspirations was greatest in the general population for finance, and in the SMI group for family; the effect of resignation was greatest in the CMD group for finance, and in the general population for health[91].

It appears therefore that aspiration and resignation do not necessarily lead to response shift, and where adaptation of this type does occur, the magnitude of the effect is relatively small.

KEY MESSAGES FOR MENTAL HEALTH SERVICES

- QOL improvements are achievable and measurable.
- Subjective changes in life-quality cannot be expected unless certain life changes (e.g. in health and/or other aspects of life) have occurred or have been brought about by relevant service interventions.
- Domain-specific evaluations of subjective life-quality are necessary as well as global assessments of QOL, if intervention effects are to be understood properly.
- Service interventions need to be targeted on specific domains to produce the most noticeable effects.
- Changes in depression and anxiety need to be assessed and controlled for.
- Respondents need to be asked about their aspirations and feelings of resignation, and have these variables controlled for, as they may mediate relationships with QOL.
- Assumptions that stable subjective QOL ratings are the result of adaptation effects are not evidence based.

REFERENCES

1. World Health Organisation. World Health Report Mental Health: new understanding, new hope. WHO 2001
2. Kingdon DG. The Care Programme Approach. Psych Bulletin 1994; 18:68–70
3. Department Of Health. The National Service Framework for Mental Health: Modern Standards and Service Models. DH, London; 1999
4. Huxley PJ, Evans S, Munroe M et al. Fair access to care services in integrated health and social care teams. Draft final report to the DH. London: Institute of Psychiatry; 2006
5. Wang PS, Demler PH, Olga MS et al. Adequacy of treatment for serious mental illness in the United States. Am J Public Health 2002; 92(1):92–98

6. Rosen A. The Australian experience of deinstitutionalization: interaction of Australian culture with the development and reform of its mental health services. Acta Psychiatr Scand 2006; 113 (Suppl. 429):81–89
7. World Health Organisation Mental Health Declaration for Europe and a Mental Health Action Plan for Europe. WHO: 2005. Available at: http://www.euro.who.int/mentalhealth2005. Accessed August 2006
8. European Communities Green Paper: Improving the mental health of the population: Towards a strategy on mental health for the European Union. Brussels 2005: 14.10.2005. COM(2005)484
9. Commonwealth of Australia National Mental Health Plan 2003-2008. Australian Government Department of Health and Ageing 2003
10. Krieble T. Towards an outcome-based mental health policy for New Zealand. Australas Psychiatry. 2003; 11 (Supplement 1):S78–S82
11. Daniels AS, Adams N. From Study to Action: A Strategic Plan for transformation of mental health care. Health Care Change Organisation 2006. Available at http://www.healthcarechange.org, accessed August 2006
12. Wilton R. Putting policy into practice? Poverty and people with serious mental illness. Soc Sci Med 2004; 58(1):25–39
13. European Social Fund Project. Inclusion Europe, Included in Society (2003-2004). 2004 Available at: http://europa.eu.int/comm/employment_social/index/socinc_en.pdf. Accessed August 2006
14. Bergner M. Quality of life, health-status and clinical research. Med Care 1989; 27(suppl):S148–156
15. Gill TM, Feinstein AR. A critical appraisal of the quality of quality of life instruments. JAMA 1994: 272:619–626
16. O'Connor R. Issues in the measurement of health-related quality of life. National Centre for Health Program Evaluation [online] 1993; Available: http://_www.rodoconnorassoc.com/issues_in_the_measurement_of_qua.htm, accessed January, 2003)
17. Namjoshi MA, Buesching DP. A review of health-related quality of life literature in bipolar disorder. Qual Life Res 2001; 10:105–115
18. Shumaker SA, Berzon R. The international assessment of health-related quality of life: theory, translation, measurement and analysis. Oxford: Rapid Communications; 1995
19. Ware JE, Sherbourne CD. The MOS 36 item short form health survey (SF36): 1. Conceptual framework and item selection. Med Care 1992; 30:473–483
20. Leidy NK, Revicki DA, Geneste B. Recommendations for evaluating the validity of quality of life claims for labelling and promotion. Value Health 1999; 2:113–127
21. Andrews F, Withey SB. Social indicators of well-being: Americans perceptions of quality of life. New York: Plenum Press; 1976
22. Oliver JPJ, Huxley PJ, Bridges K et al. Quality of life and mental health services. London: Routledge; 1996
23. Campbell A, Converse PE, Rodgers WL. The quality of American life: perceptions, evaluations and satisfactions. New York: Russell Sage Foundation; 1976
24. Lehman AF. The well-being of chronic mental patients: assessing their quality of life. Arch Gen Psychiatry 1983; 40:369–373
25. van Nieuwenhuizen C, Schene AH, Koeter MWJ et al. The Lancashire Quality of Life Profile: modification and psychometric evaluation. Soc Psychiatry Psychiatr Epidemiol 2001; 36:36–44
26. Angermeyer MC, Holzinger A, Kilian R et al. Quality of life as defined by schizophrenic patients and psychiatrists. Int J Soc Psychiatry 2001; 47:34–42
27. Lawton MP. The varieties of well-being. In: Maltatesta CZ, Izard CE. (eds) Emotion in adult development. Beverly Hills: Sage; 1984
28. Bradburn N. The structure of psychological well-being. Chicago: Aldine Publishing Co; 1969
29. Allardt E. Dimensions of welfare in a comparative Scandinavian study. Acta Sociologica 1976; 19:227–239
30. Atkinson T. Stability and validity of quality of life measures. Soc Indicators Res 1982; 10:113–132
31. Headey B. The quality of life in Australia. Soc Indicators Res 1981; 9:155–181
32. Sorensen T. The intricacy of the ordinary. Br J Psychiatry 1994; 23(suppl):108–114

33. Priebe S, Kaiser W, Huxley P et al. Do different subjective evaluation criteria reflect distinct constructs? J Nerv Men Dis 1998; 186:385–392
34. Cantril H. The pattern of human concerns. New Brunswick, N.J.: Rutgers University Press; 1965
35. Zautra A, Goodhart D. Quality of life indicators: a review of the literature. Community Men Health Rev 1979; 4:2–10
36. Skantze K, Malm U. A new approach to facilitation of working alliances based on patients' quality of life goals. Nord J Psychiatry 1994; 48:37–55
37. Angermeyer MC, Kilian R. Theoretical models of quality of life for mental disorders. In: Katschnig H, Freeman H, Sartorius N. (eds). Quality of life in mental disorders. Chichester: John Wiley & Sons Ltd; 1997
38. Barry MM. Well-being and life satisfaction as components of quality of life in mental disorders in Katschnig H, Freeman H, Sartorius N. (eds). Quality of life in mental disorders. Chichester: John Wiley & Sons Ltd; 1997
39. Baker F, Intagliata J. Quality of life in the evaluation of community support systems. Eval Program Plann 1982; 5:69–79
40. Bigelow DA, Brodsky G, Stewart L et al. The concept and measurement of quality of life as a dependent variable in the evaluation of mental health services. In: Stahler GJ, Tash WR (eds). Innovative approaches to mental health evaluation:. New York: Academic Press Inc; 1982:345–366
41. Zissi A, Barry MM & Cochrane R. A mediational model of QOL for individuals with severe mental health problems. Psychol Med 1998; 28:1221–1230
42. Evans S. Quality of Life and Mental Health in the Community. PhD Thesis. London: University of London, Institute of Psychiatry; 2004
43. Trauer T, McKinnon A. Why are we weighting? The role of importance ratings in quality of life measurement. Qual Life Res 2001; 10:579–585
44. Lehman AF A quality of life interview for the chronically mentally ill. Eval Prog Plann 1988; 11: 51–62
45. Bengtsson-Tops A, Hansson L. Subjective quality of life in schizophrenic patients living in the community: relationship to clinical and social characteristics. Eur Psychiatry 1999; 14:256–263
46. UK700 Group. Predictors of quality of life in people with severe mental illness: Study methodology with baseline analysis in the UK700 trial. Br J Psychiatry 1999; 178:426–432
47. Taylor RE, Leese M, Clarkson P et al. Quality of life outcomes for intensive versus standard community mental health services. Br J Psychiatry 1998; 173:416–422
48. Bowling A. What things are important in people's lives? A survey of the public's judgements to inform scales of health-related quality of life. Soc Sci Medicine 1995; 41:1447–1462
49. Bowling A, Windsor J. Towards the good life: a population survey of dimensions of quality of life. J Happiness Studies 2001; 2:55–81
50. van Dongen CJ. Quality of life and self-esteem in working and non-working persons with mental illness. Comm Ment Health J 1996; 32:535–548
51. Ruggeri M, Nose M, Bonetto C et al. Changes and predictors of change in objective and subjective quality of life: multi-wave follow-up study in community psychiatric practice. Br J Psychiatry 2005; 187:121–30
52. de Souza LA, & Coutinho ESF. The quality of life of people with schizophrenia living in community in Rio de Janeiro, Brazil. Soc Psychiatry Psychiatr Epidemiol 2006; Vol. 41(5):347–356
53. Lehman AF, Ward NC, Linn LS. Chronic mental patients: the quality of life issue. Am J Psychiatry_(1982) 139:1271–1276
54. Evans S, Banerjee S, Leese M et al. The impact of mental illness on quality of life: A comparison of severe mental illness, common mental disorder and healthy population samples. Qual Life Res 2006, in press
55. Michalos AC, Zumbo BD & Hubley A. Health and the quality of life. Soc Indicators Res 2000; 51:245–286
56. United Nations Office on Drugs and Crime, Centre for International Crime Prevention (1998-2000) United States Crime Statistics. www.nationmaster.com/red/country/us-united-states/ Crime. Accessed August 2006

57. Lehman AF. The effects of psychiatric symptoms on quality of life assessments among the chronic mentally ill. Eval Prog Plann 1983; 6:143–151
58. Huxley P, Evans S, Burns T et al. Quality of life outcome in a randomized controlled trial of case-management. Soc Psychiatry Psychiatr Epidemiol 2001; 36:249–255
59. Becker T, Leese M, Clarkson P et al. Links between social networks and quality of life: an epidemiologically representative study of psychotic patients in South London. Soc Psychiatry Psychiatr Epidemiol 1998; 33:299–304
60. Schmidt K, Staupendahl A, Vollmoeller W. Quality of life of schizophrenic psychiatric outpatients as a criterion for treatment planning in psychiatric institutions. Int J Soc Psychiatry 2004; 50 (3)
61. Boyd AS & Bentley KJ. The relationship between the level of personal empowerment and quality of life among psychosocial clubhouse members and consumer-operated drop-in center participants. Soc Work Ment Health. 2005; 4(2):67–93
62. Pukrop R, Moller HJ, Steinmeyer EM. Quality of life in psychiatry: a systematic contribution to construct validation and the development of the integrative assessment tool 'modular system for quality of life'. Eur Arch Psychiatry Clin Neurosci 2000; 250:120–132
63. Frost-Gaskin M, O'Kelly R, Henderson C et al. A welfare benefits outreach project to users of community mental health services. Int J Soc Psychiatry 2003; 49(4):251–63
64. McCrone P, Thornicroft G. Credit where credit's due. Community Care 1997; September 18-24:23
65. Goldney RD, Fisher LJ, Wilson DH et al. Major depression and its associated morbidity and quality of life in a random, representative Australian community sample. Aust NZJ Psychiatry 2000; 34:1022–1029
66. Schyns P. Income and satisfaction in Russia. J Happiness Studies 2001; 2:173–204
67. Stein LI, Test MA. Alternatives to mental hospital treatment: conceptual model, treatment program and clinical evaluation. Arc Gen Psychiatry 1980; 37:392–397
68. Atkinson JM, Coia DA, Harper Gilmour W et al. The impact of education groups for people with schizophrenia on social functioning and quality of life. Br J Psychiatry 1996; 168:199–204
69. Barry MM, Crosby C. Quality of life as an evaluative measure in assessing the impact of community care on people with long-term psychiatric disorders. Br J Psychiatry 1996; 168:210–216
70. Browne S, Roe N, Lane A et al. A preliminary report on the efficacy of a psychosocial and educative rehabilitation programme on quality of life and symptomatology in schizophrenia. Eur Psychiatry 1996; 11:386–389
71. Lam JA, Rosenheck RA. Correlates of improvement in quality of life among homeless persons with serious mental illness. Psychiatr Serv 2000; 51:116–118
72. Ruggeri M, Bisoffi G, Fontecedro L et al. Subjective and objective dimensions of quality of life in psychiatric patients: a factor analytical approach - The South Verona Outcome Project 4. Br J Psychiatry 2001; 178: 268–275
73. Frisch WB. Improving mental and physical health care through quality of life therapy and assessment. In: Diener E, Rahtz DR. Soc Indicators Res 4: Advances in Quality of Life Theory and Research. Amsterdam: Kluwer Academic Publishers; 2000
74. Baker F, Jodrey D, Intagliata J. Social support and quality of life in community support clients. Comm Ment Health J 1992; 28: 397–411
75. Evenson RC, Vieweg BW. Using a quality of life measure to investigate outcome in outpatient treatment of severely impaired psychiatric clients. Compr Psychiatry 1998; 39:57–62
76. Skantze K. Subjective quality of life and standard of living: a 10-year follow-up of out-patients with schizophrenia. Acta Psychiatr Scand 1998; 98:390–399
77. Bengtsson Tops A, Hansson L. Quantitative and qualitative aspects of the social network in schizophrenic patients living in the community: relationship to socio-demographic characteristics, and clinical factors and quality of life. Int J Soc Psychiatry 2001; 47:67–77
78. Bystritsky A, Liberman RP, Sun-Hwang MS et al. Social functioning and quality of life comparisons between obsessive-compulsive and schizophrenic disorders. Depress Anxiety 2001; 14:214–218
79. Marshall M, Lockwood A & Gath D. Social services case-management for long-term mental disorders: a randomised controlled trial. Lancet 1995; 345:409–412

80. Tempier R, Mercier C, Leouffre P et al. Quality of life and social integration of severly mentally ill patient: a longitudinal study. J Psychiatry Neurosc 1997; 22:249–255
81. Priebe S, Roeder-Wanner UU, Kaiser W. Quality of life in first admitted schizophrenia patients - a follow up study. Psychol Med 2000; 30:225–30
82. Franz M, Meyer T, Spitznagel A et al. Responsiveness of subjective QOL assessment in schizophrenic patients: a quasi-experimental pilot study. Eur Psychiatry 2001; 16:99–103
83. Schneider J, Wooff D, Carpenter J et al. Community mental healthcare in England: associations between service organisation and quality of life. Health Social Care Comm 2002; 10:423–434
84. Lasalvia A, Bonetto C, Malchiodi F et al. Listening to patients' needs to improve their subjective quality of life. Psychol Med 2005; 35(11):1655–65
85. Slade M, Leese M, Cahill S et al. Patient-rated mental health needs and quality of life improvement. Br J Psychiatry 2005; 187:256–261
86. Phelan M, Slade M, Thornicroft G et al. The Camberwell Assessment of Need: the validity and reliability of an instrument to assess the needs of people with severe mental illness. Br J Psychiatry 1995; 167:589–595
87. Ritsner M. Predicting Changes in Domain-Specific Quality of Life of Schizophrenia Patients. J Nerv Men Dis 2003; 191:287–294
88. Olfson M. Assertive community treatment: an evaluation of the experimental evidence. Hosp Community Psychiatry 1990; 41:634–641
89. Marshall M, Gray A, Lockwood A et al. Case-management of people with severe mental disorders. Cochrane Database Systematic Rev 2 (CD000050); 2000
90. Huppert JD, Smith TE. Longitudinal analysis of subjective quality of life in schizophrenia: anxiety as the best symptom predictor. J Nerv Ment Dis 2001; 189:669–675
91. Evans S, Huxley P. Adaptation, response shift and quality of life ratings in mentally well and unwell groups. Qual Life Res 2005; 14(7):1719–1732

CHAPTER 19

SUBJECTIVE QUALITY OF LIFE IN RELATION TO PSYCHIATRIC REHABILITATION AND DAILY LIFE

MONA EKLUND

Department of Health Sciences, Lund University, Lund, Sweden

Abstract: Due to the great variation in approaches to psychiatric rehabilitation, the literature does not provide a uniform description of these methods. In this review, outcomes in terms of subjective quality of life are discussed in relation to vocational training, activity-based rehabilitation, case management, and social skills training. Relationships between aspects of daily life and subjective quality of life are also illuminated. In all the rehabilitation approaches studied, clients have shown an improved quality of life during the rehabilitation period, but no more than the respective comparison group receiving some other form of intervention. Concerning daily life, having employment, being engaged in and satisfied with daily activities, having a supportive social network, and living in the community have consistently been shown to be related to a better quality of life. The reasons why subjective quality of life has not been shown to improve to any substantial degree as a result of psychiatric rehabilitation are probably manifold, and research so far has left many questions unanswered. More, well-designed studies, including multi-methodological approaches and long follow-up periods, are needed to elucidate how subjective quality of life is affected by various rehabilitation strategies

Keywords: Subjective quality of life, Psychiatric rehabilitation, Vocational training, Case management, Occupational therapy, Daily activity, Social network

INTRODUCTION

This chapter focuses on psychiatric rehabilitation, daily life, and subjective quality of life. Subjective quality of life has been defined as perceived well-being, not deducible from objective indicators[1]. One area that will be explored is the outcome of psychiatric rehabilitation in terms of subjective quality of life. According to Elizur[2], rehabilitation has three main aims: (1) clinical aims, in terms of symptom reduction, (2) functional aims, such as strengthening skills and adjustment to independent life, and (3) systemic aims, such as reducing environmental demands and promoting social support. Subjective quality of life pertains to the functional

M.S. Ritsner and A.G. Awad (eds.), Quality of Life Impairment in Schizophrenia, Mood and Anxiety Disorders, 355–372.

aims, according to this taxonomy. Besides rehabilitation, another topic that will be examined is people's daily life in relation to quality of life. Daily life is closely linked to the functional aims of rehabilitation, although it does not focus on interventions, but on naturally occurring everyday situations. Finally, issues pertaining to rehabilitation and daily life that need to be further developed and investigated in relation to subjective quality of life are discussed.

PSYCHIATRIC REHABILITATION AND SUBJECTIVE QUALITY OF LIFE

It has been proposed that quality of life should be the focus of the rehabilitation of persons with severe and persistent mental illness, and that the client's subjective assessment of quality of life should be the main measure of rehabilitation outcome[3]. Improved quality of life is included in the definition of psychosocial rehabilitation adopted by the NIMH[2]. Mueser and Bond[4] expected the development in psychosocial treatments to result in improved quality of life for people with severe mental illness. Despite this, a vast number of studies have indicated that self-rated quality of life does not seem to reflect the changes expected to result from psychiatric rehabilitaiton[5]. The outcome of psychiatric rehabilitation is an under-investigated area, however, and studies exploring whether rehabilitation promotes quality of life are limited in number and often restricted methodologically. Psychiatric rehabilitation constitutes a broad area, including several types of interventions, and various approaches reported in the literature can be identified as successful[2,4]. Moreover, there is no clear-cut limit between rehabilitation and treatment. Kopelowicz and Liberman[6] considered them to be integrated, seamless approaches, a standpoint that has been adopted as a guiding principle in this review. This chapter distinguishes some of the approaches identified in the literature, namely *vocational rehabilitation, activity-based rehabilitation* (such as occupational therapy), *case management*, and *social skills training*. Although the goals and methods used may overlap, these approaches can be identified as specific in psychiatric rehabilitation, and in all areas there are examples of outcome studies employing subjective quality of life as an indicator of effectiveness. However, the number and quality of outcome studies vary greatly between these areas. Finally, some additional rehabilitation approaches are addressed, in which quality of life could be a potentially interesting and relevant outcome, but where research has not yet really touched upon quality of life.

Vocational Rehabilitation

Relatively, vocational rehabilitation has been extensively evaluated regarding outcomes. The most common outcome criteria are work related, for example indicators of paid employment, proportion of consumers who obtain any kind of gainful employment, quality of work performance, and satisfaction with work situation[7,8]. Successful effects have been demonstrated, especially in programs focusing on placement in real workplaces and individually designed support, so-called Individual

Placement and Support (IPS)[9–12]. Patients participating in IPS have been shown to improve more in vocational domains, in terms of number of working hours and the proportion of persons having a job, than those receiving prevocational training, training according to the Fountain House model, or standard treatment[10]. However, the results regarding non-vocational outcomes, such as subjective quality of life, have not revealed any differences between those receiving IPS and those participating in other types of vocational training or standard care. In a number of reviews[7] quality of life has not been among the studied outcomes. In cases where quality of life has been used as an outcome variable in controlled studies, no significant differences were found between the comparison groups[8,10]. When improvements in non-vocational outcomes do occur, they often take place in both the IPS and the comparison group[8]. Prevocational training has been shown to be more effective when combined with a technique aimed at increasing the participant's motivation, such as payment or adding some psychological intervention[10]. Just as in IPS, however, the outcomes in which participants differ tend to be work-related, with no differences having been reported concerning quality of life. Indeed, Drake and associates[8] concluded that it is probably unrealistic to expect that promoting competitive employment through IPS or any other vocational rehabilitation program will have strong positive effects on non-vocational areas of functioning.

Activity-based Rehabilitation

Another alternative in psychiatric rehabilitation is occupational therapy and other activity-based rehabilitation programs. Research within this area is limited. A recent review showed that the evidence for the efficacy of occupational therapy in dealing with mental health issues is insufficient[13]. In one of the few outcome studies reported, no improvement in quality of life was observed during the rehabilitation period, although improvements were identified in areas such as psychiatric symptoms and communication skills. Furthermore, the occupational therapy group improved more regarding psychosocial functioning than a comparison group receiving standard care[14]. However, at a one-year follow up when further improvement was noted, improvement in quality was also noted[15]. This was corroborated by a professional's rating of patients' quality of life[16]. The authors suggested that changes in quality of life might not be immediately discernible following rehabilitation, but that for improvement to become felt patients might need a period of consolidation and of trying new skills in a daily life context[15]. Another study using occupational therapy as the comparison treatment showed that the subjective quality of life of the persons participating in the occupational therapy program improved during treatment, as did that of the experimental group who received social skills training[17]. Yet another study found that satisfaction with the living situation improved as a result of involvement in self-chosen meaningful daily activities[18].

A study carried out within a consumer-run, activity-based program showed that the quality of life of the participants improved, as indicated by a combined qualitative and quantitative inquiry[19]. Psychiatric day-care centers constitute another activity-based rehabilitation alternative, but in an attempt to review the existing literature regarding the effects of day-care center programs, Catty et al.[20] did not find any study meeting the criteria for a randomized controlled trial. They found that non-randomized comparative studies gave conflicting results, and concluded that people with severe mental illness and their carers have to take a pragmatic decision on which type of unit best meets their needs.

Many health care professionals spend significant parts of their working time training people with severe mental illness in the area of life skills. Life skills training is focused on the activities required in daily living, and aims at self-care functioning at personal and domestic level. A recent review did not find any clear effects[21]. However, the review was based only on two small randomized studies, and the authors concluded that more studies in this area are urgently needed.

Case Management

Quite a few studies concerning case management have reported encouraging positive results regarding effectiveness. A systematic review[22] indicated that Assertive Community Treatment (ACT), an outreach and more intense type of case management, with improved subjective quality of life as one of the rehabilitation goals, is effective in preventing dropout from psychiatric care and in reducing the frequency, length, and cost of hospitalization. Moreover, ACT patients showed a better outcome than those receiving standard community care regarding accommodation status, employment, and patient satisfaction. However, regarding quality of life, ACT and outreach case management in general have not been shown to yield better outcomes than other types of rehabilitation[5].

Some studies, less rigorous than the ones included in the Cochrane reviews by Marshall and associates[22,23], have reported a more positive picture regarding subjective quality of life outcomes of case management. Lafave et al.[24] showed that patients in ACT reported better quality of life than clients in hospital-based programs after one year of treatment. McGrew and associates[25] found that the quality of life of persons participating in ACT improved, as measured by both patient and staff ratings. In two literature reviews based on a large number of original studies, Mueser and associates[4,26] concluded that ACT moderately improved subjective quality of life. Thus, findings regarding subjective quality of life as the outcome of ACT are inconclusive, indicating the need for more refined research.

According to a recent systematic review[23], less intensive forms of case management than ACT are not superior to standard care in any respect. A randomized study comparing patients in case management with those in standard care exemplifies this conclusion by showing that there was no difference between the groups at a 36-month follow-up[27]. The superiority of ACT compared to less intensive forms of case management was illustrated in an individual study examining the relationship between type of case management and quality of life. The results showed that monitoring, an aspect of assertive outreach, was effective in terms of being associated with better quality of life, while other forms of case management were not[28].

Social Skills Training

The rehabilitation approaches described above target a broad array of skills, but since social skills have been identified as a crucial factor, shown to be deficient in people with a disability following severe mental illness, social skills training has been developed as a specific rehabilitation method. However, outcome studies of this type of training have their natural focus on independent living skills, as apparent in several reports[4,17]. Moreover, investigations of the effects of social skills training tend to rely on outcome measures that are close approximations of the training exercises used, often performed in the training setting, and this type of measurement generally yields greater effect sizes (moderate to strong) than assessments based on measures that differ in form from the training exercises and are performed outside the training setting[29,30]. The lasting effects of social skills training have been questioned[8], and patients' ability to generalize learnt skills to new areas seems to be limited[29]. However, adding training in real-life situations, aiming at bridging the gap between clinic-based skills training and the use of skills in everyday life, seems to promote social adjustment[31]. A review confined to randomized studies[30] indicted less positive effects of social skills training than previous reviews based on a broader range of studies, and concluded that the evidence so far is unconvincing.

A focus on quality of life seems to be lacking in most outcome studies of social skills training. An early review that included quality of life as outcome reported no effects[32]. In a more recent review, Pilling et al.[30] concluded that there was only one randomized study that has used quality of life as an outcome measure. That study showed that the subjective quality of life improved after social skills training, but not significantly more than after participation in the comparison treatment, which was occupational therapy[17]. Considering the strong emphasis placed on social skills training in clinical practice and the state of research over the past three decades, it appears that clarifying the role of social skills training in promoting quality of life is a matter of urgency.

Other Approaches

In addition to the rehabilitation methods discussed above, a few other approaches deserve mentioning. Rosenfield[33] showed that a program based on empowerment promoted overall quality of life, possibly mediated by perceptions of mastery. Empowerment- or mastery-based rehabilitation programs[34,35] have not been extensively evaluated regarding their impact on quality of life. However, perceived empowerment has been shown to be related to subjective quality of life[36,37]. Thus, empowerment and mastery potentially offer a new area of research.

Family interventions have been shown to be effective in terms of preventing relapse, re-admission, and treatment dropout[4,38,39], but quality of life does not seem to be considered an important outcome in this area. In a review by Pilling et al.[39] none of the studies reviewed used quality of life as an outcome measure, while only 2 out of 28 studies in a Cochrane review by Pharoah et al.[38] had addressed quality of life. The conclusion in the latter study was that the overall quality of life of family members may increase as a result of family intervention.

Among people with severe mental illness, schizophrenia is the prevailing diagnosis, and researchers have consistently found that people with schizophrenia score more poorly than others in a wide array of cognitive tasks[40]. Thus, cognitive rehabilitation techniques have been developed to remedy such impairments. Cognitive rehabilitation has not been shown to be more effective than comparison treatments regarding cognitive functioning, however, but the fact that clients in cognitive rehabilitation have been shown to improve more than a comparison group receiving occupational therapy regarding self-esteem[41] indicates that subjective quality of life could be a relevant outcome measure in any future trials addressing cognitive rehabilitation.

DAILY LIFE AND SUBJECTIVE QUALITY OF LIFE IN PEOPLE WITH SEVERE MENTAL ILLNESS

Life is characterized by involvement in different types of daily occupations, which may be defined in several ways. The definition given by the Canadian Association of Occupational Therapists[42] includes productive occupations (such as engaging in paid work, studies, and household chores), self-care, and leisure occupations. For people with mental illness, the social network is a crucial aspect of daily life[43,44,45]. Although included in daily occupations, which often occur in social contexts[42], social interaction may be distinguished as a specific area of daily life. A third perspective of daily life is people's objective life conditions. Thus, in examining the links between daily life and subjective quality of life, this section will focus on daily occupations, the social network, and objective life conditions.

Daily Occupations

Studies on persons with severe mental illness have shown that satisfaction with daily occupations and engagement in tasks of daily life are consistently related to a better subjective quality of life[46–50]. Research findings also suggest that daily activities need to be perceived as satisfying, valuable and meaningful in order to promote a good quality of life[18,46,51,52]. A study of activity-related life roles showed that having valued roles as a worker, a friend, a family member, or a hobbyist was associated with better subjective quality of life[53]. Although no direct causal relationships has been demonstrated in these studies, the consistent relationship found suggests that engagement in different types of everyday tasks might be an important rehabilitation goal in promoting quality of life.

Paid employment on the competitive labor market is the daily occupation most often investigated regarding its relationship to quality of life. Several studies have shown that people who have employment are more satisfied with their quality of life than those who are unemployed[47,54–59]. This was also the conclusion of a review by Marwaha and Johnson[12]. This conclusion was also reported in a study that controlled for different clinical characteristics, such as diagnosis, psychopathology, and history of illness[57]. Although the direction of such relationships has not been established[12], findings such as these could be regarded as being in conflict with the findings described above, that vocational training does not seem to result in a better quality of life. Results like these may also be seen as indirect evidence that vocational rehabilitation has the potential of stimulating clients towards better quality of life, although the quality-of-life-enhancing features of vocational rehabilitation have not yet been fully identified. Rüesch et al.[57] showed that social support mediated the impact of employment on quality of life, and they also suggested that social support should be regarded as an influential factor when designing vocational training programs. Thus, social support may constitute one such not yet sufficiently identified and evaluated aspect of vocational rehabilitation. Another factor could be the job satisfaction associated with the vocational rehabilitation, since job satisfaction has been shown to be related to quality of life[60,61].

Kager et al.[54] found that, in a mixed group of mentally ill patients, those who had competitive employment scored higher than those who were unemployed or were early retired regarding the quality of life domains of subjective well-being, leisure, social relationships, and general quality of life. On the other hand, an investigation comparing three groups representing different types of daily occupations, work/studies, visiting day-care centers, and no regular daily occupation, found no differences between these groups regarding subjective quality of life[62]. Thus, there are conflicting results in this area as well. However, aggregated quality of life ratings may conceal differences in certain domains, and re-analysis of two samples [49,62] showed that people who were employed rated their satisfaction significantly higher in the quality of life domains of work, living, security, physical health, psychological health, and family relations. No differences were found regarding financial situation, friends, or leisure (M Eklund, manuscript in preparation).

Some attempts have been made to develop explanatory models of quality of life in which daily occupation or activity were one of the predictors tested. Rüesch et al.[57] used type of daily occupation, in terms of work or work-like occupations versus no work-like occupation, as one of the predictors. Other predicators used were diagnosis and degree of psychopathology, while living conditions and social support were defined as mediators. Occupation was related to both mediators, and directly related to one of two subjective quality of life aspects, namely satisfaction with social ties. The other quality of life variable, satisfaction with health, was not directly related to occupation. In conclusion, in this model, occupation was an important determinant of a socially loaded aspect of quality of life, further mediated by the availability of social support. In another study, actual participation in activities and events in the community (denoted physical integration) was found to explain a minor proportion of the variance in subjective quality of life[63]. A well-being variable, namely perceived symptom distress, was found to be the most important predictor. Furthermore, a theoretical model has been developed[64] based on previous findings from multi-factorial model testing and correlational studies. This model, presented in Figure 1, includes some factors not previously incorporated in multi-factorial models, namely activity level, satisfaction with daily activities, and satisfaction with medial care (medication and contact with the psychiatrist). Previously well-tested

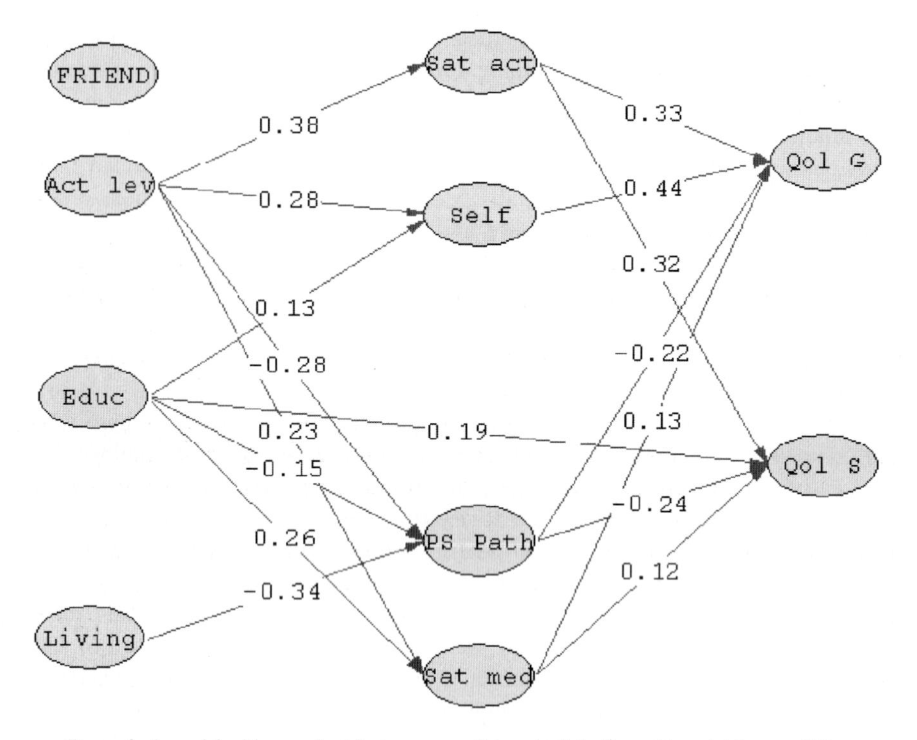

Figure 1. A model with appraised factors as mediators (satisfaction with activities, a self factor, interviewer assessed psychopathology, and satisfaction with medical care) and actual life circumstances as predictors (activity level, educational level, and living accommodation). Subjective quality of life was best explained by two factors, a general factor and a security-related factor. (Previously published in Eklund & Bäckström[64], p. 1164, Figure 2. Reprinted with kind permission of Springer Science and Business Media.)

factors such as psychopathology and self-related variables were also included. Objective life circumstances were defined as factors and subjective estimates as mediators. The analysis indicated that two quality of life factors best represented the data: a general factor and a factor related to security. Satisfaction with daily activities was related to both quality of life factors and also mediated the relationship between actual doing, in terms of activity level, and the quality of life factors. Although, as shown in Figure 1, a self-factor was found to be the most significant factor in explaining the general quality of life factor, satisfaction with activities was the most important factor regarding the quality of life factor related to security.

Social Networks

Research has shown that social networks are important for good quality of life [65–67]. This is especially true for qualitative aspects of the social network, such as perceived support, which tend to exhibit higher correlations with quality of life than quantitative aspects do, such as the size of the network[50,66,68]. Regarding the size of the network, Becker and associates[69] showed that an increase in the social network up to about 15-20 people was associated with increasing levels of an aggregated measure of subjective quality of life. Regarding satisfaction with the quality of life domain of social relations, the peak was reached at a network of 10-12 people. Thus, the size of the network seems to be important for subjective quality of life, and the optimal number of contacts seems to be 10-20.

Recent guidelines for individuals with schizophrenia and related disorders identified availability of social support and access to meaningful social roles as essential components in recovery and quality of life[11]. Social support has also been shown to mediate the relationship between daily occupation and quality of life[57]. A study comparing the subjective quality of life of patients treated within the community health care system and those in long-term hospital care indicated better quality of life among those treated in the community[70]. When social support and other potentially influencing variables were included in the analysis, the location of treatment was, however, insignificant, and social support explained most of the variation in subjective quality of life. A recent study of the social network among people with persistent mental illness indicated that higher levels of quality of life and self-esteem and being a cohabitant were related to higher scores regarding different aspects of the social network[68]. The authors concluded that this effect was probably dynamic, and, just as much as a more developed social network may lead to better quality of life, perceptions of a good quality of life probably promote a more favorable situation regarding the social network.

Yanos et al.[71] investigated the nature of the relationship between social relations and quality of life. They distinguished negative and supportive social relations, stigma, and subjective and objective quality of life and controlled for diagnosis and demographic variables. The findings indicated that negative social relations were associated with lower ratings of subjective quality of life, but not with objective quality of life. Supportive social relations were related to both subjective and objective quality of life, but more consistently to the objective aspect. Furthermore, perceived stigma mediated the relationship between negative social interactions and quality of life.

An important part of the social environment is the family. Sullivan and associates[72] found that better family relations, together with fewer side effects and fewer depressive symptoms, were related to better self-rated quality of life. Summarizing results from studies including family relations, Prince and Prince[5] concluded that positive family relations seemed to have a favorable influence and negative relations an adverse effect on quality of life and adjustment.

Objective Life Conditions

Although sometimes seen as indicators of quality of life in themselves, in terms of objective quality of life[73], objective life circumstances may also be regarded as aspects of daily life of potential interest in the perception of quality of life. Among such life conditions are standards of living, including the housing situation and the financial situation of the client. Research has generally failed to show any strong or consistent relationships between objective life circumstances and subjective quality of life[4,5,75,76]. However, living in the community is an objective condition that does seem to promote people's quality of life, as shown both in studies comparing groups with different types of housing conditions[77,78] and in studies of discharged patients, assessing changes in quality of life related to the move from hospital to community living[79,80]. Some research, however, indicates that the social contacts and support rather than where the client lives per se constitute the more important factor[70,80,81].

Studies have shown that neither the standard of living nor the patient's economic situation seems to be related to subjective quality of life[5,56]. In fact, a lower income may be related to a better quality of life when combined with worse psychopathology[64]. This is probably due to the fact that patients with severe psychiatric symptoms receive a disability allowance affording them a stable, but low, income, which in turn makes them feel secure. Thus, stability and security regarding their financial situation may be more important than the size of the income.

Although correlational studies generally have failed to identify any links between objective life circumstances and subjective quality of life, findings from multi-factorial research indicate that different types of objective life conditions may account for minor proportions of the variance in subjective quality of life when mediated through self-related factors or autonomy[82–84]. This indicates that the appraisal process, suggested to be important by Barry et al.[74], does have an impact on ratings of subjective quality of life and might explain part of the very low or lack of associations between objective life circumstances and subjective quality of life.

The non-findings regarding a relationship between objective life conditions and subjective quality of life should not be taken as indicating that this is an uninteresting area of research that should be abandoned. Curiously, in studies of persons with severe mental illness there seems to be a relationship between higher education and worse ratings of subjective quality of life. This is a common finding in studies that have included educational level in the analysis[58,64,75], although there are also reports of no relationship[85]. It has been speculated that the gap between expectations and how life actually is, is wider for those who have higher education, but as yet there has been no thorough investigation or discussion regarding how this association may be explained. Since education is an ingredient often used in rehabilitation programs, the nature of this relationship and how it may be influenced is of interest.

Is there a General Satisfaction Factor?

The frequently repeated finding of a relationship between different satisfaction variables and quality of life raises the question of whether there is a general satisfaction factor that manifests itself in different measures of satisfaction, such as quality of life, general well-being, satisfaction with care, satisfaction with daily occupations, and job satisfaction. This has previously been suggested by Priebe et al.[86]. The model developed by Rüesch et al.[57] indicated some unexplained error correlation, which they interpreted as reflecting a general satisfaction factor, not made explicit in their model. Similarly, a study of satisfaction with care showed that although a certain proportion of the variance seemed to be specific, part of it could be explained by a general satisfaction factor, also expressing itself in quality of life[87]. Speaking against the hypothesis that a general satisfaction factor could explain the variance in subjective quality of life are some findings in a study by Eklund and Hansson[16], in which longitudinal data based on both the patients' self-rated quality of life and a professional's rating of the patients' quality of life were compared. The professional's rating was based on an interview carried out by another professional regarding the patient's subjective quality of life. The rater was blind to whether the interview was performed before or after the intervention. From both the patient's and the professional's perspectives, the patient's quality of life improved during occupational therapy in two areas, interpersonal relationships and internal psychological state, but not in a third area that concerned satisfaction with external life conditions. The external life conditions were not influenced by the intervention. They de facto did not change, and neither did the satisfaction concerning these areas. Despite this indication of specific satisfaction in different areas, when interpreting quality of life models that include satisfaction variables as determinants and mediators of quality of life, one should bear in mind that some of this variation could be due to a common satisfaction factor.

GENERAL DISCUSSION

Potential Explanations of Non-findings Regarding Effects of Psychiatric Rehabilitation on Subjective Quality of Life

Despite the fact that some progress has been demonstrated in several studies of psychiatric rehabilitation regarding objective life circumstances, this has not been accompanied by improved subjective quality of life[5], and one may speculate on the reasons why. Six explanations can be gleaned from the scientific work so far.

i) Quality of life measures have been criticized for not adequately reflecting the views of patients, since the quality of life domains are typically defined by professionals, sometimes on the basis of interviews with samples from the general population. Thus, the reason why the outcomes of psychiatric rehabilitation are not reflected in changes in subjective quality of life could be methodological problems, such as a misfit between the measures used in quality of life assessments and the needs and experiences of the patients[5], or the fact that the appraisal process is complex and may lead to a positive response bias[74].

ii) It could also be that subjective quality of life simply does not change as a result of psychiatric rehabilitation, as proposed by some researchers[5,8].

iii) Personal factors, such as personality, a sense of self, and self-esteem, explain substantial proportions of the variance in subjective quality of life[50,64,82–84,88,89], and factors such as these tend to be fairly stable over time. A seven-year follow-up study of patients with schizophrenia illustrated this, by showing that the participants had the same scores as at baseline[90].

iv) Another likely reason is that people accommodate to their life situation and recalibrate their expectations and personal goals along with changes in their life circumstances [1,5]. Being exposed to a condition for a long time, e.g. in terms of receiving a certain type of psychiatric care, is often associated with better quality of life [78], and the effects of negative events diminish over time [75], which renders support to this assumption.

v) Moreover, factors that are common to different rehabilitation approaches, such as the ward atmosphere, the client-staff relationship, or social support, shown to be related to quality of life outcomes [50,66,70,91], may be of greater importance than the method of rehabilitation per se.

vi) Finally, longer follow-up periods than those employed in most longitudinal studies might be needed to detect quality of life changes. A two-year follow-up of return to the community by long-term psychiatric patients indicated stability or improvements in subjective quality of life over the follow-up period [92]. These results support the suggestion that changes in subjective quality of life might not take place immediately after rehabilitation, but evolve during a subsequent consolidation period [15].

In essence, the above suggested explanations relate to different phenomena. Issues i) and ii) have to do with insufficient or inadequate measurement. Issues iii) and iv) are concerned with factors extraneous to both the rehabilitation process and the measurement issue and pertain to naturally occurring phenomena that interact with perceptions of quality of life, namely personal factors and adjustment over time. Issue v) concerns possibly obscured mechanisms in rehabilitation, while issue vi) has bearing on the research design chosen when investigating the effects of psychiatric rehabilitation on subjective quality of life.

Needs for Future Research

Issues iii) and iv) above may be seen as conflicting hypotheses. On the one hand, a person's perception of his or her quality of life is supposed to improve over time together with adaptation to the situation at hand. This suggests that the immediate outcome after an intervention is important. On the other hand, it is proposed that a certain period of consolidation is needed in order for quality of life improvements to emerge, which indicates that follow-ups might reveal interesting results that were not seen immediately after completed rehabilitation. On the basis of current research it is difficult to say to what extent any changes after a follow-up period are due to accommodation to life circumstances or to increased satisfaction as a result of having practiced newly acquired skills in the real-life context. Future research will be needed to elucidate this. The proposed explanations are in fact hypotheses, implying that a number of studies would have to be undertaken to ascertain whether any of them are valid.

Not only well-established rehabilitation approaches with potential influence on subjective quality of life are presented in this chapter, but also less investigated ones, yielding tentative results that may serve as inspiration for future research to confirm or refute. In general, the literature reviewed revealed inconsistent findings regarding quality of life outcomes of different types of psychiatric rehabilitation, and in most areas few controlled studies could be identified that had used quality of life as an indicator of effect. Furthermore, longitudinal research was scarce. Since rehabilitation is a very common alternative for many persons with severe mental illness this situation is far from satisfactory. Psychiatric rehabilitation places a tremendous economic burden on society, and the lack of evaluation means that we do not know if resources are being used in a cost-effective way. Quality of life is a justified indicator of successful rehabilitation[3–5]. If external needs are met, such as a meaningful daily occupation, this is of course very valuable, and such external factors may be regarded as objective indicators of quality of life[73]. However, if such external aspects of quality of life are not accompanied by increased satisfaction in the individual, then a vital aspect of the desired outcome is missing.

Thus, more and better-designed research is urgently needed in the area of psychiatric rehabilitation and subjective quality of life. The research cited in this chapter varies considerably in methodological respects. Regarding vocational rehabilitation, ACT, and social skills training, the research is fairly broad in scope; several methodologically sound studies have been performed, and outcomes closely related to the rehabilitation focus have been investigated. However, quality of life has been inadequately addressed, and therefore more studies are needed that target subjective quality of life. In other areas, such as rehabilitation based on activities and mastery/empowerment, the body of research is limited and the outcomes are largely uninvestigated. In these areas, methodologically sound evaluation studies that include quality of life aspects must be designed and carried out; otherwise it is difficult to justify these types of intervention at all.

Future outcome studies addressing quality of life should include a follow-up of at least 12 months, preferably longer, since previous research has indicated that changes in quality of life may require some time to become established. If current forms of psychiatric rehabilitation are shown to be ineffective from a quality of life perspective in studies with adequate follow-up periods, psychiatric rehabilitation needs to be reappraised and new methods developed. There are indications that existing types of psychiatric rehabilitation, including vocational rehabilitation, place too little emphasis on the quality of the social support associated with rehabilitation. Furthermore, in relation to vocational rehabilitation, the issue of job satisfaction needs to be investigated regarding its influence on non-vocational outcomes such as subjective quality of life.

CONCLUSIONS

"Everybody has won and all must have prizes..." This quote from Alice in Wonderland has often been used in psychotherapy research to illustrate the notion that no single theoretical school is superior to any other, but that all types of well-administered interventions seem effective[93]. Regarding psychiatric rehabilitation, the reverse could be said: Nobody has won and nobody is entitled to a prize.

However, since research has shown that several approaches are associated with quality of life amelioration during the rehabilitation period, this is probably too pessimistic a standpoint. The fact still remains that quality of life has not been shown to improve to any substantial degree as a result of psychiatric rehabilitation. The reasons for this are probably manifold. The quality of life measures used may not be sufficiently sensitive to change, or the follow-up period may be too short. Moreover, quality of life is a complex phenomenon, influenced to a great deal by factors related to personality and the self. This makes quality of life relatively stable over time, and it may therefore not be suitable as a measure of outcome. On the other hand, a better life is the ultimate goal of psychiatric rehabilitation, and therefore quality of life cannot be abandoned as an indicator of outcome. Several indicators of relationships between factors characterizing people's daily life further underscore this. Such associations, e.g. between meaningful daily activities and subjective quality of life, and between social support and quality of life, suggest that there might be undiscovered potentials in psychiatric rehabilitation. Thus, more, well-designed studies, including multi-methodological approaches and long follow-up periods, are needed to clarify how quality of life is affected by different rehabilitation strategies.

REFERENCES

1. Cheng S-T. Subjective quality of life in the planning and evaluation of programmes. Eval Progr Plann 1988; 11:123–34.
2. Elizur A. Rehabilitation of the disabled mentally ill in the community. Isr J Psychiatry Relat Sci 2004; 41:248–58.
3. Sartorius N. Rehabilitation and quality of life. Hosp Comm Psychiatry 1992; 43:1180–1.
4. Mueser KT, Bond GR. Psychosocial treatment approaches for schizophrenia. Curr Opin Psychiatry 2000; 13:27–35.
5. Prince PN, Prince CR. Subjective quality of life in the evaluation of programs for people with serious and persistent mental illness. Clin Psychol Rev 2001; 7:1005–36.
6. Kopelowicz A, Liberman RP. Integrating treatment with rehabilitation for persons with major mental illnesses. Psychiatr Serv 2003; 54:1491–8.
7. Cook JA, Razzano L. Vocational rehabilitation for persons with schizophrenia: Recent research and implications for practice. Schizophr Bull 2000; 26:87–103.
8. Drake RE, Becker DR, Clark RE, Meuser KT. Research on the individual placement and support model of supported employment. Psychiatr Q 1999; 70:289–301.
9. Bond GR. Supported employment: Evidence for an evidence-based practice. Psychiatr Rehab J 2004; 27:345–59.
10. Crowther R, Marshall M, Bond G, Huxley P. Vocational rehabilitation for people with severe mental illness (Review). The Cochrane Library 2006, Issue 2. Chichester: John Wiley & Sons, 2006.
11. Royal Australian and New Zealand College of Psychiatrists Clinical Practice Guidelines Team for the Treatment of Schizophrenia and Related Disorders. Royal Australian and New Zealand College of Psychiatrists clinical practice guidelines for the treatment of schizophrenia and related disorders. Aust N Z J Psychiatry 2005; 39:1–30.
12. Marwaha S, Johnson S. Schizophrenia and employment. Soc Psychiatry Psychiatr Epidemiol 2004; 39:337–49.

13. Steultjens EMJ, Dekker J., Bouter LM, Leemrijse CJ, van den Ende CHM. Evidence of the efficacy of occupational therapy in different conditions: an overview of systematic reviews. Clin Rehab 2005; 19:247–54.
14. Eklund M. Outcome of occupational therapy in a psychiatric day care unit for long-term mentally ill patients. Occupational Therapy in Mental Health 1999; 14(4):21–45.
15. Eklund M, Hansson L. Stability of improvement in patients receiving psychiatric occupational therapy: A 1-year follow-up. Scand J Occup Ther 1997; 4:15–22.
16. Eklund M, Hansson L. Subjectively and independently assessed quality of life for the long-term mentally ill. Nord J Psychiatry 1998; 52:285–94.
17. Liberman RP, Wallace CJ, Blackwell G, Kopelowicz A, Vaccaro JV, Mintz J. Skills training versus psychosocial occupational therapy for persons with persistent schizophrenia. Am J Psychiatry 1998; 155:1087–91.
18. Champney TF, Cox Dzurec L. Involvement in productive activities and satisfaction with living situation among severely mentally disabled adults. Hosp Community Psychiatry 1992; 43:899–903.
19. Rebeiro KL, Day DG, Semeniuk B, O'Brien MC, Wilson B. Northern Initiative for Social Action: an occupation-based mental health program. Am J Occup Ther 2001; 55:493–500.
20. Catty J, Burns T, Comas A. Day centers for severe mental illness (Review). The Cochrane Library 2006, Issue 2. Chichester: John Wiley & Sons, 2006.
21. Robertson L, Connaughton J, Nicol M. Life skills programmes for chronic mental illnesses (Review). The Cochrane Library 2006, Issue 2. Chichester: John Wiley & Sons, 2006.
22. Marshall M, Lockwood A. Assertive community treatment for people with severe mental disorders (Review). The Cochrane Library 2006, Issue 2. Chichester: John Wiley & Sons, 2006.
23. Marshall M, Gray A, Lockwood A, Green R. Case management for people with severe mental disorders (Review). The Cochrane Library 2006, Issue 2. Chichester: John Wiley & Sons, 2006.
24. Lafave HG, de Souza HR, Gerber GJ. Assertive community treatment of severe mental illness: a Canadian experience. Psychiatr Serv 1996; 47:757–9.
25. McGrew JH, Bond GR, Dietzen L, McKasson M, Miller LD. A multisite study of client outcomes in assertive community treatment. Psychiatr Serv 1995; 46:696–701.
26. Mueser KT, Bond GR, Drake RE, Resnick SG. Models of community care for severe mental illness: a review of research on case management. Schizophr Bull 1998; 24:37–74.
27. Björkman T, Hansson L, Sandlund M. Outcome of case management based on the strengths model compared to standard care. A randomised controlled trial. Soc Psychiatry Psychiatr Epidemiol 2002; 37:147–52.
28. Huxley P, Warner R. Case management, quality of life, and satisfaction with services of long-term psychiatric patients. Hosp Community Psychiatry 1992; 43:799–802.
29. Dilk MN, Bond GR. Meta-analytic evaluation of skills training research for individuals with severe mental illness. J Consult Clin Psychol 1996; 64:1337–46.
30. Pilling S, Bebbington P, Kuipers E, Garety P, Geddes J, Martindale B, Orbach G, Morgan C. Psychological treatments in schizophrenia: II. Meta-analyses of randomized controlled trials of social skills training and cognitive remediation. Psychol Med 2002; 32:783–91.
31. Liberman RP, Glynn S, Blair KE, Ross D, Marder SR. In vivo amplified skills training: promoting generalization of independent living skills for clients with schizophrenia. Psychiatry 2002; 65:137–55.
32. Wallace CJ, Nelson CJ, Liberman RP, Aitchison RA, Lukoff D, Elder JP, Ferris C. A review and critique of social skills training with schizophrenic patients. Schizophr Bull 1980; 6:42–63.
33. Rosenfield S. Factors contributing to the subjective quality of life of the chronic mentally ill. J Health Soc Behav 1992; 33:299–315.
34. Gunatilake S, Ananth J, Parameswaran S, Brown S, Silva W. Rehabilitation of schizophrenic patients. Curr Pharm Des 2004; 10:2277–88.
35. Starkey D, Flannery RB. Schizophrenia, psychiatric rehabilitation, and healthy development: A theoretical framework. Psychiatr Q 1997; 68:155–66.
36. Crane-Ross D, Lutz WJ, Roth D. Consumer and case manager perspectives of service empowerment: relationship to mental health recovery. J Behav Health Serv Res 2006; 33:142–55.

37. Hansson L, Björkman T. Empowerment in people with a mental illness: reliability and validity of the Swedish version of an empowerment scale. Scand J Caring Sci 2005; 19:32–8.
38. Pharoah FM, Rathbone J, Mari JJ, Streiner D. Family intervention for schizophrenia (Review). The Cochrane Library 2006, Issue 3. Chichester: John Wiley & Sons, 2006.
39. Pilling S, Bebbington P, Kuipers E, Garety P, Geddes J, Orbach G, Morgan C. Psychological treatments in schizophrenia: I. Meta-analysis of family intervention and cognitive behaviour therapy. Psychol Med 2002; 32:763–82.
40. Hayes RL, McGrath JJ. Cognitive rehabilitation for people with schizophrenia and related conditions (Review). The Cochrane Library 2005, Issue 4. Chichester: John Wiley & Sons, 2005.
41. Wykes T, Reeder C, Corner J, Williams C, Everitt B. The effects of neurocognitive remediation on executive processing in patients with schizophrenia. Schizophr Bull 1999; 25:291–307.
42. Canadian Association of Occupational Therapists. Enabling occupation: An occupational therapy perspective. Ottawa, Canada: CAOT Publications, 2002.
43. Becker T, Albert M, Angermeyer MA, Thornicroft G. Social networks and service utilization in patients with severe mental illness. In Tansella M (ed), Making Rational Mental Health Services (pp. 113–125). Rome: Il Pensiero Scientifico Editore, 1997.
44. Clinton M, Lunney P, Edwards H, Weir R, Barr J. Perceived social support and community adaptation in schizophrenia. J Adv Nurs 1998; 27:955–65.
45. Howard L, Leese M, Thornicroft G. Social networks and functional status in patients with psychosis. Acta Psychiatr Scand 2000; 102:376–85.
46. Aubin G, Hachey R, Mercier C. Meaning of daily activities and subjective quality of life in people with severe mental illness. Scand J Occup Ther 1999; 6:53–62.
47. Eklund M, Hansson L, Bejerholm U. Relationships between satisfaction with occupational factors and health-related variables in schizophrenia outpatients. Soc Psychiatry Psychiatr Epidemiol 2001; 36:79–85.
48. Goldberg B, Britnell ES, Goldberg J. The relationship between engagement in meaningful activities and quality of life in persons disabled by mental illness. Occupational Therapy in Mental Health 2002; 19(2):17–44.
49. Laliberte-Rudman D, Yu B, Scott E, Pajouhandeh P. Exploring of the perspectives of persons with schizophrenia regarding quality of life. Am J Occup Ther 2000; 54:137–47.
50. Prince PN, Gerber GJ. Subjective well-being and community integration among clients of assertive community treatment. Qual Life Res 2005; 14:161–9.
51. Eklund M, Erlandsson L-K, Persson D. Occupational value among individuals with long-term mental illness. Can J Occup Ther 2003; 70:276–84.
52. Eklund M. Satisfaction with daily occupations – a tool for client evaluation in mental health care. Scand J Occup Ther 2004; 11:136–42.
53. Eklund M. Psychiatric patients' occupational roles: Changes over time and associations with self-rated quality of life. Scand J Occup Ther 2001; 8:125–30.
54. Kager A, Lang G, Berghofer G, Henkel H, Schmitz M. Die Bedeutung von Arbeit für die Lebensqualität psychisch kranker Menschen [The impact of work on quality of life for persons with severe mental illness]. Nervenheilkunde 2000; 19:560–5.
55. Mueser KT, Becker DR, Torrey WC, Xie H, Bond GR, Drake RE, Dain BJ. Work and nonvocational domains of functioning in persons with severe mental illness: A longitudinal analysis. J Nerv Ment Dis 1997; 185:419–26.
56. Priebe S, Warner R, Hubschmid T, Eckle I. Employment, attitudes towards work, and quality of life among people with schizophrenia in three countries. Schizophr Bull 1998; 24:469–77.
57. Rüesch P, Graf J, Meyer PC, Rössler W, Hell D. Occupation, social support and quality of life in persons with schizophrenia or affective disorders. Soc Psychiatry Psychiatr Epidemiol 2004; 39:686–94.
58. Skantze K, Malm U, Dencker SJ, May PRA, Corrigan P. Comparison of quality of life with standard of living in schizophrenic out-patients. Br J Psychiatry 1992; 161:797–801.
59. Van Dongen CJ. Quality of life and self-esteem in working and nonworking persons with mental illness. Community Ment Health J 1996; 32:535–48.

60. Romney DM, Evans DR. Toward a general model of health-related quality of life. Qual Life Res 1996; 5:235–41.
61. Tsang HWH, Wong A. Development and validation of the Chinese version of Indiana Job Satisfaction Scale (CV-IJSS) for people with mental illness. Int J Soc Psychiatry 2005; 51:177–91.
62. Eklund M, Hansson L, Ahlqvist C. The importance of work as compared to other forms of daily occupations for wellbeing and functioning among persons with long-term mental illness. Community Ment Health J 2004; 40:465–77.
63. Chan PS, Krupa T, Stuart Lawson J, Eastabrook S. An outcome in need of clarity: Building a predictive model of subjective quality of life for persons with severe mental illness living in the community. Am J Occup Ther 2005; 59:181–90.
64. Eklund M, Bäckström M. A model of subjective quality of life for outpatients with schizophrenia and other psychoses. Qual Life Res 2005; 14:1157–68.
65. Barry MM, Zissi A. Quality of life as an outcome measure in evaluating mental health services: a review of the empirical evidence. Soc Psychiatry Psychiatr Epidemiol 1997; 32:38–47.
66. Bengtsson-Tops A, Hansson L. Quantitative and qualitative aspects of the social network in schizophrenic patients living in the community. Relationships to sociodemographic characteristics and clinical factors and subjective quality of life. Int J Soc Psychiatry 2001; 47:67–77.
67. Brunt D, Hansson L. The social networks of severely mentally ill persons in inpatient settings and sheltered community settings. Journal of Mental Health 2002; 11:611–21.
68. Eklund M, Hansson L. Social network among people with persistent mental illness: Associations with sociodemographic, clinical and health-related factors. Manuscript submitted for publication, 2006.
69. Becker T, Leese M, Clarkson P, Taylor RE, Turner D, Kleckham J, Thornicroft, G. Links between social networks and quality of life: an epidemiologically representative study of psychotic patients in South London. Soc Psychiatry Psychiatr Epidemiol 1998; 33:299–304.
70. Rössler W, Salize HJ, Cucciaro G, Reinhard I, Kernig C. Does the place of treatment influence the quality of life of schizophrenics? Acta Psychiatr Scand 1999; 100:142–48.
71. Yanos PT, Rosenfield S, Horwitz AV. Negative and supportive social interactions and quality of life among persons diagnosed with severe mental illness. Community Ment Health J 2001; 37:405–19.
72. Sullivan G, Wells KB, Leake B. Clinical factors associated with better quality of life in a seriously mentally ill population. Hosp Community Psychiatry 1992; 43:794–8.
73. Post MWM, de Witte LP, Schrijvers AJP. Quality of life and the ICIDH: Towards an integrated conceptual model for rehabilitation outcomes research. Clin Rehabil 1999; 13:5–15.
74. Barry MM, Crosby C, Bogg J. Methodological issues in evaluating the quality of life of long-stay psychiatric patients. Journal of Mental Health 1993; 2:43–56.
75. Chan GWL, Ungvari GS, Shek DTL, Leung JJP. Hospital and community-based care for patients with chronic schizophrenia in Hong Kong. Soc Psychiatry Psychiatr Epidemiol 2003; 38:196–203.
76. Corten P, Mercier C, Pelc I. "Subjective qol": clinical model for assessment of rehabilitation treatment in psychiatry. Soc Psychiatry Psychiatr Epidemiol 1994; 29:178–83.
77. Brunt D, Hansson L. The quality of life of persons with severe mental illness across housing settings. Nord J Psychiatry 2004; 58:293–8.
78. Lehman AF, Possidente S, Hawker F. The quality of life of chronic mental patients in a state hospital and in community residences. Hosp Community Psychiatry 1986; 37:901–7.
79. Gibbons JS, Butler JP. Quality of life for 'new' long-stay psychiatric in-patients: the effects of moving to a hospital. Br J Psychiatry 1987; 151:347–54.
80. Leff J, Dayson D, Gooch C, Thornicroft G, Wills W. Quality of life of long-stay patients discharged from two psychiatric institutions. Psychiatr Serv 1995; 47:62–7.
81. Levitt AJ, Hogan TP, Bucosky CM. Quality of life in chronically mentally ill patients in day treatment. Psychol Med 1990; 20:703–10.
82. Eklund M, Bäckström M, Hansson L. Personality and self-variables: Important determinants of subjective quality of life in schizophrenia outpatients. Acta Psychiatr Scand 2003; 108:134–43.
83. Holloway F, Carson J. Quality of life in severe mental illness. Int Rev Psychiatry 2002; 14:175–84.

84. Zissi A, Barry MM, Cochrane R. A mediational model of quality of life for individuals with severe mental health problems. Psychol Med 1998; 28:1221–30.
85. Ritsner M, Gibel A, Ratner Y. Determinants of change in perceived quality of life in the course of schizophrenia. Qual Life Res 2006; 15:515–26.
86. Priebe S, Kaiser W, Huxley PJ, Röder-Wanner U-U, Rudolf H. Do different subjective evaluation criteria reflect distinct constructs? J Nerv Ment Dis 1998; 186:385–92.
87. Eklund M, Hansson L. Determinants of satisfaction with community based psychiatric services: A cross-sectional study among schizophrenia outpatients. Nord J Psychiatry 2001; 55:413–8.
88. Hansson L, Eklund M, Bengtsson-Tops A. The relationship of personality dimensions as measured by the TCI and quality of life in individuals with schizophrenia or schizoaffective disorder living in the community. Qual Life Res 2001; 10:133–9.
89. Ritsner M, Farkas H, Gibel A. Satisfaction with quality of life varies with temperament types of patients with schizophrenia. J Nerv Ment Dis 2003; 191:668–74.
90. Tempier R, Mercier C, Leoffre P, Caron J. Quality of life and social integration of severely mentally ill patients: A longitudinal study. J Psychiatry Neurosci 1997; 22:249–55.
91. Eklund M, Hansson L. Relationships between characteristics of the ward atmosphere and treatment outcome in a psychiatric day care unit based on occupational therapy. Acta Psychiatr Scand 1997; 95:329–35.
92. Gerber GJ, Coleman GE, Johnston L, Lafave HG. Quality of life of people with psychiatric disabilities 1 and 3 years after discharge from hospital. Qual Life Res 1994; 3:379–83.
93. Lambert MJ, Bergin AE. The effectiveness of psychotherapy. In: Bergin AE, Garfield SL eds. Handbook of psychotherapy and behavior change, 4th ed. New York: John Wiley & Sons, 1994:143–89.

CHAPTER 20

COST-UTILITY ANALYSIS

Research and practical applications

MOJCA Z. DERNOVSEK[1,2], VALENTINA PREVOLNIK-RUPEL[3] AND ROK TAVCAR[1,4]

[1]*University Psychiatric Hospital, Studenec 48, SI-1260 Ljubljana-Polje, Slovenia*
[2]*Institute of Public Health of Republic of Slovenia, Ljubljana, Slovenia*
[3]*Ministry of Health of Republic of Slovenia, Ljubljana, Slovenia*
[4]*University of Ljubljana, School of Medicine, Chair of Psychiatry, Ljubljana, Slovenia*

Abstract: Cost-utility analysis is a method which is most often used when benefits cannot be expressed in monetary (profit) or metric values (days of sick leave). The utilities in cost-utility analyses are in fact preferences of each person, a selected group, or the whole population. Since quality of life is one of the preferences, estimation of quality of life is frequently used in health economics. The results of cost-utility analyses are expressed in QALY – quality adjusted life years. QALY indicates the average number of years of quality life which a person with a defined health status will be able to live in a case that a certain intervention is carried out. This indicator therefore shows the cost of intervention with regard to a specific outcome, life in quality. In psychiatry the measurements of health outcome like survival, disability, sick leave, quality of life, satisfaction of clients, etc., were traditionally important but in the recent years the awareness of costs associated with any health intervention has grown. In studying mental disorders the situation seems somewhat specific since traditional outcome indicators do not always reflect the many (complex) faces of mental disorder. Although the methodology is appropriate to compare health economic issues of different mental disorders these studies are scarce, probably due to a complicated design. It is much easier to conduct an outcome study with a single diagnostic category. On the other hand there were many studies that were comparing costs and other outcomes of different preventive and therapeutic interventions in a single diagnostic category. The results of cost-utility analyses are useful in many situations: planning of service development, resource allocations, to find out the best available intervention for persons with a certain health status, etc.

This chapter provides a short introduction to economic analysis giving a special emphasis to a cost-utility analysis and defining its place among other methods in health economics. Beside theoretical considerations, several practical applications of cost-utility analysis using QALYs are discussed

Keywords: Cost-utility analysis, effectiveness, QALY, health states

M.S. Ritsner and A.G. Awad (eds.), Quality of Life Impairment in Schizophrenia, Mood and Anxiety Disorders, 373–384.

INTRODUCTION

Health related quality of life represents an important measure of therapeutic effectiveness and has hence become an important element in the economic evaluations. Economic evaluations are important to facilitate the choices which must be made concerning the deployment of the resources. Therefore each type of economic analysis compares the costs and the outcome of the therapeutic procedure. Costs are relatively easy to estimate, whereas many debates are going on at the outcome side. In order to be able to compare the different options for the use of common resources the quantification of health outcomes using a common measurement unit is required[1]. Cost-utility analysis was developed to address the problem of comparing interventions with different measures of primary effectiveness. It provides a method through which the various disparate outcomes can be combined into a single composite summary outcome, allowing broad comparisons across widely differing interventions. Besides, cost-utility analysis is based on people's utilities or preferences, through which a higher value is attached to the better or higher-quality (HRQoL) outcomes. The utility of QoL in clinical psychiatric research, drug trials, and economic analyses could be enhanced by appropriate conceptual models which endorse subject's perception of the outcome of an interaction between severity of clinical symptoms, side effects including subjective responses to psychoactive drugs, and the level of psychosocial performance[2].

COST-UTILITY ANALYSIS

Decades ago as economics entered medicine, it was perceived as a threat likely to lead threat to dehumanization of health care. Insistence on quality of life, its measurement and making its improvements as a desirable outcome of health care lifts health care from the market place to the level of equity, where it belongs[3].

So what do Economic Evaluation and Quality of Life have in Common?

Economic analyses have become a requirement in some countries as a result of the introduction of spending limits which prompted a search for higher efficiency. Treatments or drugs for which the highest benefit per unit of costs is proven has become a priority. This latter does not mean that only positive effects of a particular treatment should be considered in QoL analyses, but one must be aware also of possible negative effects of treatment like adverse events[4].

Though cost measurement seems to be similar among studies – it is the measurement of benefits, which takes many approaches. Most prospective treatment studies include several determinants of outcome such as (health-related QoL, clinical outcome, medical costs, adverse events which then are used to perform) cost-effectiveness analysis. However, it is not clear why most authors stop at this point and do not perform cost-utility analyses. The reasons for this are not clear but may include sample size, inappropriate instruments, complicated procedure of cost-utility analysis, limited research resources or simply satisfaction with good results of cost-effectiveness analyses.

Two Examples were Selected to Show the Usefulness of Cost Utility Analysis

Namjoshi et al.[5] studied economic, clinical and QoL outcomes in a 3-weeks randomized controlled clinical trial of olanzapine treatment in mania followed by 49-weeks open phase treatment. The results showed that olanzapine was effective in reducing clinical symptoms, improving QoL of patients and was cost-effective (lower treatment costs in study year than in previous year) but there are no data on utilities based on patients' preferences. In-patient hospital stay was significantly reduced under olanzapine treatment resulting in lower treatment costs in the study year. Shorter hospitalization could be related to several different issues: more stable course of illness, better compliance or better tolerance of olanzapine in such bipolar population. It would enhance our understanding of results and increase the impact of the study if patients' preferences were elicited. This additional data would also allow the conducting of cost utility analysis.

On the other hand Peveler et al.[6] conducted one-year open randomized controlled trial which compared three classes of antidepressants (tricyclics, SSRIs, and lofepramine) as first-choice treatment of depression in primary care. Comprehensive economic analysis included both cost-utility and cost-effectiveness approach in addition to profound sensitivity analysis. Outcome variables in cost-effectiveness analysis were the number of depression-free weeks and direct costs. Maximum acceptable costs per depression-free week were used in comparisons aiming to determine the most cost-effective strategy for the amount of money that health system can pay. For example, if the most a health system could afford for additional depression-free week was £10, prescribing tricyclics as the first line treatment would be the cost-effective option approximately 20% of the time, SSRIs 15% of the time, and lofepramine 65% of the time. On the other hand, if the health system could afford up to a just over £50 for a depression-free week, a first-line treatment of lofepramine is most likely to be the most cost-effective strategy. Above the sum of £50 per depression free week, SSRIs would be the most cost-effective treatment strategy. In cost-utility analysis EQ-5D tariff scores were used to weight survival and calculate QALYs. The pattern of results of cost-utility analysis was similar as in previously mentioned cost-effectiveness analysis. The authors found that due to the low probability of significant differences in cost-effectiveness it is appropriate to base the first choice between these three classes of antidepressants in primary care on doctor and patient preferences.

Cost-utility analysis is an adaptation of cost-effectiveness analysis which measures the effect of treatment on quantitative (mortality) as well as qualitative (morbidity) aspect of health. It takes into account the principle, laid by WHO: the goals of healthcare are to add years to life and to add life to years[7]. The relative efficiency of the intervention in cost-utility analysis is expressed by cost-utility ratio (Box 1). By comparing and ranking cost-utility ratios of different interventions (the lower the ratio, the higher a priority of the intervention) the priorities are defined.s

In cost – utility ratios the health benefits are assessed in terms of quality adjusted survival, measured in quality adjusted life years (QALYs). The cost-utility ratios will thus be expressed in monetary units (dollars) per QALY (Box 2).

Box 1 Cost-utility ratio calculation

Cost-utility ratio = (costs of treatment A − costs of treatment B) / (QALYs produced by treatment A − QALYs produced by treatment B)

Box 2 Example of cost-utility calculations

Treatment	Costs ($)	Life years gained	Quality weight	QALYs
A	15,000	4	0.90	3.60
B	30,000	5	0.85	4.25
C	35,000	5.3	0.85	4.51

QALYs in the table are the product of life years gained and quality weight for each treatment. Marginal cost-utility ratios give incremental costs (in dollars) for each quality adjusted life year gained, they adjust life years gained for quality weight. The most expensive alternative C produces QALY at a lower marginal cost than does the less expensive treatment, B.

Marginal cost-utility ratios:

A to B: (30,000 − 15,000)/(4.25 − 3.60) = 23,077 per QALY gained

A to C: (35,000 − 15,000)/(4.51 − 3.60) = 21,978 per QALY gained

The average costs per QALY are higher for C than for B treatment, but average costs are not the one to be looked at in economic analyses. The benefits gained from health care expenditures will be maximized by basing decisions on gains at the margin. Cost-utility ratios have a meaning therefore only when compared to other ratios.

MEASURING QUALITY OF LIFE IN ECONOMIC ANALYSIS

Almost 60 years ago Karnofsky et al.[8] started the measurement of health status in cancer patients. At the same time, von Neumann and Morgenstern developed the foundations of assessing utilities, which can be defined as strength of preferences for various health states. Estimation of preferences is a very active area of research in economic evaluation. The term preferences is used for values and utilities – however, there is a distinction between them[19]. **Utilities** are numbers that represent the strength of an individual's preferences for different health outcomes under the conditions of uncertainty.

Values, on the other hand, are the numbers people assign to different health outcomes in certainty conditions – when measurement is not undertaken under the conditions of uncertainty [10].

Whatever they are called, the key is to show that the domains measured by health-related QoL instrument must include important patient's preferences [11]. On the other hand not enough is known about preferences related to individual domains of well-known, standardized and frequently used QoL instruments. One may wonder that this may be one of the reasons that relatively few cost-utility analyses are published. Franz et al. [12] tried to overcome the shortcomings of previous studies by taking individual preferences into account and measuring the importance of each domain of subjective QoL. They found very high correlations between the pure and the weighted satisfaction scores. In mentioned study pure scores were those that were actually measured while weighted scores of each domain were calculated by adding weights to pure scores corresponding to the importance of the domain for each individual.

Preferences can be generated by investigating various populations like members of general population, patients, their caregivers or physicians using different methods like visual analogue scale (VAS), standard gamble method (SG), paired comparisons, person trade-off (PTO) or time trade-off method (TTO). Depending on whether the condition of uncertainty is fulfilled in the method, the preferences elicited are called values or utilities (Table 1).

SG Produces Utilities While VAS and TTO Produce Values

With VAS the respondent (e.g. patient) locates the health state to be assessed between two fixed points indicating death and full health. Death is ascribed a value of 0 and full health is ascribed a value of 1. VAS is very easy to administer and use, but it has also important drawbacks like it does not involve choice and it is not

Table 1. Methods for measuring preferences, taken from ([19])

Response method	Question framing	
	Certainty (values)	Uncertainty (utilities)
Scaling	Rating scale Category scaling Visual analogue scale Ratio scale	
Choice	Time trade-off Paired comparison Equivalence Person trade-off	Standard gamble

clear whether it really is an interval scale (meaning that change from 0.2-0.3 does not have the same utility as a change from 0.8-0.9)[13].

SG method compares a specific number of years in the health state to a gamble with a probability p of achieving full health for the same number of years and a complementary probability $1 - p$ of immediate death. The probability of full health (p) is varied until the individual is indifferent between the alternatives, and the *quality weight equals p*. SG involves uncertainty and choice for the respondent, but is very difficult to understand and use.

TTO method compares X years in the health state to Y years in full health. The number of years in full health (Y) is varied until the individual is indifferent between the alternatives and the quality of weight of the health state equals Y/X. In this method the individual trades off survival for a higher health-related quality of life[14].

In terms of methods for describing health states, some researchers have measured quality of life weights based on direct, holistic utility assessment (asking patients in clinical studies a time trade-off question), whereas most of them base weights on prespecified health state classification systems. The latter systems are designed to be complete and general enough to apply across many different types of conditions and treatments. They provide an indirect means of obtaining preference weights: a respondent is assigned a health state classification, which is based on responses to health status questionnaire and prespecified preference weights for this health state, obtained from other population beforehand[15]. Although it is sometimes debated whether patient or general population preferences should be used in cost-effectiveness studies, patients with experience in particular disease may be the best sources of health state preference data[16].

For example, Revicki et al.[17] studied preferences for schizophrenia–related health states and demonstrated that patient preferences were correlated with clinician and caregiver preferences for current health. The cited study used SG method and categorical rating scales but found that SG method may be too demanding for patients with schizophrenia and therefore the TTO or paired comparison was used instead. This is in line with previous reports that SG method may not be possible with patients with significant thought disturbances and cognitive impairment[18]. Another study assessed health utilities in schizophrenia and found that rating scales, TTO and Willingness-to-Pay techniques were the preferred methods of utility evaluation[19]. Study design included test and retest (after one week) approach and in addition a staff member assessed patients' health state utility rating. The authors concluded that clinically stabilized patients with schizophrenia can provide accurate health state descriptions and assign them utilities with a fair degree of reliability and validity.

However, patients with depression were able to participate in SG method[20] and clearly distinguished 11 health states related to depression severity. In this study the patients estimated utility for severe depression 0.30, which was lower than the utilities reported for chronic medical diseases (0.63 for chronic renal disease[21], 0.87 for severe angina[22]). The same study ([20]) also explored the utilities of the current health of patients after two months of treatment for depressive episode.

Table 2. Some characteristics of the most used instruments with predefined utility weights

Instrument	Attributes	Method	Values generated on
Euroqol EQ-5D	Mobility, Self care, Usual activity, Pain/discomfort, Anxiety/depression	Time trade-off	Random sample of the general public
HUI	Physical functioning, Role function, Social-emotional function, Health problem	Time trade-off	Random sample of the general public
QWB	Mobility, Physical activity Social activity Symptom problem complex	Category scaling	Random sample of the general public

The utility was 0.74, which was comparable to utilities for schizophrenia (0.74)[17], depressive and anxiety disorders (0.69-0.78)[23], severe angina (0.83)[22], and chronic renal disease (0.63)[21]). There is a specially designed utility measure instrument (McSad) for patients with depression aiming to obtain directly measured utilities for depression health states[24].

The other question in measuring preferences is the validity of subjects' responses. Lenert et al.[25] recommend the use of computerized interview for VAS, paired comparison, and SG method and special post tests to identify potentially invalid preference ratings.

The most used systems with predefined utility weights are EuroQol EQ-5D (Table 2)[26], Health Utilities Index HUI[27], Quality of well-being QWB[28]. There are also two other instruments: 15D[29], and SF-6D[30]. As quality of life is a multi-dimensional entity, it includes a number of domains, which has been differently identified by different authors and different instruments. However, they can be translated into physical health, occupational functioning, psychological functioning, social role performance and somatic sensation[38].

Each of the five dimensions has three levels (no problems, some problems, extreme problems), which can result in a total of 243 possible health states (rank 1 on all five dimensions 11111 would interpret like having no problems in any of five defined dimensions). Health states unconsciousness and death are added to make 245 health states in total. A quality weight from previously conducted study is known for each of 245 so defined health states and by placing a patient into one of health states we know immediately his quality of life weight.

Combination of any method of eliciting health preferences/utilities (VAS, SG, TTO, PTO) and any method of description of health states in a quality weight

(EQ-5D, HUI, QWB), can be later incorporated into QALY and used on a benefit side in a cost-utility analysis.

WHAT ARE QALYS?

QALY is the arithmetic product of life expectancy and a quality weight obtained for a defined health state (Box 20.2). Each year of life is multiplied by a weight reflecting quality of life- QALY places a weight on time in different health states. A year of perfect health is worth 1; a year of less than perfect health is worth less than 1. Death is equivalent to 0. Some health states may be considered worse than death and have a negative quality weight.

The main advantage of QALY is that it is a common measure of benefits and can be used across a variety of interventions and measure health-related quality of life and survival for the patient.

The idea for QALY league tables or Off-the-Shelf Weights were to aggregate preferences of the diseases and disorders, chronic states or procedures and treatments in a way to enable comparisons. Their construction and use has generated considerable discussion and debate and they should be used with caution. The studies included in such tables were carried out at different times and in different locations and settings and have used different quality of life measures [15].

There are three uses of QALY. First approach adjusts survival for the QoL during years prior to death. Others calculate QALYs by explicitly making both QoL and time part of the utility judgment [31] or prospectively evaluate QoL and combine it with time to estimate QALYs [32]. The last-mentioned method has two variants of calculation which have to be considered in designing of prospective studies.

The QALY has been proposed as a standard outcome measure for cost-effectiveness analyses [33], [34]. Resource allocation decisions between patient groups competing for medical care can be taken on the basis of QALYs. Although QALY provides an indication of the benefits gained from various treatments in terms of quality of life and life expectancy of the patients, it is far from being perfect as a measure of outcome. There is a debate whether utilities, which always belong to individuals, can be aggregated. A gain of 0.5 QALY for one person should be equal as gains of 0.25 QALY for two persons. A gain of 0.5 QALY for one year should be then equal to gain of 0.25 QALY for two years (without taking discounting into account) [13].

In their cost-utility study Dernovsek et al. [35] studied the QoL and costs in 200 outpatients with schizophrenia, treated with depot neuroleptics. EQ-5D was used. In analysis it was literally assumed that the patient would not survive without treatment. Such an assumption was necessary because the study measured health states' quality status in one time point and there was no control group. The authors found out that the patients in a sample can expect to survive on average 14.75 QALY.

With data relating to both health-related quality of life and survival, it is then possible to chart the impact of a healthcare intervention of an individual patient. For example, in research setting, it is possible to compare the health profile of a patient receiving the intervention with that of a patient who does not receive the intervention (or similarly, comparing the health profile of a patient receiving the intervention A with that of a patient receiving intervention B).

The Washington Panel of Cost-Effectiveness in Health and Medicine[36] distinguished between two broad approaches of the assignment of preference weights to health states in computing QALYs – those based on expected utility theory and those derived from psychological or psychophysical scaling methods. It was noted that the diversity in how preference weights are gathered markedly constrains the ability to credibly compare analyses where the effectiveness is presented in QALYs. National Institute for Clinical Excellence NICE takes a slightly different position: in favoring cost-utility analysis as a means of providing a comparative context for judging the relative value of health benefits from interventions in different disease areas NICE accords equal statues to SG and TTO utilities. Some authors like Drummond[19] set out subtle but important difference between utilities and values and the role of uncertainty and methods of choice. The fact is that each method yields different results and that there is no consensus which method is the "golden standard". There is no consensus whether the role of uncertainty conditions and the role of choice is important for gaining utilities. In the three sets, gained through three methods, it is still impossible to decide which numbers are utilities and which are proper to use in cost-utility analysis. Kind[37], one of leading health economists, stresses the absence of the recognized standard and proposes that in such a situation the multiple measurement methods must be tolerated as having some claim for legitimacy. Recent attention given to this question suggests that in many studies, quality adjustment had relatively little effect on the final cost-utility ratio. Its impact was important in moving ratios across a $50,000/QALY cost-utility threshold in only 20% of investigated cases[37]. Nearly half a century since the beginnings of health outcomes measurement the enquiry into the central question of how to value health in QALY calculations and which methods to use to obtain utilities remains still largely unresolved.

The QALY represents one of the most elegant forms of health outcome measurement, combining information on both intensity and duration. QALY has changed a lot overtime. The early clearly effectiveness issue had been surrounded by issues such as distribution and equity that are not yet resolved, but discover many disadvantages surrounding QALY.

Yet another concern of QALY is its subjectivity: the subjects who are experiencing a disease will assign higher utilities to decreased health states as healthy individuals due to disability adjustment process – this will result in lesser possible improvement in the treatment process, lower QALY and higher cost-utility ratio, ultimately resulting in a discriminating policy decision[38]. Some researchers[15] have argued for methods other than QALYs. Proposed alternatives include healthy-years equivalents (HYE), saved young life equivalents and disability-adjusted life years (DALYs). These approaches, similar to QALYs, are not perfect and have their own limitations and disadvantages.

In resource allocation decisions, mental health service providers compete for limited resources with providers in other areas of medicine. Although there are special problems connected to the use of utility-based measures in mental health care, connected to psychopathological symptoms (affective symptoms, cognitive decline and reality distortion)[3] it is necessary to somehow confirm that mental health services provide a benefit to consumers. As the goal in mental health is the same as in other areas of medicine – to increase the length of life and improve its quality – QALY is a metric that allows a comparison among different types of services and is valuable as outcome measure in any type of health care interventions[39].

CONCLUSIONS

To make judgment about efficiency, economic evaluation has to compare health outcomes, however measured, with costs. Measures of quality of life which go beyond both clinical and mortality endpoints are becoming more common. Quality of life measures that are based on preferences tend to use general population to obtain them. There are different methods through which preferences can be obtained and it is still not clear, which method represents the gold standard and is most suitable for that purpose. Until this issue is resolved, all of them are used, which adds some confusion, non-transparency and non-comparability in the calculation of cost-utility ratios. It is precisely the question of the measurement of benefit which lies at the heart of any evaluation.

All methodological shortcomings should not divert interested researchers from conducting research in this area.

Our recommendations:

1. In a case that a cost-utility analysis is included in the research plan, we recommend to add one generic preference-based instrument to the selected set of instruments for quality of life assessment.
2. The person with skills and knowledge in health economics should be a member of a team in a study that includes economic analysis.
3. We do not recommend using holistic measurements of preferences. As predetermined preferences scales are widely used it is simpler to use them, the results will be more comparable, and the scales have determined validity.
4. More research in mental health economics will provide us with more data on instruments, disorders, interventions and health care determinants. There is no golden standard for complete procedure of cost utility analysis. There are numerous research methods, each with its valid aspects but also with shortcomings. The only postulated methodological determinant at the moment is the use of QALY.

ACKNOWLEDGEMENT

We are especially grateful to Miss Eva Sajovic for her help with managing of references.

REFERENCES

1. Pyne JM, Patterson TL, Kaplan RM, et al. Preliminary longitudinal assessment of quality of life in patients with major depression. Psychopharmacol Bull 1997; 33: 23–29
2. Awad AG, Voruganti LN, Heslegrave RJ. A conceptual model of quality of life in schizophrenia: description and preliminary clinical validation. Qual Life Res 1997; 6: 21–26
3. Katschnig H. How useful is the concept of quality of life in Psychiatry? In: Katschnig H, Freeman H, Sartorius N, eds. Quality of Life in Mental Disorders. Chichester: Wiley; 1997: 3–16
4. Wiklund I, Karlberg J. Evaluation of quality of life in clinical trials: selecting quality-of-life measures. Control Clin Trials 1991; 12. 204S-216S
5. Namjoshi MA, Rajamannar G, Jacobs T, et al. Economic, clinical, and quality-of-life outcomes associated with olanzapine treatment in mania: results from a randomized controlled trial. J Affect Disord 2002; 69: 109–118
6. Peveler R, Kendrick T, Buxton M, et al. A randomised controlled trial to compare the cost-effectiveness of tricyclic antidepressants, selective serotonin reuptake inhibitors and lofepramine. Health Technol Assess 2005; 9: 1–134
7. WHO. Health Promotion. A discussion document on the concepts and principles. Copenhagen: WHO Regional Office of Europe, 1984
8. Weeks J. Taking quality of life into account in health economic analyses. J Natl Cancer Inst Monogr, 1996; 20: 23–27
19. Drummond FM, O'Brien B, Stoddart GL, Torrance GW. Methods for the economic evaluation of health care programmes. 2nd edition. Oxford: Oxford Medical Publications; 1997: 1–305
10. Revicki DA, Kaplan RM. Relationship between psychometric and utility-based approaches to the measurement of health-related quality of life. Qual Life Res 1993; 2: 477–487
11. Revicki DA, Osoba D, Fairclough D, et al. Recommendations on health- related quality of life research to support labeling and promotional claims in the United States. Qual Life Res 2000; 9: 887–900
12. Franz M, Meyer T, Gallhofer B. Are importance ratings useful in the assessment of subjective quality of life in schizophrenic patients? Int J Methods Psychiatr Res 1999; 8: 204–211
13. Schwartz S, Richardson J, Glasziou PP. Quality-adjusted life years: origins, measurements, applications, objections. Aust J Public Health 1993; 17: 272–278
14. Johannesson M, Jönsson B, Karlsson G. Outcome measurement in economic evaluation. Health Econ 1996; 5: 279–296
15. Neumann PJ, Goldie SJ, Weinstein MC. Preference-based measures in economic evaluation in health care. Annu Rev Public Health 2000; 21: 587–611
16. Jenkins DC. Assessment of outcomes of health intervention. Soc Sci Med 1992; 35: 367–375
17. Revicki DA, Shakespeare A, Kind P. Preferences for schizophrenia – related health states: a comparison of patients, caregivers and psychiatrists. Int Clin Psychopharmacol 1996; 11: 101–118
18. Revicki DA, Wu AW, Murray MI. Change in clinical status, health status, and health utility outcomes in HIV-infected patients. Med Care. 1995; 33 (4 Suppl):AS173–182
19. Voruganti LNP, Awad AG, Oyewumi LK, Cortese L, Zirul S, Dhawan R. Assessing health utilities in schizophrenia. A feasibility study. Pharmacoeconomics 2000; 17:273–286
20. Revicki DA, Wood M. Patient-assigned health state utilities for depression-related outcomes: differences by depression severity and antidepressant medications. J Affect Disord 1998; 48: 25–36
21. Revicki DA. Relationship between health utility and psychometric health status measures. Med Care 1992; 30 (5 Suppl): MS274–282
22. Nease RF, Kneeland T, O'Connor GT, et al. Variation in patient utilities for outcomes of the management of chronic stable angina. Implications for clinical practice guidelines. Ischemic Heart Disease Patient Outcomes Research Team. JAMA 1995; 273: 1185–1190
23. Patrick DL, Mathias SD, Elkin EP, et al. Health state preferences of persons with anxiety. San Francisco: Technology Assessment Group, 1995

24. Benett KJ, Torrance GW, Boyle MH, Guscott R, Moran LA. Development and testing of a utility measure for major, unipolar depression (McSad). Qual Life Res 2000; 9: 109–120
25. Lenert LA, Morss S, Goldstein MK, Bergen MR, Faustman WO, Garber AM. Measurement of the validity of utility elicitations performed by computerized interview. Med Care 1997; 35: 915–920
26. http://www.euroqol.org
27. Torrance GW, Feeny DH, Furlong WJ, Barr RD, Zhang Y, Wang Q. Multiattribute utility function for a comprehensive health status classification system: Health Utilities Index Mark 2. Med Care, 1996; 34: 702–722
28. Kaplan RM, Anderson JP. The general health policy model: an integrated approach. In: Spilker B (ed.): Quality of life and pharmacoeconomics in clinical trials. Philadelphia: Lippincott-Raven; 1996: 309–322
29. Sintonen H. The 15D instrument of health-related quality of life: properties and applications. Ann Med 2001; 33: 328–336
30. http://www.shef.ac.uk/scharr/sections/heds/mvh/sf-6d
31. Mehrez A, Gafni A. The healthy-years equivalents: how to measure them using the standard gamble approach. Med Decis Making 1991; 11: 140–146
32. Ganiats TG, Browner DK, Kaplan RM. Comparison of two methods of calculating Quality-adjusted Life Years. Qual Life Res 1996; 5: 162–164
33. Kaplan RM, Anderson JP, Ganiats TG. The quality of well-being scale: rationale for a single quality of life index. In: Walker SR, Rosser RM, eds. Quality of life assessment. Key issues in the 1990s. London: MTM Press, 1993: 65–94
34. Gold MR, Siegal JE, Russell LB, Weinstein MC, eds. Cost-effectiveness in health and medicine. New York: Oxford University Press, 1996: 21–39
35. Dernovsek MZ, Prevolnik Rupel V, Rebolj M, Tavcar R. The quality of life of schizophrenic outpatients, treated with depot neuroleptics. Eur Psychiatry 2001; 16: 474–482
36. Gold MR, Patrick DL, Torrance GW, et al. Identifying and valuing outcomes. In: Gold MR, Russell LB, Weinstein MC, eds. Cost-effectiveness in Health and Medicine. New York: Oxford University Press; 1996: 82–134
37. Kind P. Putting the "Q" into QALYs. Paper, presented at a conference Be reasonable: Following the Williams way – a conference to celebrate the work of Alan Williams. York, July 25–26, 2006
38. Raisch DW. Understanding Quality-adjusted Life Years and their application to pharmacoeconomic research. Ann Pharmacother 2000; 34: 906–914
39. Patterson TL, Kaplan RM, Jeste DV. Measuring the effect of treatment in quality of life in patients with schizophrenia: focus on utility-based measures. CNS Drugs 1999; 12: 49–64

INDEX